Advances in Multivariate Approximation

Proceedings of the 3rd International Conference
on Multivariate Approximation Theory
held at Witten-Bommerholz, Germany,
September 27–October 2, 1998

edited by
Werner Haußmann
Kurt Jetter
Manfred Reimer

WILEY-VCH

Berlin · Weinheim · New York · Chichester
Brisbane · Singapore · Toronto

Editors:

Prof. Dr. Werner Haußmann, Gerhard-Mercator-University of Duisburg
Prof. Dr. Kurt Jetter, University of Hohenheim
Prof. Dr. Manfred Reimer, University of Dortmund

With 19 figs. and 4 tabs.

1st edition

Die Deutsche Bibliothek – CIP-Einheitsaufnahme

Advances in multivariate approximation : proceedings of the 3rd International Conference
on Multivariate Approximation Theory, held at Witten-Bommerholz, Germany,
September 27–October 2, 1998 / ed. by Werner Haußmann ... – 1. ed. – Berlin ; Weinheim ;
New York ; Chichester ; Brisbane ; Singapore ; Toronto : WILEY-VCH, 1999
   (Mathematical research ; Vol. 107)
   ISBN 3-527-40236-5

ISSN 0138-3019

© WILEY-VCH Verlag Berlin GmbH, Berlin (Federal Republic of Germany), 1999

Printed on non-acid paper.
The paper used corresponds to both the U.S. standard ANSI Z.39.48 – 1984
and the European standard ISO TC 46.

Printing: GAM Media GmbH, Berlin

Bookbinding: Druckhaus „Thomas Müntzer", Bad Langensalza

Printed in the Federal Republic of Germany

WILEY-VCH Verlag Berlin GmbH
Bühringstr. 10
D-13086 Berlin
Federal Republic of Germany

Preface

The Third International Conference on Multivariate Approximation at "Haus Bommerholz", the guest–house of the University of Dortmund, took place during the week of September 27 through October 2, 1998. It was attended by 50 participants from 18 countries. The program included 10 invited lectures, 25 contributed talks, and two problem sessions. In these Proceedings we collect most of the lectures, suitably prepared by the authors for publication, and carefully peer–refereed.

As was the case for the preceding conferences, the meeting was intended to advance several selected topics from modern aspects of Multivariate Approximation Theory. This time, emphasis was put on the following subjects: Interpolation and Approximation on Spheres and Balls, Approximation by Solutions of Partial Differential Equations, Construction of Node Systems, and Scattered Data Techniques. Leading experts in these areas and quite a number of young researchers made the meeting a stimulating event, with lively discussions and scientific interactions to support and initiate future research.

The editors are grateful to all who have contributed to the success of the conference and to the production of this Proceedings volume. In particular, our thanks go to the Deutsche Forschungsgemeinschaft whose funding made the conference possible, and to the Ministerium für Schule, Weiterbildung, Wissenschaft und Forschung of the Land Nordrhein–Westfalen for sponsoring this volume. We acknowledge the help of the authors and the referees to guarantee publications of high scientific standard, and we thank the WILEY–VCH publishers for their kind cooperation. In addition, we acknowledge the extensive help of Dr. Hermann Hoch and Dipl.–Math. Jörg Zabold (Duisburg) during the production of this Proceedings volume. Finally, we would especially like to thank the conference secretary, Mrs. Lamprecht, and the staff of Haus Bommerholz, for all their help.

May 1, 1999

Werner Haußmann

Kurt Jetter

Manfred Reimer

Acknowledgement

We are deeply indebted to the

Deutsche Forschungsgemeinschaft (DFG)

whose funding made the conference possible.

The production of this volume was sponsored by the

Ministerium für Schule und Weiterbildung,
Wissenschaft und Forschung
des Landes Nordrhein–Westfalen.

We are grateful for the generous support.

The Editors

Advances in
Multivariate Approximation

MATHEMATICAL RESEARCH

Volume 107

Table of Contents

Participants

David Armitage, *Department of Pure Mathematics, Queen's University, Belfast BT7 1NN, Northern Ireland*
email: d.armitage@qub.ac.uk

Marek Beśka, *Department of Mathematics, Technical University of Gdansk, Narutowicza 11/12, 80-952 Gdansk, Poland*
email: beska@mif.pg.gda.pl

Peter Binev, *Institute of Mathematics, Bulgarian Academy of Sciences, Sofia 1090, Bulgaria*
email: binev@fmi.uni-sofia.de

Len Bos, *Department of Mathematics and Statistics, University of Calgary, Calgary, Alberta T2N 1N4, Canada*
email: lpbos@math.ucalgary.ca

Dietrich Braess, *Institut für Mathematik, Ruhr-Universität Bochum, D–44780 Bochum, Germany*
email: braess@num.ruhr-uni-bochum.de

Zbigniew Ciesielski, *Mathematical Institute PAN, ul. 23 Abrahama 18, 81–825 Sopot, Poland*
email: Z.Ciesielski@impan.gda.pl

Stephan Dahlke, *Institut für Geometrie und Praktische Mathematik, RWTH Aachen, Templergraben 55, D–52062 Aachen, Germany*
email: dahlke@igpm.rwth-aachen.de

Wolfgang Dahmen, *Institut für Geometrie und Praktische Mathematik, RWTH Aachen, Templergraben 55, D–52062 Aachen, Germany*
email: dahmen@igpm.rwth-aachen.de

Oleg Davydov, *Fachbereich Mathematik, Universität Dortmund, D–44221 Dortmund, Germany*
email: davydov@math.uni-dortmund.de

Franz–Jürgen Delvos, *Fachbereich 6 – Mathematik, Universität–GH–Siegen, Hölderlinstraße 3, D–57076 Siegen, Germany*

Stefano de Marchi, *Dipartimento di Matemàtica e Informàtica, Università di Udine, via della Scienze 206, I–33100 Udine, Italy*
email: demarchi@dimi.uniud.it

Uwe Depczynski, *Institut für Angewandte Mathematik und Statistik, Universität Hohenheim, D–70593 Stuttgart, Germany*
email: depczyns@uni-hohenheim.de

Anita Faul, *DAMTP, University of Cambridge, Silver Street, Cambridge CB3 9EW, Great Britain*
email: A.C.Faul@damtp.cam.ac.uk

Michael Felten, *Fachbereich Mathematik, Universität Dortmund, D–44221 Dortmund, Germany*
email: felten@zx2.hrz.uni-dortmund.de

Willi Freeden, *AG Geomathematik, Universität Kaiserslautern, Postfach 3049, D–67653 Kaiserslautern, Germany*
email: freeden@mathematik.uni-kl.de

Stephen J. Gardiner, *Department of Mathematics, University College Dublin, Dublin 4, Ireland*
email: gardiner@acadamh.ucd.ie

Werner Haußmann, *Fachbereich Mathematik, Universität Duisburg, D–47048 Duisburg, Germany*
email: haussmann@math.uni-duisburg.de

Kurt Jetter, *Institut für Angewandte Mathematik und Statistik, Universität Hohenheim, D–70593 Stuttgart, Germany*
email: kjetter@uni-hohenheim.de

Lavi Karp, *Department of Applied Mathematics, Ort Braude College,*
P.O. Box 78, Karmiel 21982, Israel
email: karp@techunix.technion.ac.il

John Klinkhammer, *Fachbereich Mathematik, Universität Duisburg,*
D–47048 Duisburg, Germany
email: klinkhammer@math.uni-duisburg.de

Hermann König, *Mathematisches Seminar, Universität Kiel,*
Ludewig–Meyn–Str. 4, D–24098 Kiel, Germany
email: nms22@rz.uni-kiel.d400.de

Ognyan Kounchev, *Institute of Mathematics, Bulgarian Academy of*
Sciences, Acad. G. Bonchev Str. 8, BG–1113 Sofia, Bulgaria
email: kounchev@cblink.net

Angela Kunoth, *Institut für Geometrie und Praktische Mathematik,*
RWTH Aachen, Templergraben 55, D–52062 Aachen, Germany
email: kunoth@igpm.rwth-aachen.de

Alain Le Méhauté, *Département de Mathématiques, Université de Nantes,*
2 Rue de la Houssinière, F–44072 Nantes Cedex, France
email: alm@math.univ-nantes.fr

Detlef Mache, *Fakultät für Mathematik, Ludwig–Maximilians–Universität,*
D–80333 München, Germany
email: mache@rz.mathematik.uni-muenchen.de

Ulrike Maier, *Fachbereich Mathematik, Universität Dortmund,*
D–44221 Dortmund, Germany
email: umaier@math.uni-dortmund.de

Marie-Laurence Mazure, *LMC – IMAG, Université Joseph Fourier,*
BP 53 X, F–38041 Grenoble, France
email: mazure@imag.fr

Volker Michel, *AG Geomathematik, Universität Kaiserslautern,*
Postfach 3049, D–67653 Kaiserslautern, Germany
email: michel@mathematik.uni-kl.de

H. Michael Möller, *Fachbereich Mathematik, Universität Dortmund,*
D–44221 Dortmund, Germany
email: moeller@math.uni-dortmund.de

Manfred W. Müller, *Fachbereich Mathematik, Universität Dortmund,*
D–44221 Dortmund, Germany
email: mueller@math.uni-dortmund.de

Günther Nürnberger, *Fakultät für Mathematik und Informatik,*
Universität Mannheim, D–68131 Mannheim, Germany
email: nuernberger@math.uni-mannheim.de

Fernando Pérez–Gonzaléz, *Departamento de Analisis Matematico,*
Universidad de Laguna, E–28271 La Laguna, Tenerife, Spain
email: fpergon@ull.es

Gerlind Plonka, *Fachbereich Mathematik, Universität Duisburg,*
D–47048 Duisburg, Germany
email: plonka@math.uni-duisburg.de

Jürgen Prestin, *Institut für Biomathematik und Biometrie,*
GSF-Forschungszentrum, Postfach 1129, D–86758 Neuherberg, Germany
email: prestin@gsf.de

Manfred Reimer, *Fachbereich Mathematik, Universität Dortmund,*
D–44221 Dortmund, Germany
email: reimer@math.uni-dortmund.de

Sherman D. Riemenschneider, *Department of Mathematical Sciences,*
University of Alberta, Edmonton, Alberta T6G 2G1, Canada
email: sherm@approx.math.ualberta.ca

Robert Schaback, *Institut für Numerische und Angewandte Mathematik,*
Universität Göttingen, D–37083 Göttingen, Germany
email: schaback@math.uni-goettingen.de

Jochen W. Schmidt, *Institut für Numerische Mathematik,*
Technische Universität Dresden, D–01062 Dresden, Germany
email: jschmidt@math.tu-dresden.de

Jacob J. Seidel, *Vesalinslaan 26, 5644 HK Eindhoven, Netherlands*
email: secdw@win.tue.nl

Ian H. Sloan, *School of Mathematics, University of New South Wales,*
Sydney, Australia 2052
email: i.sloan@unsw.edu.au

Frauke Sprengel, *CWI, Kruislaan 413, P.O.Box 94079,*
1090 GB Amsterdam, Netherlands
email: Frauke.Sprengel@cwi.nl

Gabriele Steidl, *Fakultät für Mathematik und Informatik,*
Universität Mannheim, D–68131 Mannheim, Germany
email: steidl@mathematik.uni-mannheim.de

Joachim Stöckler, *Department of Mathematics and Computer Science,*
University of Missouri, St. Louis, MO 63121–4499, USA
email: stoeckler@arch.umsl.edu

Arne Stray, *Matematisk Institutt, Universitetet i Bergen,*
Johs. Bruns Gt. 12, 5020 Bergen, Norway
email: Arne.Stray@mi.uib.no

Manfred Tasche, *Fachbereich Mathematik, Universität Rostock,*
D–18051 Rostock, Germany
email: manfred.tasche@mathematik.uni-rostock.de

Robert F. Tichy, *Institut für Mathematik A, Technische Universität Graz,*
Steyrergasse 30, A–8010 Graz, Austria
email: tichy@weyl.math.tu-graz.ac.at

Joseph D. Ward, *Department of Mathematics, Texas A&M University,*
College Station, TX 77843–3368, USA
email: jward@math.tamu.edu

Pei–Cai Xuan, *Department of Computer Sciences, Shaoxing College of*
Arts and Sciences, Shaoxing, Zhejiang 312 000, P.R.China
email: xuanj@public.sxptt.zj.cn

Vladimir A. Yudin, *Moscow Institute of Power Engineering, Department of*
Higher Mathematics, Krasnokazarmennaya 14, Moscow 105 835, Russia
email: andreev@nw.math.msu.su

Georg Zimmermann, *Institut für Angewandte Mathematik und Statistik,*
Universität Hohenheim, D–70593 Stuttgart, Germany
email: gzim@uni-hohenheim.de

Scientific Program of the Conference

Monday, September 28:

Opening Session: Chair: *K. Jetter*

9:00	Opening of the Conference
9:30 – 10:30	*R. F. Tichy*: Point Distributions, Discrepancy and Applications

Morning Session: Chair: *R. F. Tichy*

11:00 – 11:30	*V. A. Yudin*: Extremal Dispositions of Points on the Sphere and the Torus
11:30 – 12:00	*J. J. Seidel*: Spherical Designs
12:00 – 12:30	*U. Maier*: Numerical Calculation of Spherical Designs

Afternoon Session: Chair: *A. Stray*

16:00 – 17:00	*F. Pérez–Gonzalez*: Approximation in the L^1–Norm by Element of a Uniform Algebra
17:00 – 17:30	*S. J. Gardiner*: Best One–Sided L^1–Approximation by Harmonic and Subharmonic Functions
17:30 – 18:00	*D. H. Armitage*: Universal Harmonic Functions

Tuesday, September 29:

1st Morning Session: Chair: *J. W. Schmidt*

9:00 – 10:00   *G. Plonka*: A New Factorization of Mask Symbols of Scaling Vectors

10:00 – 10:30   *J. Prestin*: Polynomial Wavelets

2nd Morning Session: Chair: *H. König*

11:00 – 11:30   *H. M. Möller*: H–Bases: A Tool for Polynomial Interpolation and System Solving

11:30 – 12:30   *L. Karp*: Generalized Newtonian Potential and its Applications

Afternoon Session: Chair: *J. D. Ward*

16:00 – 17:00   *W. Freeden*: Inverse Multiscale Modelling of the Geopotential

17:00 – 17:30   *V. Michel*: A New Way of Reconstructing the Density of the Earth from Geodetic Measurements

17:30 – 18:00   *J. W. Schmidt*: Scattered Data Interpolation with Rational Splines on PS Refinements

Evening Session: Chair: *G. Plonka*

19:30 – 20:00   *A. Faul*: Proof of Convergence of an Iterative Technique for Thin Plate Splines in Two Dimensions

20:00 – 20:30   *A. Le Méhauté* : Inf–Convolution and Radial Basis Functions

20:30 – 21:00   *F.-J. Delvos*: Optimal Interpolation in Multivariate Periodic Hilbert Spaces

21:00 – 21:30   1st Problem Session

Wednesday, September 30:

1st Morning Session: Chair: *S. Riemenschneider*

9:00 – 10:00   *W. Dahmen*: Adaptive Wavelet Methods for Elliptic Problems – Convergence Rates

10:00 – 10:30   *A. Kunoth*: Multiscale Galerkin Methods for Elliptic PDEs – Appending Boundary Conditions by Lagrange Multipliers and Fictitious Domains

2nd Morning Session: Chair: *D. Braess*

11:00 – 11:30   *S. Dahlke*: Recent Results on Besov Regularity for Elliptic Boundary Value Problems

11:30 – 12:00   *P.-C. Xuan*: The Equivalence Between the Weighted K–Functional and Moduli of Smoothness on the Simplex and Their Applications

12:00 – 12:30   *F. Sprengel*: Interpolation on Sparse Grids and Nikol'ski–Besov Spaces of Dominating Mixed Smoothness

Thursday, October 1:

1st Morning Session: Chair: *Z. Ciesielski*

9:00 – 10:00   *M. Reimer*: Spherical Polynomial Approximation: A Survey

10:00 – 10:30   *L. Bos*: On the Asymptotics of Fekete Points for Radial Basis Functions

2nd Morning Session: Chair: *L. Bos*

11:00 – 11:30   *S. de Marchi*: On the Limit under Scaling of Polynomial Lagrange Interpolation on Spheres and Analytic Manifolds

11:30 – 12:30   *A. Stray*: Simultaneous Approximation in the Dirichlet Space

Afternoon Session: Chair: *G. Nürnberger*

16:00 – 17:00   *H. König*: Cubature Formulas on Spheres

17:00 – 17:30   *M. Beśka*: The Saturation Problem for some Classical
Operators

17:30 – 18:00   *S. D. Riemenschneider*: Cardinal Interpolation by Gaussian
Radial Basis Functions: Behaviour as $\lambda \to 0+$

Evening Session: Chair: *M. Reimer*

19:30 – 20:00   *M. Tasche*: How Stable are Fast Cosine Transforms?

20:00 – 21:00   2nd Problem Session

Friday, October 2:

1st Morning Session: Chair: *I. H. Sloan*

9:00 – 10:00   *J. D. Ward*: Approximation from Spaces of Shifts and of a
Positive Definite Kernel

10:00 – 10:30   *P. Binev*: Using Three–Scale Equation in Adaptive
Refinement

2nd Morning Session: Chair: *W. Haußmann*

11:00 – 11:30   *O. Davydov*: Bases of Bivariate Piecewise Polynomials

11:30 – 12:00   *G. Nürnberger*: Spline Interpolation on Triangulations

12:00 – 12:30   *I. H. Sloan*: Uniform Constructive Approximation by
Polynomials on the Sphere

12:30   Closing of the Conference

Problems of Approximation Theory in Discrete Geometry

Nikolay N. Andreev and Vladimir A. Yudin

Dedicated to Professor Manfred Reimer on the occasion of his 65th birthday

Abstract

Relations between discrete geometry problems and extremal problems from the theory of functions on a sphere and on a torus are studied.

Introduction

There is a list of old interesting problems in $\mathbb{R}^3$ on the best arrangement of points on a sphere. Such problems arise in different fields of science and go back to I. Newton, I. Tamm, J. J. Thomson and others [22, 17, 19, 14, 11]. The multidimensional analogues of these problems are of prominent interest in coding theory. Essential progress in this field has been achieved during the last 20–30 years. To solve such a problem, one should make two steps: First find an extremal point distribution, then prove that it is the best possible. In the second step, i.e. in finding estimates for optimal parameters of the arrangement, analytical methods which use extremal properties of polynomials and functions are used. This paper is devoted to problems where optimal parameters are obtained using Fourier analysis.

Let $x = (x_1, \ldots, x_d) \in \mathbb{R}^d$ and let $xy = x_1 y_1 + \ldots + x_d y_d$ be the inner product of x and y; $|x| = \sqrt{xx}$. Further, let $S^{d-1} = \{x \in \mathbb{R}^d \,:\, |x| = 1\}$ be the unit sphere of $\mathbb{R}^d$. For a fixed dimension $d \geq 2$ we denote by $\{P_\nu^d(t)\}_{\nu=0}^\infty$ the system of Gegenbauer polynomials, which are orthogonal on $[-1,1]$ with the weight $(1 - t^2)^{\frac{d-3}{2}}$ and normalized such that $P_\nu^d(1) = 1$, (see [8]):

$$P_0^d(t) = 1, \quad P_1^d(t) = t, \quad P_2^d(t) = \frac{dt^2 - 1}{d - 1}, \quad P_3^d(t) = \frac{(d+2)t^3 - 3t}{d - 1}, \ldots,$$

$$(k + d - 2)P_{k+1}^d(t) = (2k + d - 2)tP_k^d(t) - kP_{k-1}^d(t).$$

Advances in Multivariate Approximation; W. Haußmann, K. Jetter and M. Reimer (eds.)
Mathematical Research, Vol. 107, pp. 19–32, ISBN 3-527-40236-5
© WILEY-VCH, Berlin 1999

They are positive–definite polynomials [8], i.e. *for any finite subset* $X = \{x^{(k)}\}_{k=1}^N \subset S^{d-1}$, *for any* $\xi_k \in \mathbb{C}$ *and for any* $\nu \in \mathbb{N}$

$$\sum_{k,l=1}^N \xi_k \overline{\xi_l}\, P_\nu^d(x^{(k)}x^{(l)}) \geq 0.$$

1   Fourier Characteristics of Finite Subsets of S^{d-1}

Let $X \neq \phi$ be a finite set of points on S^{d-1}. By $|X|$ we denote the cardinality of X.

The next two concepts are well–known and very important in coding theory.

The set X is called a τ–code if and only if

$$\forall x, y \in X, \ x \neq y \ \text{ we have } \ xy \leq \tau.$$

The set X is called a design of order q (this is one of some equivalent definitions) iff

$$\sum_{x,y \in X} P_\nu^d(xy) = 0 \quad \text{for} \quad \nu = 1, 2, \ldots, q.$$

The inner products set of X is defined by

$$S(X) = \{xy \ : \ x, y \in X, \ x \neq y\}.$$

It is obvious that if X is a τ–code then $\tau = \max\{t \ : \ t \in S(X)\}$.

The next definition seems to be new.

Definition. *We call the set*

$$Z(X) = \{\nu \in \mathbb{N} \ : \ \sum_{x,y \in X} P_\nu^d(xy) = 0\}$$

the zero set of X.

This characteristic is close to the other well–known characteristic – the order of a design. We shall see, that sometimes all the set Z(X) is needed, not only its first part, which is the design characteristic. For an antipodal set X the zero set $Z(X)$ is infinite: all odd natural numbers belong to it because the Gegenbauer polynomials of odd degree are odd functions. At the same time the number of even numbers in the set $Z(X)$ is always finite.

In this paper we make this more concrete for $d = 4$, as follows. Let

$$L(X) = \sup\{2k \ : \ 2k \in Z(X)\}.$$

Lemma. *Assume $d = 4$. If X is an antipodal spherical τ-code in S^3, then*

$$L(X) \leq \frac{|X| - 2}{2\sqrt{1 - \tau^2}} - 1. \tag{1.1}$$

Proof. Since $d = 4$, the Gegenbauer polynomials are the Chebyshev polynomials of the second kind [20]:

$$P_\nu^4(t) = \frac{\sin(\nu + 1)\varphi}{(\nu + 1)\sin\varphi}, \quad \cos\varphi = t.$$

Hence we obtain

$$|P_\nu^4(t)| \leq \frac{1}{(\nu + 1)\sin\varphi} = \frac{1}{(\nu + 1)\sqrt{1 - t^2}}, \quad -1 < t < 1;$$

we shall use the estimate

$$P_\nu^4(t) \geq -\frac{1}{(\nu + 1)\sqrt{1 - t^2}} \tag{1.2}$$

only. Now let $\nu \in Z(X)$ be even. Then

$$0 = \sum_{\substack{x,y \in X}} P_\nu^4(xy) = \sum_{\substack{x,y \in X \\ x=y}} P_\nu^4(xy) + \sum_{\substack{x,y \in X \\ x=-y}} P_\nu^4(xy) + \sum_{\substack{x,y \in X \\ x \neq y}} P_\nu^4(xy) =$$

$$= 2|X| + \sum_{\substack{x,y \in X \\ |xy| \leq \tau}} P_\nu^4(xy).$$

By (1.2) we conclude that

$$0 \geq 2|X| - \frac{|X|^2 - 2|X|}{(\nu + 1)\sqrt{1 - \tau^2}},$$

and since $|X| > 0$ this completes the proof of the Lemma. $\square$

In general this bound cannot be improved. For instance we can take X to be the set of vertices of the octahedron $\pm(1,0,0,0)$, ..., $\pm(0,0,0,1)$. It is a well-known fact that this is a design of order 3, and therefore

$$L(X) \geq 2.$$

At the same time this is a 0-code. So the Lemma gives us

$$L(X) \leq \frac{8 - 2}{2 \cdot 1} - 1 = 2.$$

In some particular cases one can prove better bounds for $L(X)$ using Gram matrix information.

2   Coding Problem

It is a very interesting but also very difficult problem to find, for fixed d and τ, a τ–code with the largest number of points. For example, for $\tau = 1/2$ this is a well–known problem of kissing numbers [8].

Upper estimates for the cardinality of spherical codes can be derived from the positive–definiteness property of Gegenbauer polynomials. Here the cone of positive–definite continuous real functions

$$K := \left\{ h(t) \in C[-1,1] \; : \; h(t) = \sum_{\nu=0}^{\infty} \widehat{h}_\nu \, P_\nu^d(t), \quad \widehat{h}_0 > 0, \; \widehat{h}_\nu \geq 0, \; \nu \in \mathbb{N} \right\}$$

plays an important role. In proving upper bounds for a code we follow the Delsarte scheme [8, 9, 10, 15, 18]. Let

$$y_\tau(t) = \begin{cases} 0, & -1 \leq t \leq \tau; \\ 1, & \tau < t \leq 1, \end{cases}$$

and assume $h \in K$ is satisfying the inequality

$$h(t) \leq y_\tau(t), \qquad -1 \leq t < 1.$$

Then, if there are N points $\{x^{(k)}\}_{k=1}^{N} \subset S^{d-1}$ forming a τ–code, then we get

$$I = \sum_{k,l=1}^{N} h(x^{(k)} x^{(l)}) \leq \sum_{k,l=1}^{N} y_\tau(x^{(k)} x^{(l)}) = N \, y_\tau(1) = N.$$

On the other hand

$$I = \sum_{k,l=1}^{N} \sum_{\nu=0}^{\infty} \widehat{h}_\nu \, P_\nu^d(x^{(k)} x^{(l)}) = \sum_{\nu=0}^{\infty} \widehat{h}_\nu \sum_{k,l=1}^{N} P_\nu^d(x^{(k)} x^{(l)}) \geq N^2 \widehat{h}_0.$$

Combining these inequalities we get

$$N \leq \frac{1}{\widehat{h}_0}. \tag{2.1}$$

To realize the upper estimate of N we can take, for instance, the solution of the following extremal problem of approximation theory: Find

$$\inf_{h \in K, \, h(t) \leq y_\tau(t)} \int_{-1}^{1} (y_\tau(t) - h(t)) \, (1 - t^2)^{\frac{d-3}{2}} \, dt. \tag{2.2}$$

Many results were based on this idea. Sometimes [8, 10, 15, 18, 7] inequality (2.1) gives the lowest upper bound for the cardinality of a spherical τ–code. It is interesting that in all these cases the solution turns out to be a polynomial.

The extremal problem for $\tau = 1/2$ and $d = 4$ is solved in [4] and it gives the bound $N \leq 25$, which is already known.

At the same time, in [3] the exact solution was found for $d = 4$, $\tau = \cos \pi/5$. The problem was solved by a principle well–known in function theory: First guess the extremal configuration, then prove that it is really the best one. The first step is simple. For instance, we can take the set X^*, $|X^*| = 120$: eight points
$$(\pm 1, 0, 0, 0),\ (0, \pm 1, 0, 0),\ (0, 0, \pm 1, 0),\ (0, 0, 0, \pm 1);$$
16 points
$$(\pm 1/2, \pm 1/2, \pm 1/2, \pm 1/2);$$
and 96 points which are the result of all even permutations of coordinates of the 8 points
$$(\pm(\sqrt{5} + 1)/4, \pm 1/2, \pm(\sqrt{5} - 1)/4, 0).$$
These points are the vertices of one of the regular polytopes in $\mathbb{R}^4$ [5]. The inner products set of X^* is

$$S(X^*) = \left\{ -1, \cos \frac{4\pi}{5}, \cos \frac{2\pi}{3}, \cos \frac{3\pi}{5}, \cos \frac{\pi}{2}, \cos \frac{2\pi}{5}, \cos \frac{\pi}{3}, \cos \frac{\pi}{5} \right\}.$$

So, this is a $\cos \frac{\pi}{5}$–code. The zero set is

$$Z(X^*) =$$

$$\{\text{odd natural numbers}\} \bigcup \{2, 4, 6, 8, 10, 14, 16, 18, 22, 26, 28, 34, 38, 46, 58\}.$$

This can be verified as follows. Odd natural numbers belong to $Z(X^*)$ by the remark in Section 1. If $\nu \in Z(X^*)$ is even then $\nu \leq 99$ by (1.1), and the numbers from 1 to 99 can be verified by a computer.

In order to realize the upper bound for the cardinality of a $\cos \frac{\pi}{5}$–code we take the polynomial

$$h(t) =$$

$$\frac{1}{505800} t^2 (t + 1)^2 (t^2 - 1/4)^2 (t^2 - (3 - \sqrt{5})/8)^2 (t + (1 + \sqrt{5})/4)^2 (t - (1 + \sqrt{5})/4) \times$$

$$\times (65536(232\sqrt{5} - 303)t^2 + 8192(3079 - 2990\sqrt{5})t + 2048(4255\sqrt{5} - 1777)).$$

As the last factor has no real zeros, we get $h(t) \leq y_{\cos(\pi/5)}(t)$, $-1 \leq t \leq 1$. All the Fourier coefficients in the expansion with respect to Gegenbauer polynomials are non–negative (all of them are positive, but $h_{12} = h_{13} = 0$) and $\widehat{h}_0 = 1/120$. That is why by (2.1) we have the sharp estimate

$$N \leq 120.$$

An important obeservation is that the spectrum of h is a subset of the zero set of the extremal configuration. This is necessary for the bound to be minimal. It follows from how we obtained the lower estimate of I above.

We note also that in this case there are many extremal polynomials of degree 17. All of them differ by the last quadric factor.

3 Minimum of the Energy

A very interesting problem of discrete geometry is to find an arrangement of N equal charges on a unit sphere S^{d-1}, $d \geq 3$, which minimizes the potential energy, i.e. which minimizes

$$W_d(N, x^{(1)}, \ldots, x^{(N)}) = \sum_{\substack{k,l=1 \\ k \neq l}}^{N} \frac{1}{|x^{(k)} - x^{(l)}|^{d-2}}.$$

(We deal with Newton's potential, only.) So, we have to find the values

$$W_d(N) = \inf_{\{x^{(k)}\}_{k=1}^N} W_d(N, x^{(1)}, \ldots, x^{(N)})$$

and the corresponding best arrangements of the points. The history of this problem and the cases where the solutions are known are described in [22, 17, 6, 2]. Now we demonstrate how the methods mentioned above work in this case. We take

$$y(t) = \{2(1 - t)\}^{1-d/2}, \quad -1 \leq t < 1.$$

Then, as in Section 1, we can find

$$W_d(N) \geq N^2 \widehat{h}_0 - N\, h(1) \tag{3.1}$$

with $h \in K$ satisfying the inequality

$$h(t) \leq y(t), \quad -1 \leq t < 1.$$

Again we get a problem of one–sided approximation in a weighted L–space, and similarly to the problem of maximal codes we guess the extremal configuration X, first. Then we calculate the inner products set

$$L(x) = \{xy \; : \; x, y \in X, \; x \neq y\} = \{t_0, t_1, \ldots, t_q\}$$

and define the polynomial $h = h_{2q+1}$ with degree $2q+1$ or $h = h_{2q}$ with degree $2q$, respectively, by the Hermite interpolatory conditions

A) $h_{2q+1}(t_k) = y(t_k)$, $h'_{2q+1}(t_k) = y'(t_k)$, $k = 0, 1, \ldots q$, if $t_0 > -1$,

B) $h_{2q}(t_k) = y(t_k)$, $k = 0, 1, \ldots, q$; $h'_{2q}(t_k) = y'(t_k)$, $k = 1, \ldots q$, if $t_0 = -1$.

Adapting Lagrange's formula for the interpolation error by a slight change in the proof of the Hermite case we get the following theorems:

Theorem A. *Let $(t_0, t_1, \ldots, t_q) \subset [a, b]$, and let $y \in C[a, b]$ be $2q + 2$–times continuously differentiable on (a, b). Then for all $t \in [a, b]$ there exists $\tau \in (a, b)$ such that*

$$y(t) - h_{2q+1}(t) = \frac{y^{(2q+2)}(\tau)}{(2q+2)!}(t - t_0)^2(t - t_1)^2 \ldots (t - t_q)^2.$$

Theorem B. *Let $(t_0, t_1, \ldots, t_q) \subset [a, b]$, and let $y \in C[a, b]$ be $2q + 1$–times continuously differentiable on (a, b). Then for all $t \in [a, b]$ there exists $\tau \in (a, b)$ such that*

$$y(t) - h_{2q}(t) = \frac{y^{(2q+1)}(\tau)}{(2q+1)!}(t - t_0)(t - t_1)^2 \ldots (t - t_q)^2.$$

Now we take $a = t_0 = -1$, while b is an arbitrary point from $(-1, 1)$, and let $y(t) := k(1 - t)^{-\alpha}$, $k > 0$, $\alpha > 0$. Obviously, y is absolutely monotone on $(-\infty, 1)$, i.e. $y(t), y'(t), y''(t), \ldots$ are non–negative on $(-\infty, 1)$. So it follows from Theorem A or B, respectively, that the interpolation polynomial satisfies the inequality

$$h(t) \leq y(t), \quad -1 \leq t < b.$$

This holds for an arbitrary $b < 1$ and hence for all $t \in [-1, 1)$.

Whether the constructed polynomials belong to the cone K has to be checked in each case separately.

Let us consider some examples. For $d \in \mathrm{I\!N}$, $d \geq 3$, $N = 2d$ the vertices of the octahedron $\pm(1, 0, \ldots, 0, 0)$, ... , $\pm(0, 0, \ldots, 0, 1)$ provide the extremal configuration [13]. The set of inner products is $\{-1, 0\}$. So, according to the method we discussed above, the extremal polynomial must be found from the Hermite conditions

$$h(-1) = y(-1), \quad h(0) = y(0), \quad h'(0) = y'(0).$$

It is possible to calculate the explicit form of $h(t)$ and its expansion with respect to the Gegenbauer polynomials. The result is

$$h(t) = 2^{-d/2}\{2 + (d-2)t + (2^{2-d/2} + d - 4)t^2\} =$$

$$= \frac{2^{-d}(4 + 2^{d/2}(3d-4))}{d} P_0^d(t) + 2^{-d/2}(d-2)P_1^d(t) +$$

$$+ \frac{2^{-d}(d-1)(4 + 2^{d/2}(d-4))}{d} P_2^d(t).$$

All Fourier coefficients are positive, hence $h \in K$. By Theorem B we get $y(t) \geq h(t)$, $-1 \leq t < 1$. So we obtain for the vertices of the octahedron, from (3.1) with $N = 2d$, that

$$W_d(2d) \geq 8d\{2^{-d} + (d-1)2^{-d/2}\}.$$

In a similar way the problem is solved in some other cases [1, 2]. In particular, for $d = 3$ and $N = 12$ the polynomial of degree 4 defined by the equations

$$h(-1) = y(-1), \quad h(\pm 1/\sqrt{5}) = y(\pm 1/\sqrt{5}), \quad h'(\pm 1/\sqrt{5}) = y'(\pm 1/\sqrt{5})$$

gives the estimate

$$W_3(12) \geq 6 + 15\sqrt{10 - 2\sqrt{5}} + 15\sqrt{10 + 2\sqrt{5}},$$

which is attained on the vertices of the icosahedron. For $d = 24$, $N = 196560$ the extremal configuration consists of the minimal vectors of the Leech lattice where

$$W_{24}(196560) = \frac{21162133021638766235178 1}{207360000000000}.$$

At the same time we cannot solve this problem even for $d = 3$ and $N = 5$, though the solution is known from numerical calculations.

Another interesting problem is to find the asymptotics of $W_d(N)$ when $N \to \infty$. One can take

$$h(t) = r^{d-2}(1 - 2rt + r^2)^{\frac{2-d}{2}}, \quad 0 < r < 1. \tag{3.2}$$

The function h belongs to K for any $r \in (0,1)$ (see [20]), whilst

$$h(t) \le y(t) = \{2(1-t)\}^{1-d/2}, \quad -1 \le t < 1,$$

for any $r \in (0,1)$. Taking r in (3.1) and (3.2) in the best way, $r = 1 - N^{-\frac{1}{d-1}}$, we get the lower estimate

$$W_d(N) \ge N(N^{1/(d-1)} - 1)^{d-1}.$$

We do not know how to construct a good upper bound for $W_d(N)$. By averaging, i.e. by using the mean value

$$MW = \frac{1}{(\text{meas } S^{d-1})^N} \int_{S^{d-1}} \cdots \int_{S^{d-1}} \frac{dx^{(1)} \ldots dx^{(N)}}{|x^{(k)} - x^{(l)}|^{d-2}} = N^2 - N,$$

we obtain

$$W_d(N) \le N^2 - N.$$

This yields

$$W_d(N) = N^2\{1 + O(N^{-1/(d-1)})\}, \quad N \to \infty.$$

But the remainder term here is poor. Best estimates can be found in [21], for example.

4 Other Manifolds

The problems discussed above have very interesting generalisations when we consider some other compact manifolds instead of the sphere. In coding theory the most popular is the Hamming space H^d. In this case there are also connections with approximation theory. For more information see [12].

In this section we are interested in extremal problems on the torus

$\mathbb{T}^d = [-1/2, 1/2)^d$. It is obvious that the trigonometric system

$\{e^{2\pi i \nu x}\}$, $\nu \in \mathbb{Z}^d$, is positive–definite. This means: for all $N \in \mathbb{N}$, $\nu \in \mathbb{Z}^d$, for all $\{\xi_k\}_{k=1}^N \subset \mathbb{C}$, $\{x^{(k)}\}_{k=1}^N \subset \mathbb{R}^d$ we get

$$\sum_{k,l=1}^N \xi_k \overline{\xi_l}\, e^{2\pi i \nu (x^{(k)} - x^{(l)})} = \left| \sum_{k=1}^N \xi_k\, e^{2\pi i \nu x^{(k)}} \right|^2 \ge 0.$$

Next let us denote by K the cone of continuous functions on $\mathbb{T}^d$ satisfying the condition

$$f(x) \sim \sum_\nu \widehat{f}_\nu e^{2\pi i \nu x}, \qquad \widehat{f}_\nu = \int_{\mathbb{T}^d} f(x) e^{-2\pi i \nu x}\, dx \geq 0,$$

where ν ranges over $\mathbb{Z}^d$.

The coding problem. Let $0 < \varepsilon \leq \sqrt{d}/2$. How many non–overlapping balls of radius ε can be displayed in $\mathbb{T}^d$? The maximum number of this kind we denote by $N_d(\varepsilon)$. In order to get upper estimates for $N_d(\varepsilon)$ we have to solve, as in Section 1, the following extremal problem from the theory of functions: Find a best one–sided approximation to the function

$$y_\varepsilon(x) = \begin{cases} 1, & |x| < 2\varepsilon; \\ 0, & |x| \geq 2\varepsilon,\ x \in \mathbb{T} \end{cases}$$

in the space $L(\mathbb{T}^d)$ by functions from the cone K, i.e. find

$$E(\varepsilon) = \inf_{f \in K,\, f(x) \leq y_\varepsilon(x)} \int_{\mathbb{T}^d} (y_\varepsilon(x) - f(x))\, dx. \tag{4.1}$$

The inequality

$$N_d(\varepsilon) \leq \frac{1}{\widehat{f}_0} \tag{4.2}$$

holds again.

For $d \geq 2$ the exact solutions of (4.1) are unknown. Using (4.2) and the function $f \in K$ constructed in [23] on the base of eigenfunctions of the Laplace operator we get the estimate

$$N_d(\varepsilon) \leq \frac{q_d\, \omega_d^2}{(4\pi\varepsilon)^d},$$

where ω_d is the volume of the unit ball in $\mathbb{R}^d$ and q_d is the first positive zero of the Bessel function $J_{d/2}(z)$. If $d = 1$ then the estimate (4.2) is best possible whenever $\varepsilon = 1/q$, $q \in \mathbb{N}$. However even in the case of a single variable the value of (4.1) is unknown for general ε. We shall demonstrate the computation of (4.1) for $\varepsilon = 1/q$, $q \in \mathbb{N}$. Since

$$f(0) \leq 1, \quad \text{and} \quad \forall j \not\equiv 0 \pmod{q} : \qquad f\left(\frac{j}{q}\right) \leq 0,$$

we have

$$J = \frac{1}{q} \sum_{j=0}^{q} f\left(\frac{j}{q}\right) \leq \frac{1}{q}.$$

On the other hand $f \in K$ and

$$J = \sum_{\substack{\nu=-\infty \\ \nu \equiv 0 \ (\mathrm{mod}\ q)}}^{\infty} \widehat{f}_\nu \geq \widehat{f}_0.$$

So $\widehat{f}_0 \leq 1/q$, and we obtain the lower estimate

$$E\left(\frac{1}{q}\right) = \inf_{f(X) \in K} (\widehat{(y_\varepsilon)}_0 - \widehat{f}_0) \geq \frac{2}{q} - \frac{1}{q} = \frac{1}{q}.$$

Two different functions are known to be extremal in this problem, namely

$$f_1(x) = \begin{cases} 1 - |qx|, & |x| \leq 1/q, \\ 0, & |x| > 1/q, \end{cases}$$

and

$$f_2(x) = \frac{1 - \cos(2\pi/q)}{q^2} \cdot \frac{\sin^2 \pi q x}{\sin^2 \pi x (\cos(2\pi x) - \cos(2\pi/q))}.$$

These functions belong to K and their mean values on the torus are equal to $1/q$. The first function has an infinite spectrum, the second function is a trigonometrical polynomial of degree $q - 2$. Moreover, the whole "segment" $\{\alpha f_1(x) + (1 - \alpha)f_2(x),\ 0 \leq \alpha \leq 1\}$ consists of extremal functions. The description of the set of all extremal functions for the problem (4.1) is an interesting but unsolved problem.

The problem of best one–sided approximation from the cone of positive definite functions is interesting not only with respect to step functions. For example the old problem (it was solved by different methods) of displaying N points on the unit circle with the greatest possible product of all possible pairwise distances can also be solved by our method. In this case we have to consider N arbitrary points $\{x^{(k)} = (\cos \varphi_k, \sin \varphi_k)\}_{k=1}^{N} \subset S^1$. As $d = 2$, the fundamental solution of the Laplace equation is $\ln |x|^{-1}$. The distance between $x^{(k)}$ and $x^{(l)}$ is equal to $2|\sin \frac{\varphi_k - \varphi_l}{2}|$. Therefore the energy functional is

$$\sum_{\substack{k,l=1 \\ k \neq l}}^{N} \ln \left| 2 \sin \frac{\varphi_k - \varphi_l}{2} \right|^{-1} = \ln \prod_{\substack{k,l=1 \\ k \neq l}}^{N} \left| 2 \sin \frac{\varphi_k - \varphi_l}{2} \right|^{-1}.$$

The problem of finding lower estimates for the energy is equivalent to the problem of finding upper estimates for the product of distances. Using the function

$$y(\varphi) = \ln |2\sin(\varphi/2)|^{-1}, \quad 0 < |\varphi| \le \pi,$$

we now have to find the value of

$$E = \inf_{f \in K,\, f(0)=\lambda \in \mathbb{R},\, f(\varphi) \le y(\varphi),\, 0 < |\varphi| \le \pi} \frac{1}{2\pi} \int_{-\pi}^{\pi} (y(\varphi) - f(\varphi))\, d\varphi.$$

We can minimize the integral by an interpolatory polynomial f with respect to the Hermite conditions

$$f\left(\frac{2k\pi}{N}\right) = y\left(\frac{2k\pi}{N}\right); \quad f'\left(\frac{2k\pi}{N}\right) = y'\left(\frac{2k\pi}{N}\right), \qquad k = 1, \ldots, N-1.$$

Again we do not know all the extremal functions.

So, we have seen how methods from approximation theory work in discrete geometry, and that there are, vice versa, many still unsolved extremal problems in the theory of functions, whose solutions would help to solve beautiful problems of discrete geometry.

Acknowledgement. The authors wish to express their thanks to M. Feigin, M. Reimer and V. Temlyakov for fruitful discussions.

References

[1]   N. N. Andreev: *An extremal property of the icosahedron*, East J. Approx. **2**, 4 (1996), 459–462.

[2]   N. N. Andreev: *Disposition of points on a sphere with minimum of energy*, Trydy Matematicheskogo Instityta Imeni V.A. Steklova **219** (1997), 27–31 (Russian); Proceedings of the Steklov Institute of Mathematics **219** (1997), 20–24.

[3]   N. N. Andreev: *A spherical code*, Russian Math. Surveys **54**, 1 (1999), 255–256 (Russian).

[4]   V. V. Arestov, A. G. Babenko: *On Delsarte scheme of estimating the contact numbers*, Trydy Matematicheskogo Instityta Imeni V. A. Steklova **219** (1997), 44–73 (Russian); Proceedings of the Steklov Institute of Mathematics **219** (1997), 44–73.

[5]   M. Berzhe: *Géométrie*, Nathan, Paris 1978.

[6]   G. Björck: *Distributions of positive mass, which maximize a certain generalized energy integral*, Ark. Math. **3** (1956), 255–269.

[7]   P. Boyvalenkov: *Extremal polynomials for obtaining bounds for spherical codes and designs*, Discrete Comput. Geom. **14** (1995), 167–183.

[8]   J. Conway, N. J. A. Sloane: *Sphere Packing, Lattices and Groups*, Springer, Berlin–New York 1988.

[9]   P. Delsarte: *Bounds for unrestricted codes by linear programming*, Philips Res. Rep. **2** (1972), 272–289.

[10]   P. Delsarte, J. M. Goethals, J. J. Seidel: *Spherical codes and designs*, Geom. Dedicata **6** (1977), 363–388.

[11]   L. Fejes Tóth: *Lagerungen in der Ebene auf der Kugel und im Raum*, Springer, Berlin–Göttingen–Heidelberg 1953.

[12]   V. I. Ivanov, O. I. Smirnov: *Jackson's constants in the space $l_2(\mathbb{Z}_2^n)$*, Trydy Matematicheskogo Instityta Imeni V. A. Steklova **219** (1997), 183–210 (Russian); Proceedings of the Steklov Institute of Mathematics **219** (1997), 183–210.

[13]   A. V. Kolushov, V. A. Yudin: *Extremal dispositions of points on the sphere*, Anal. Math. **23**, 1 (1997), 25–34.

[14]   J. Korevaar, J. L. H. Meyers: *Spherical Faraday cage for the case of equal point charges and Chebyshev-type quadrature on the sphere*, Integral Transforms and Spherical Functions, **1**, 2 (1993), 105–117.

[15]   V. I. Levenstein: *On bounds for packing of metric spaces and some of their applications*, Problemy Kibernetiki **40** (1983), 43–110 (Russian).

[16]   A. A. Markov: *Collected Papers*, Moscow 1948.

[17]   T. W. Melnyk, O. Knop, W. R. Smith: *Extremal arrangements of points and unit charges on a sphere: equilibrium configurations revised*, Canadian J. Chemistry **55**, 10 (1977), 1745–1761.

[18]   V. M. Sidelnikov: *On extremal polynomials used in bounds of the cardinality of a code*, Problemy Perdachi Informachii (Problems of Information Transmission) **16** (1980), 17–30 (Russian).

[19]   N. J. A. Sloane: *Sphere packing*, Scientific American, **250**, 1 (1984).

[20]   G. Szegö: *Orthogonal Polynomials*, Amer. Math. Soc., Providence, Rhode Island, 1959.

[21]   G. Wagner: *On means of distances on the surface of a sphere (lower bounds)*, Pacific J. Math. **144** (1990), 389–398.

[22]   L. L. Whyte: *Unique arrangements of points on a sphere*, Amer. Math. Monthly **59**, 9 (1952), 606–611.

[23]   V. A. Yudin: *Sphere packing in Euclidean space and extremal problems for trigonometrical polynomials*, Discrete Math. Appl. **1** (1991), 69–72.

[24]   V. A. Yudin: *The minimum of potential energy of a system of point charges*, Discrete Math. Appl. **3** (1993), 75–82.

Addresses:

NIKOLAY N. ANDREEV
Department of Mathematics and Mechanics
Moscow State University
Main Building
Vorob'evy Gory, Moscow 119899
Russia

VLADIMIR A. YUDIN
Department of Higher Mathematics
Moscow Institute of Power Engineering
Krasnokazarmennaya 14
Moscow 105835
Russia

Dense Vector Spaces of Universal Harmonic Functions

David H. Armitage

For Manfred Reimer on his 65th birthday, November 22nd, 1998

Abstract

A theorem of G. R. MacLane asserts that the space $\mathcal{E}$ of all (holomorphic) entire functions on $\mathbb{C}$ contains elements f having the property that the sequence $(f^{(n)})$ of derivatives is dense in $\mathcal{E}$ with the topology of local uniform convergence. We call such a function f a universal (entire) function. Recently, M. P. Aldred and the author introduced analogous universal functions in the space $\mathcal{H}_N$ of all harmonic functions on $\mathbb{R}^N$. This article briefly summarises what is known about the abundance and the growth rates of universal functions in both the holomorphic and harmonic contexts. It also includes a proof of a new result about the existence of a dense vector subspace of $\mathcal{H}_N$ all of whose elements, except 0, are universal harmonic functions of prescribed growth.

Introduction and Survey of Results

Let X, Y be topological spaces and let $\{T_\alpha : \alpha \in J\}$ be a collection of continuous mappings from X into Y. An element ξ of X is called *universal (with respect to the family $\{T_\alpha\}$)* if the set $\{T_\alpha(\xi) : \alpha \in J\}$ is dense in Y.

A classical example of universality is due to G. R. MacLane [18]. Let $\mathcal{E}$ denote the vector space of all (holomorphic) entire functions on $\mathbb{C}$, equipped with the topology of local uniform convergence, and let $D^n : \mathcal{E} \to \mathcal{E}$ be the continuous mapping given by $D^n(g) = g^{(n)}$, the nth derivative of g. MacLane showed that there exist functions f that are universal with respect to $\{D^n : n \in \mathbb{N}\}$, where $\mathbb{N} = \{0, 1, 2, \ldots\}$. We call such a function f a *universal entire function*. Thus, by definition, a function $f \in \mathcal{E}$ is a universal entire function if and only if for every $g \in \mathcal{E}$, every compact set $K \subset \mathbb{C}$, and every $\epsilon > 0$ there exists $n \in \mathbb{N}$ such that $|f^{(n)} - g| < \epsilon$ on K. A succinct proof, essentially the same as MacLane's, of the existence of universal entire functions is given by Blair and

Advances in Multivariate Approximation; W. Haußmann, K. Jetter and M. Reimer (eds.)
Mathematical Research, Vol. 107, pp. 33–42, ISBN 3-527-40236-5
© WILEY–VCH, Berlin 1999

Rubel [6] who unwittingly rediscovered it. MacLane showed in fact that there exist universal entire functions f of exponential type 1; that is to say,

$$M_\infty(f, r) := \max_{|z|=r} |f(z)| = O(e^{(1+\epsilon)r}) \quad (r \to +\infty)$$

for every $\epsilon > 0$. This result was refined by G. Herzog [16] and perfected by K.-G. Große–Erdmann [14], who proved the following theorem: if $\phi : (0, +\infty) \to (0, +\infty)$ is a function such that $\phi(t) \to +\infty$ as $t \to +\infty$, then there exists a universal entire function f such that

$$M_\infty(f, r) = O(\phi(r)r^{-1/2}e^r), \tag{0.1}$$

and this result fails if $\phi \equiv 1$. Although no explicit example of a universal entire function has ever been exhibited, such functions are very abundant: Duyos Ruiz [10] showed that in the Baire space $\mathcal{E}$ the universal entire functions form a residual set (that is, their complementary set is of the first category).

For a bibliographic survey of work on universality, see [15].

Recently, M. P. Aldred and the author [1, 2] obtained similar results in the space $\mathcal{H}_N$ of all harmonic functions on $\mathbb{R}^N$, where $N \geq 2$. To describe these results we use standard multi–index notation. Thus, if $\alpha = (\alpha_1, \ldots, \alpha_N) \in \mathbb{N}^N$, then we write

$$|\alpha| = \alpha_1 + \ldots + \alpha_N, \quad \partial^\alpha = \frac{\partial^{|\alpha|}}{\partial x_1^{\alpha_1} \ldots \partial x_N^{\alpha_N}}.$$

Taking $\mathcal{H}_N$ to be equipped with the topology of local uniform convergence, we say that a function $h \in \mathcal{H}_N$ is a *universal harmonic function* if $\{\partial^\alpha h : \alpha \in \mathbb{N}^N\}$ is dense in $\mathcal{H}_N$. In [1] we showed that the universal harmonic functions form a residual subset of $\mathcal{H}_N$ and proved an analogue, which we now describe, of Große–Erdmann's growth theorem. Let $M_2(h, r)$ denote the L^2–norm of a function $h \in \mathcal{H}_N$ on the sphere of radius r centred at the origin; that is,

$$M_2(h, r) = \left(\int_S (h(rx))^2 d\sigma(x) \right)^{1/2},$$

where σ is $(N-1)$–dimensional measure on the unit sphere S, normalized so that $\sigma(S) = 1$. The main result in [1] asserts that if $\phi : (0, +\infty) \to (0, +\infty)$ is such that $\phi(r) \to +\infty$ as $r \to +\infty$, then there exists a universal harmonic function h such that

$$M_2(h, r) = O(\phi(r)r^{-(N-1)/2}e^r), \tag{0.2}$$

and this result fails if $\phi \equiv 1$. In this paper we show that there is a large class of universal harmonic functions satisfying (0.2):

Theorem 0.1. *Let* $\phi : (0, +\infty) \to (0, +\infty)$ *be such that* $\phi(r) \to +\infty$ *as* $r \to +\infty$. *There exists a dense subspace* V *of* $\mathcal{H}_N$ *such that every element* h *of* $V \backslash \{0\}$ *is a universal harmonic function and satisfies the growth condition* (0.2).

1 Proof of Theorem 0.1

1.1 We introduce some special classes of universal harmonic functions. If J is an infinite subset of $\mathbb{N}^N$, then we say that a function $h \in \mathcal{H}_N$ is J–*universal* if $\{\partial^\alpha h : \alpha \in J\}$ is dense in $\mathcal{H}_N$. The existence of J–universal functions, for arbitrary J, was established in [2], where we also considered their growth properties; some details are given in Section 2, below. We write $J_1 = \{(\alpha_1, 0, \ldots, 0) \in \mathbb{N}^N : \alpha_1 \in \mathbb{N}\}$. If $h \in \mathcal{H}_N$, then V_h denotes the subspace of $\mathcal{H}_N$ spanned by $\{\partial^\alpha h : \alpha \in J_1\}$. The main step in the proof of Theorem 0.1 is the following result which may be of some independent interest and which is given in a little more generality than is actually required.

Theorem 1.1. *If* h *is a* J–*universal function for some infinite subset* J *of* $\mathbb{N}^N$, *then every element of* $V_h \backslash \{0\}$ *is also* J–*universal.*

1.2 For the proof of Theorem 1.1, we need some more notation and preliminary results. Let $\mathcal{H}_{j,N}$, where $j \in \mathbb{N}$, denote the space of all homogeneous harmonic polynomials of degree j on $\mathbb{R}^N$. It is well known that there is a unique element $I_{j,N}$ of $\mathcal{H}_{j,N}$ such that $I_{j,N}$ is x_1–axial (that is, $I_{j,N}(x_1, \ldots, x_N)$ depends only on x_1 and $x_2^2 + \ldots + x_N^2$) and $I_{j,N}(1, 0, \ldots, 0) = 1$; (see [7] or [4, Chapter 5]). We write $\partial_1 = \partial/\partial x_1$. Note that if $j \geq 1$, then $\partial_1 I_{j,N}$ is an x_1–axial element of $\mathcal{H}_{j-1,N}$ taking the value j at $(1, 0, \ldots, 0)$. Hence, by the uniqueness of $I_{j-1,N}$,

$$\partial_1 I_{j,N} = j I_{j-1,N}. \tag{1.1}$$

For each $p \in \mathbb{N}$, let $I_{j,N+2p}^*$ be the (not necessarily harmonic) polynomial on $\mathbb{R}^N$ given by

$$I_{j,N+2p}^*(x_1, \ldots, x_N) = I_{j,N+2p}(x_1, \ldots, x_N, 0, \ldots, 0)$$

(the number of zeros on the right–hand side being $2p$). Using (1.1) and induction, we obtain

$$\partial_1^k I_{j,N+2p}^* = \frac{j!}{(j-k)!} I_{j-k,N+2p} \quad (k = 0, 1, \ldots, j). \tag{1.2}$$

Let

$$\mathcal{H}^*_{p,N} = \{F \in \mathcal{H}_{p,N} : \partial_1 F = 0\}.$$

Kuran [17, Theorems 2, 3] showed that if $j, p \in \mathrm{I\!N}$ and $F \in \mathcal{H}^*_{p,N}$, then $FI^*_{j,N+2p}$ is a harmonic polynomial and, moreover, every harmonic polynomial H can be written as the sum of such polynomials: if H is of degree n, then

$$H = \sum_{\nu=0}^{n} \sum_{p=0}^{\nu} F_{p,\nu} I_{\nu-p,N+2p},$$

where $F_{p,\nu} \in \mathcal{H}^*_{p,N}$.

1.3 In the proof of Theorem 1.1 we use the following lemma.

Lemma 1.2. *Let $a_0, a_1, \ldots, a_m$ be real numbers, not all zero. If H is a harmonic polynomial, then there exists a harmonic polynomial G such that*

$$\sum_{k=0}^{m} a_k \partial_1^k G = H. \tag{1.3}$$

According to the foregoing remarks, it is enough to prove Lemma 1.2 in the case where $H = FI^*_{j,N+2p}$, where $j, p \in \mathrm{I\!N}$ and $F \in \mathcal{H}^*_{p,N} \setminus \{0\}$. We work at first with the additional assumption that $a_0 \neq 0$. We shall show that there exist constants $\alpha_0, \ldots, \alpha_j$ such that the function

$$G = F \sum_{\nu=0}^{j} \alpha_\nu I^*_{\nu,N+2p}$$

satisfies (1.3). If $m < j$, then we define $a_{m+1} = a_{m+2} = \ldots = a_j = 0$. Using (1.2), we obtain

$$\begin{aligned}
\sum_{k=0}^{m} a_k \partial_1^k G &= F \sum_{k=0}^{j} \sum_{\nu=k}^{j} a_k \alpha_\nu \frac{\nu!}{(\nu-k)!} I^*_{\nu-k,N+2p} \\
&= F \sum_{\nu=0}^{j} \sum_{k=0}^{j-\nu} a_k \alpha_{k+\nu} \frac{(k+\nu)!}{\nu!} I^*_{\nu,N+2p}.
\end{aligned}$$

This expression will be equal to $FI^*_{j,N+2p} (= H)$, as required, if $\alpha_0, \ldots, \alpha_j$ satisfy

$$a_0 \alpha_j = 1, \quad \frac{1}{\nu!} \sum_{k=0}^{j-\nu} a_k (k+\nu)! \alpha_{k+\nu} = 0 \quad (\nu = 0, 1, \ldots j-1).$$

The coefficient matrix for this system is triangular with all elements on the leading diagonal equal to $a_0 \neq 0$, so the equations do indeed have a solution. This completes the proof in the case where $a_0 \neq 0$.

Now suppose that $a_0 = 0$ and let η be the smallest index for which $a_\eta \neq 0$. The preceding paragraph shows that there is a harmonic polynomial G_0 such that

$$\sum_{k=\eta}^{m} a_k \partial_1^{k-\eta} G_0 = H,$$

so it is enough to verify that there is a harmonic polynomial G such that $\partial_1^\eta G = G_0$. Earlier remarks show that it suffices to treat the case where $G_0 = F_0 I_{n,N+2q}$ with $n, q \in \mathbb{N}$ and $F_0 \in \mathcal{H}_{q,N}^*$. From (1.2) we see that

$$G = \frac{n!}{(n+\eta)!} F_0 I_{n+\eta,N+2q}^*$$

has the required property.

1.4 We now prove Theorem 1.1. Let h be a J–universal function and let $Eh = \sum_{k=0}^{m} a_k \partial_1^k h$ be an element of $V_h \backslash \{0\}$; here $a_0, \ldots, a_m$ are real numbers, not all zero. We have to show that Eh is J–universal. Let $g \in \mathcal{H}_N$ and let r, ϵ be positive numbers. It is enough to show that there exists $\alpha \in J$ such that $|\partial^\alpha Eh - g| < \epsilon$ on the open ball $B(r)$ of radius r centred at the origin of $\mathbb{R}^N$. First note that there is a constant C, depending only on E, r and N, such that

$$\sup_{B(r)} |Ef| \leq C \sup_{B(2r)} |f| \qquad (f \in \mathcal{H}_N)$$

(see e.g. [4, p. 33]). Since the space of all harmonic polynomials is dense in $\mathcal{H}_N$, there exists a harmonic polynomial H such that $|H - g| < \epsilon/2$ on $B(r)$. By Lemma 1.2, there exists a harmonic polynomial G such that $EG = H$. Since h is J–universal, there exists $\alpha \in J$ such that $|\partial^\alpha h - G| \leq \epsilon/(2C)$ on $B(2r)$ and hence $|E\partial^\alpha h - EG| \leq \epsilon/2$ on $B(r)$. It now follows that on $B(r)$

$$|\partial^\alpha Eh - g| \leq |\partial^\alpha Eh - H| + |H - g| < |E\partial^\alpha h - EG| + \epsilon/2 \leq \epsilon,$$

and this completes the proof.

1.5 Now let ϕ be as in Theorem 0.1. The proof of [1, Theorem 1(i)] shows that there exists a J_1–universal function h satisfying the growth condition (0.2), where again $J_1 = \{(\alpha_1, 0, \ldots, 0) \in \mathbb{N}^N : \alpha_1 \in \mathbb{N}\}$. Fix such a function h and consider the vector space V_h. By Theorem 1.1 every element of $V_h \backslash \{0\}$ is J_1–universal. Also, V_h is dense in $\mathcal{H}_N$, since V_h contains $\{\partial_1^j h : j \in \mathbb{N}\}$ which is dense in $\mathcal{H}_N$ by the J_1–universality of h. By replacing ϕ by a smaller function, if necessary, we can arrange that $M_2(h,r) = O(e^r)$. It will follow from the next lemma that every element g of V_h satisfies the growth condition (0.2), and the proof of Theorem 0.1 will be complete.

Lemma 1.3. *If $h \in \mathcal{H}_N\backslash\{0\}$ and $M_2(h,r) = O(e^{\lambda r})$ for some positive number λ, then every function $g \in V_h$ satisfies*

$$M_2(g,r) = O(M_2(h,r)). \tag{1.4}$$

It is enough to prove the lemma in the case where $g = \partial_1 h$, for it will then follow by induction that (1.4) holds whenever $g = \partial_1^j h$ for $j \in \mathbb{N}$ and hence for all $g \in V_h$. Since $h \in \mathcal{H}_N$, we can write $h = \sum_{j=0}^{\infty} H_j$, where $H_j \in \mathcal{H}_{j,N}$ and the series converges absolutely and locally uniformly on $\mathbb{R}^N$. Moreover, we can differentiate term by term to obtain $\partial_1 h = \sum_{j=1}^{\infty} \partial_1 H_j$ (see e.g. [4, pp. 18–27]). The spaces $\mathcal{H}_{j,N}$ are mutually orthogonal in the sense that

$$\int_S H_j(rx)H_k(rx)d\sigma(x) = 0 \qquad (H_j \in \mathcal{H}_{j,N}, H_k \in \mathcal{H}_{k,N}, j \neq k)$$

for all $r > 0$ (see e.g. [4, p. 75]). Hence

$$M_2^2(h,r) = \sum_{j=0}^{\infty} M_2^2(H_j,r)$$

and

$$M_2^2(\partial_1 h,r) = \sum_{j=1}^{\infty} M_2^2(\partial_1 H_j,r).$$

An inequality of Calderón and Zygmund [8, formula (1.5.2), Chapter I, §§7,8] (or see [17, Theorem 4 and formula (5)]) shows that

$$M_2(\partial_1 H_j,r) \leq r^{-1}j\sqrt{d_j/d_{j-1}}\,M_2(H_j,r) \quad (j \geq 1, r > 0),$$

where $d_j = \dim \mathcal{H}_{j,N}$. Writing $b_j = M_2^2(H_j,1)$, we have

$$M_2^2(h,r) = \sum_{j=0}^{\infty} b_j r^{2j}$$

and

$$M_2^2(\partial_1 h,r) \;\leq\; r^{-2}\sum_{j=1}^{\infty} j^2 b_j (d_j/d_{j-1})r^{2j}$$

$$= \; O\Big(r^{-2}\sum_{j=1}^{\infty} j^2 b_j r^{2j}\Big),$$

since $d_j/d_{j-1} = O(1)$ (see e.g. the formula for d_j in [4, p. 82]). Let ψ be the entire function defined by $\psi(z) = \sum_{j=0}^{\infty} b_j z^{2j}$. By Cauchy's estimates,

$$b_j \leq r^{-2j} M_\infty(\psi, r) = r^{-2j} \psi(r) = r^{-2j} M_2^2(h, r) \leq A r^{-2j} e^{2\lambda r}$$

for all $j \geq 1, r > 0$ and some constant A independent of j and r. Taking $r = j/\lambda$, we find that $b_j = O((\lambda e/j)^{2j})$, and hence

$$
\begin{aligned}
M_2^2(\partial_1 h, r) &\leq O\Big(r^{-2} \sum_{1 \leq j \leq 2\lambda er} j^2 b_j r^{2j}\Big) + O\Big(r^{-2} \sum_{j > 2\lambda er} j^2 (\lambda er/j)^{2j}\Big) \\
&\leq O\Big(\sum_{j=0}^{\infty} b_j r^{2j}\Big) + o(r^{-2}) \\
&= O\big(M_2^2(h, r)\big) + o(1).
\end{aligned}
$$

2 Remarks and Questions

2.1 The holomorphic analogue of Theorem 0.1 seems not to have been recorded hitherto. We state it here and briefly indicate its proof, which follows the same pattern as that of Theorem 0.1 but is technically simpler.

Theorem 2.1. *Let $\phi : (0, +\infty) \to (0, +\infty)$ be such that $\phi(r) \to +\infty$ as $r \to +\infty$. There exists a dense subspace U of $\mathcal{E}$ such that every element f of $U \backslash \{0\}$ is a universal entire function and satisfies the growth condition* (0.1).

A holomorphic analogue of Lemma 1.2 states that if P is a (holomorphic) polynomial and $a_0, \ldots, a_m$ are complex numbers, not all zero, then there exists a polynomial Q such that $\sum_{k=0}^{m} a_k Q^{(k)} = P$. This is easy to prove when $m = 0$ or 1, and the general result then follows since $\sum_{k=0}^{m} a_k D^k$ can be expressed as a product of first degree operators. Now let f be a universal entire function and let U_f be the space spanned by $\{f, f', f'', \ldots\}$. An argument mimicking the proof of Theorem 1.1 shows that every element of $U_f \backslash \{0\}$ is universal, and clearly U_f is dense in $\mathcal{E}$. Finally, if f is chosen so that

$$M_\infty(f, r) = O(\phi(r-1) r^{-1/2} e^r),$$

then Cauchy's estimates can be used to show that every element of U_f satisfies (0.1).

2.2 Here we summarise what is known about J–universal functions and mention some open questions. In [2] we showed in the case $N = 2$ that if J is an arbitrary infinite subset of $\mathbb{N}^2$, then there exist J-universal functions of exponential type 1. This is the best possible result concerning the exponential type of such functions: there is no universal harmonic function of exponential type less than 1. In all dimensions, we also have precise knowledge about growth rates in the special case where $J = J_1$: the proof of Theorem 0.1 shows that we can take $V \setminus \{0\}$ to consist of J_1-universal functions satisfying (0.2). However, there are gaps in our knowledge in the case where $N \geq 3$ and J is arbitrary. It is not always true that J-universal functions of exponential type 1 exist. For instance (see [2]) if $J = \{(\alpha, \ldots, \alpha) \in \mathbb{N}^N : \alpha \in \mathbb{N}\}$, then there is no J-universal function of exponential type less than $\sqrt{N/2}$. On the other hand, if $N \geq 3$ and J is arbitrary, then there exists a J-universal function of exponential type at most

$$c_N := N \left(\prod_{j=1}^{N-1} \frac{(2j)^{2j}}{(2j+1)^{2j+1}} \right)^{1/(2N)} .$$

We note that $0 < c_N - \sqrt{N/2} \to 0$ as $N \to \infty$ and conjecture that J-universal functions of exponential type $\sqrt{N/2}$ always exist. Another open question concerns the possible extension of Theorem 1.1 to J-universal functions: is there always a dense vector subspace V_J of $\mathcal{H}_N$ such that $V_J \setminus \{0\}$ consists of J-universal functions? – If so, can we arrange for the elements of V_J to satisfy constraints on their growth rates?

2.3 We say that an entire function f is *T-universal* if the set of translates $\{z \mapsto f(z + a) : a \in \mathbb{C}\}$ is dense in $\mathcal{E}$. Similarly, an element h of $\mathcal{H}_N$ is called *T-universal* if the set of translates $\{x \mapsto h(x + y) : y \in \mathbb{R}^N\}$ is dense in $\mathcal{H}_N$. The existence of T-universal functions was established by G. D. Birkhoff [5] and O. P. Dzagnidze [11] in the holomorphic and harmonic contexts respectively. Modern theorems about harmonic (or holomorphic) tangential approximation on unbounded sets can be used to give very short existence proofs for T-universal functions (see [13, pp. 248-249], [12, pp. 116–117], [3, pp. 5-6]), and once existence is proved, it is not difficult to deduce that the T-universal functions form residual sets in their respective spaces $\mathcal{E}$ and $\mathcal{H}_N$ (see [10] for the holomorphic case). Although T-universal entire functions, in contrast with universal functions of MacLane type, can have arbitrarily slow transcendental growth (Duyos Ruiz [9]), there appears to be nothing known about the permissible growth rates of T-universal harmonic functions.

References

[1] M. P. Aldred, D. H. Armitage: *Harmonic analogues of G. R. MacLane's universal functions*, J. London Math. Soc. (2) **57** (1998), 148–156.

[2] M. P. Aldred, D. H. Armitage: *Harmonic analogues of G. R. MacLane's universal functions II*, J. Math. Anal. Appl. **220** (1998), 382–395.

[3] D. H. Armitage, P. M. Gauthier: *Recent developments in harmonic approximation, with applications*, Results in Math. **29** (1996), 1–15.

[4] S. Axler, P. Bourdon, W. Ramey: *Harmonic Function Theory*, Springer–Verlag, New York 1992.

[5] G. D. Birkhoff: *Démonstration d'un théorème élémentaire sur les fonctions entières*, C.R. Acad. Sci. Paris **189** (1929), 473–475.

[6] C. E. Blair, L. A. Rubel: *A universal entire function*, Amer. Math. Monthly **90** (1983), 331–332; addendum 732.

[7] M. Brelot, G. Choquet: *Polynômes harmoniques et polyharmoniques*, Seconde Colloque sur les équations aux dérivées partielles, Bruxelles, 45–55, Georges Thone, Liège; Masson et Cie, Paris, 1955.

[8] A.P. Calderón, A. Zygmund: *On higher gradients of harmonic functions*, Studia Math. **24** (1964), 211–226.

[9] S. M. Duyos Ruiz: *On the existence of universal functions*, Soviet Math. Dokl. **27** (1983), 9–13.

[10] S. M. Duyos Ruiz: *Universal functions and the structure of the space of entire functions*, Soviet Math. Dokl. **30** (1984), 713–716.

[11] O. P. Dzagnidze: *The universal harmonic function in the space E_n*, (Russian), Skharth. SSR Mecn. Akad. Moambe **55** (1969), 41–44.

[12] S. J. Gardiner: *Harmonic Approximation*, London Math. Soc. Lect. Notes **221**, Cambridge Univ. Press, 1995.

[13]   P. M. Gauthier: *Uniform approximation,* Complex potential theory (Proc. NATO Adv. Study Inst., Montreal, 1993), 235–271, Kluwer Acad. Publ., Dordrecht 1994.

[14]   K.-G. Große–Erdmann: *On the universal functions of G. R. MacLane,* Complex Variables Theory Appl. **15** (1990), 193–196.

[15]   K.-G. Große–Erdmann: *Universal families and hypercyclic operators,* Bull. Amer. Math. Soc., to appear.

[16]   G. Herzog: *Universelle Funktionen,* Diplomarbeit, Universität Karlsruhe 1988.

[17]   Ü. Kuran: *On Brelot–Choquet axial polynomials,* J. London Math. Soc. (2) **4** (1971), 15–26.

[18]   G. R. MacLane: *Sequences of derivatives and normal families,* J. Analyse Math. **2** (1952), 72–87.

Address:

DAVID H. ARMITAGE
Department of Pure Mathematics
Queen's University
Belfast BT7 1NN
Northern Ireland

Best One–sided L^1–Approximation by Harmonic and Subharmonic Functions

David H. Armitage and Stephen J. Gardiner

To Manfred Reimer on his 65th birthday

Abstract

Let Ω be a bounded open set in $\mathbb{R}^n$ and let $U \subseteq C(\overline{\Omega})$. If $f \in C(\overline{\Omega})$, then we define $\mathcal{U}(f) = \{u \in U : u \leq f \text{ on } \overline{\Omega}\}$. A function u^* in $\mathcal{U}(f)$ is said to be a best one–sided L^1–approximant to f from U if $\|f - u^*\|_1 \leq \|f - u\|_1$ for every u in $\mathcal{U}(f)$, where $\| \cdot \|_1$ denotes the usual L^1–norm. This paper presents characterizations of such best approximants in the cases where U consists of functions harmonic or subharmonic on Ω. In the case of harmonic functions, Ω is assumed to be a quadrature domain.

Introduction

Let Ω be a bounded open set in Euclidean space $\mathbb{R}^n$, where $n \geq 2$, and let U be a subset of $C(\overline{\Omega})$, the collection of all real–valued continuous functions on $\overline{\Omega}$. If $f \in C(\overline{\Omega})$, then we define

$$\mathcal{U}(f) = \{u \in U : u \leq f \text{ on } \Omega\}.$$

Let

$$\|g\|_1 = \int_{\overline{\Omega}} |g|\, d\lambda \qquad (g \in C(\overline{\Omega})),$$

where λ denotes n–dimensional Lebesgue measure. A function u^* in $\mathcal{U}(f)$ is said to be a *best one–sided L^1–approximant to f from U* if

$$\|f - u^*\|_1 \leq \|f - u\|_1 \qquad \text{for every } u \in \mathcal{U}(f).$$

An introduction to this kind of L^1–approximation may be found in Pinkus [11, Chapter 5], which emphasizes mainly the case where U is a finite–dimensional

Advances in Multivariate Approximation; W. Haußmann, K. Jetter and M. Reimer (eds.)
Mathematical Research, Vol. 107, pp. 43–56, ISBN 3-527-40236-5
© WILEY–VCH, Berlin 1999

vector space. In this paper we consider the cases where U consists of functions harmonic or subharmonic on Ω. For a recent survey of the literature on best harmonic approximation in the uniform and L^1 norms we refer to [2].

We begin with a general result which will be proved in § 1. For any $g \in C(\overline{\Omega})$ we define $Z(g) = \{x \in \overline{\Omega} : g(x) = 0\}$.

Proposition 1. *Let U be a subspace of $C(\overline{\Omega})$ containing the constant functions, let $f \in C(\overline{\Omega})$ and $u^* \in \mathcal{U}(f)$, and let $Z = Z(f - u^*)$. The following are equivalent:*

(a) *u^* is a best one–sided L^1–approximant to f from U;*

(b) *$\int_{\overline{\Omega}} u \, d\lambda \leq 0$ whenever $u \in U$ and $u \leq 0$ on Z;*

(c) *there is a measure μ with support in Z such that $\int_{\overline{\Omega}} u \, d\lambda = \int_Z u \, d\mu$ for all u in U.*

In the case where $U = \mathcal{H}$, the vector space of members of $C(\overline{\Omega})$ which are harmonic on Ω, the equivalence of (b) and (c) above is essentially given in [8, Proposition 2.2]. Inspired by condition (c) we now specialize to the case where $U = \mathcal{H}$ and Ω is a quadrature domain associated with a measure ν; that is, Ω is a domain (connected open set), ν has compact support $K_\nu \subset \Omega$ and $\int_\Omega h \, d\lambda = \int_{K_\nu} h \, d\nu$ for every λ–integrable harmonic function h on Ω. Let $B(x, r)$ denote the open ball in $\mathbb{R}^n$ with centre x and radius r. The simplest example of a quadrature domain is where $\Omega = B(0, 1)$ and ν is a point measure at 0 with mass equal to $\lambda(\Omega)$. However, there are many examples: if ν is a measure with compact support K_ν contained in $B(0, r)$ and if $\nu(K_\nu) \geq \lambda(B(0, 2r))$, then there is a unique quadrature domain Ω for ν and, further, $\partial\Omega$ is smooth (see Sakai [12, Theorem 2]).

In order to state the next result we recall that the *fine topology* on $\mathbb{R}^n$ is the coarsest topology for which all subharmonic functions are continuous. It is strictly finer than the Euclidean topology. A set E is said to be *thin* at a point y if y is not a fine limit point of E, that is, if there is a fine neighbourhood of y disjoint from $E \setminus \{y\}$. An account of these concepts may be found in Helms [9, Chapter 10]. We will assume that

$$\mathbb{R}^n \setminus K_\nu \text{ is nowhere thin} \tag{1}$$

(or, equivalently, K_ν has empty fine interior) and

$$B \backslash K_\nu \text{ is connected for any open ball } B. \qquad (2)$$

Finally, if L is a compact set in $\overline{\Omega}$, we denote by $\hat{L}$ the union of L with the (connected) components of $\mathbb{R}^n \backslash L$ which are contained in Ω.

Theorem 1. *Let Ω be a quadrature domain associated with a measure ν such that $\partial \overline{\Omega} = \partial \Omega$ and such that (1) and (2) hold, let $U = \mathcal{H}$, let $f \in C(\overline{\Omega})$ and $h^* \in \mathcal{U}(f)$, and let $Z = Z(f - h^*)$. Then h^* is a best one–sided L^1–approximant to f from $\mathcal{H}$ if and only if $K_\nu \subseteq \hat{Z}$.*

The special case of Theorem 1 where $\Omega = B(0,1)$ and ν is a point measure at 0 was given in [1], which also treated the (more difficult) case of an infinite cylinder. We note from this special case that the "only if" part of Theorem 1 fails if $\Omega = B(0,1)$ and ν is either Lebesgue measure on $B(0,r)$ or surface measure on $\partial B(0,r)$, where $0 < r < 1$, normalized in each case so that $\nu(K_\nu) = \lambda(\Omega)$. Thus neither (1) nor (2) can be dropped. However, the proof of the "if" part of the theorem does not depend on these conditions. Further discussion of (1) and (2) may be found in § 5.

Example 1. Let $a > 0$, let Ω denote the prolate ellipsoid

$$\left\{ (x', t) \in \mathbb{R}^{n-1} \times \mathbb{R} : \frac{|x'|^2}{a^2} + \frac{t^2}{1 + a^2} < 1 \right\},$$

and let ν be the measure with support $K_\nu = \{0\}^{n-1} \times [-1, 1]$ defined by

$$\int_{K_\nu} f \, d\nu = \lambda(\Omega) \left\{ \int_{-1}^{1} (1 - t^2)^{(n-1)/2} \, dt \right\}^{-1} \int_{-1}^{1} f(0', t)(1 - t^2)^{(n-1)/2} \, dt$$

for $f \in C(K_\nu)$. Then Ω is a quadrature domain associated with ν. (This can be proved by taking the limit as $\kappa \to -a^2$ in the equation

$$\frac{1}{\lambda(\Omega)} \int_{\Omega} h \, d\lambda = \frac{1}{\lambda(\omega_\kappa)} \int_{\omega_\kappa} h \, d\lambda$$

where $h \in \mathcal{H}$ and ω_κ is the confocal ellipsoid given by

$$\omega_\kappa = \left\{ (x', t) \in \mathbb{R}^{n-1} \times \mathbb{R} : \frac{|x'|^2}{a^2 + \kappa} + \frac{t^2}{1 + a^2 + \kappa} < 1 \right\} \qquad (-a^2 < \kappa < 0);$$

see [4, p. 753]). When $n \geq 3$, conditions (1) and (2) hold, so Theorem 1 applies. When $n = 2$, condition (2) fails: it is an open question whether (with the notation of Theorem 1) the inter–focal line segment K_ν is necessarily contained in $\widehat{Z}$ whenever h^* is a best one–sided L^1–approximant from $\mathcal{H}$ to a function $f \in C(\overline{\Omega})$.

We now consider the case where $U = \mathcal{S}$, the collection of members of $C(\overline{\Omega})$ which are subharmonic on Ω. Thus U is a convex cone rather than a subspace of $C(\overline{\Omega})$. We will deal with arbitrary bounded open sets Ω, and we define

$$\check{\Omega} = \{x \in \mathbb{R}^n : B(x, r)\backslash\Omega \text{ is a polar set for some } r > 0\}.$$

(We recall that a set E is called polar if there is a subharmonic function on $\mathbb{R}^n$ which is valued $-\infty$ on E.) Clearly $\check{\Omega}$ is open, $\Omega \subseteq \check{\Omega}$ and $\check{\Omega}\backslash\Omega$ is a polar set. In order to state the next result we recall that Ω is said to be *regular* for the Dirichlet problem if, for every f in $C(\partial\Omega)$, there exists h in $\mathcal{H}$ such that $h = f$ on $\partial\Omega$. This is equivalent to the condition that $\mathbb{R}^n\backslash\Omega$ is non–thin at each point of $\partial\Omega$. (See [9, Chapters 8, 10] for further details.)

Theorem 2. *Let $U = \mathcal{S}$, let $f \in C(\overline{\Omega})$ and $s^* \in \mathcal{U}(f)$, and let $Z = Z(f - s^*)$.*

(a) *The function s^* is a best one–sided L^1–approximant to f from $\mathcal{S}$ if and only if $\partial(\check{\Omega}) \subseteq Z$ and s^* is harmonic on $\Omega\backslash Z$.*

(b) *If there exists a best one–sided L^1–approximant to f from $\mathcal{S}$, then it is unique.*

(c) *The following are equivalent:*

    (i) *each member of $C(\overline{\Omega})$ has a best one–sided L^1–approximant from $\mathcal{S}$;*

    (ii) *$\check{\Omega}$ is regular for the Dirichlet problem.*

Theorems 1 and 2 will be proved in §§ 2, 3 respectively. The question of uniqueness of best approximants in Theorem 1 is discussed in § 4.

1   Proof of Proposition 1

1.1 Suppose that (c) holds and let $u \in \mathcal{U}(f)$. Then $u \leq f = u^*$ on Z, and hence $\int_{\overline{\Omega}} u^* \, d\lambda \geq \int_{\overline{\Omega}} u \, d\lambda$ by (c). So

$$\|f - u\|_1 = \int_{\overline{\Omega}} f \, d\lambda - \int_{\overline{\Omega}} u \, d\lambda \geq \int_{\overline{\Omega}} f \, d\lambda - \int_{\overline{\Omega}} u^* \, d\lambda = \|f - u^*\|_1,$$

and hence (a) holds.

1.2 Suppose that (a) holds. To prove (b), let $u \in U$ be such that $u \leq 0$ on Z and suppose, for the sake of contradiction, that $\int_{\overline{\Omega}} u \, d\lambda > 0$. If we define

$$v(x) = u(x) - \{2\lambda(\overline{\Omega})\}^{-1} \int_{\overline{\Omega}} u \, d\lambda \qquad (x \in \overline{\Omega}),$$

then $v \in U$ and $v < 0$ on Z, and

$$\int_{\overline{\Omega}} v \, d\lambda = 2^{-1} \int_{\overline{\Omega}} u \, d\lambda > 0. \tag{3}$$

Let $F = \{x \in \overline{\Omega} : v(x) \geq 0\}$. Then F is compact and disjoint from Z, so there exists a positive number δ such that $f - u^* > \delta$ on F. Let

$$w(x) = u^*(x) + \frac{\delta v(x)}{\sup_{\overline{\Omega}} v} \qquad (x \in \overline{\Omega}).$$

Then $w \leq u^* + \delta < f$ on F and $w < u^* \leq f$ on $\overline{\Omega} \backslash F$. Hence $w \in \mathcal{U}(f)$ and

$$\|f - w\|_1 = \int_{\overline{\Omega}} f \, d\lambda - \int_{\overline{\Omega}} w \, d\lambda < \int_{\overline{\Omega}} f \, d\lambda - \int_{\overline{\Omega}} u^* \, d\lambda = \|f - u^*\|_1,$$

by (3). This contradicts (a). Thus $\int_{\overline{\Omega}} u \, d\lambda \leq 0$, and (b) is established.

1.3 Suppose that (b) holds and let $V = \{u|_Z : u \in U\}$. Given $v \in V$, we define $L(v) = \int_{\overline{\Omega}} u \, d\lambda$, where u is a member of U such that $u = v$ on Z. We note that the choice of u here is immaterial since, if $u_1, u_2 \in U$ and $u_1 = v = u_2$ on Z, then (b) shows that

$$\int_{\overline{\Omega}} (u_1 - u_2) \, d\lambda \leq 0 \leq \int_{\overline{\Omega}} (u_1 - u_2) \, d\lambda.$$

Thus L is well–defined. Let $\|v\| = \sup_Z |v|$ for each v in V. Since $-\|v\| \leq v \leq \|v\|$ on Z, it follows from (b) that $|L(v)| \leq \lambda(\overline{\Omega})\|v\|$. Hence L is a bounded linear functional on V and, by the Hahn–Banach theorem, extends to a bounded linear functional $\overline{L}$, with the same norm, on $C(Z)$. Since $\|\overline{L}\| = \|L\| = \lambda(\overline{\Omega}) = L(1)$, it follows that $\overline{L}$ is positive. Hence, by the Riesz Representation Theorem, there is a measure μ with support in Z such that $\overline{L}(u) = \int_Z u\, d\mu$ for all u in $C(Z)$. In particular,

$$\int_{\overline{\Omega}} u\, d\lambda = \overline{L}(u|_Z) = \int_Z u\, d\mu \qquad \text{for all } u \text{ in } U,$$

and (c) is established.

2 Proof of Theorem 1

2.1 Suppose that $K_\nu \subseteq \hat{Z}$. If $u \in \mathcal{H}$ and $u \leq 0$ on Z, then $u \leq 0$ on K_ν by the maximum principle, and so

$$\int_{\overline{\Omega}} u\, d\lambda = \int_\Omega u\, d\lambda = \int_{K_\nu} u\, d\nu \leq 0$$

(we note from [10, pp. 18 ff.] that $\lambda(\partial\Omega) = 0$ because Ω is a quadrature domain). It follows from Proposition 1 that h^* is a best one–sided L^1–approximant to f from $\mathcal{H}$.

2.2 Conversely, suppose that h^* is a best one–sided L^1–approximant to f from $\mathcal{H}$. By Proposition 1 there is a measure μ with support in Z such that $\int_{\overline{\Omega}} h\, d\lambda = \int_Z h\, d\mu$ for all h in $\mathcal{H}$. Let

$$h_x(y) = \begin{cases} \log 1/|x - y| & (n = 2) \\ |x - y|^{2-n} & (n \geq 3) \end{cases}$$

for each $x, y \in \mathbb{R}^n$. Then, since Ω is a quadrature domain associated with ν,

$$\int_Z h_x(y)\, d\mu(y) = \int_\Omega h_x(y)\, d\lambda(y) = \int_{K_\nu} h_x(y)\, d\nu(y) \qquad (4)$$

for $x \in \mathbb{R}^n \backslash \overline{\Omega}$.

Now suppose, for the sake of contradiction, that $K_\nu\backslash\hat{Z} \neq \emptyset$. Let $x_0 \in K_\nu\backslash\hat{Z}$ and let ω be the component of $\mathbb{R}^n\backslash\hat{Z}$ which contains x_0. Then $\omega\backslash\Omega \neq \emptyset$ by the definition of $\hat{Z}$. Hence $\omega\backslash\overline{\Omega} \neq \emptyset$, since we have assumed that $\partial\overline{\Omega} = \partial\Omega$. We can now choose a finite sequence of open balls $B_1,\ldots,B_m$ in ω such that $x_0 \in B_1$, $B_i \cap B_{i+1} \neq \emptyset$ $(i = 1,\ldots,m-1)$ and $B_m\backslash\overline{\Omega} \neq \emptyset$. It follows from (1) that $(B_i \cap B_{i+1})\backslash K_\nu \neq \emptyset$ for each $i \in \{1,\ldots,m-1\}$. Also, $B_i\backslash K_\nu$ is connected for each i, in view of (2). Hence $(\cup_i B_i)\backslash K_\nu$ is a connected open subset of $\mathbb{R}^n\backslash(Z \cup K_\nu)$ which intersects $\mathbb{R}^n\backslash\overline{\Omega}$. Since the functions

$$H_1(x) = \int_Z h_x(y)\, d\mu(y) \text{ and } H_2(x) = \int_{K_\nu} h_x(y)\, d\nu(y)$$

are both harmonic on this connected set, it follows from (4) that $H_1 = H_2$ on $B_1\backslash K_\nu$, and hence on B_1, by (1). This is impossible since H_1 is harmonic on a neighbourhood of $x_0 \in K_\nu$, but H_2 is not.

3 Proof of Theorem 2

3.1 Suppose that $\partial(\check{\Omega}) \subseteq Z$ and that s^* is harmonic on $\Omega\backslash Z$. Further, let $s \in \mathcal{U}(f)$. Then, by a standard extension result (see [9, Theorem 7.7]), s and s^* are both subharmonic on $\check{\Omega}$ and s^* is harmonic on $\check{\Omega}\backslash Z$. On Z we have $s \leq f = s^*$. Since $\partial(\check{\Omega}\backslash Z) \subseteq \partial(\check{\Omega}) \cup \partial Z \subseteq Z$, it follows from the maximum principle that $s \leq s^*$ on $\check{\Omega}\backslash Z$, and thus $s \leq s^*$ on the closure of $\check{\Omega}$, that is, on $\overline{\Omega}$. Hence s^* is a (in fact, the) best one–sided L^1–approximant to f from $\mathcal{S}$.

3.2 Conversely, suppose that s^* is a best one–sided L^1–approximant to f from $\mathcal{S}$. Suppose further, for the sake of contradiction, that $\partial(\check{\Omega})$ is not a subset of Z, and let $y \in \partial(\check{\Omega})\backslash Z$. Let $r > 0$ be such that $\overline{B(y,2r)} \cap Z = \emptyset$. Then, by the definition of $\check{\Omega}$, the compact set $\overline{B(y,r)}\backslash\Omega$ is non–polar. Thus there is a non–zero measure μ on this set such that the corresponding potential u in $B(y,2r)$ is bounded above by 1. By suitably restricting μ it can be arranged that u is continuous on $B(y,2r)$ (see [9, Theorem 6.21]), and u can be extended to a continuous function on $\mathbb{R}^n$ by assigning it the value 0 on $\mathbb{R}^n\backslash B(y,2r)$. Since u is subharmonic outside $\overline{B(y,r)}\backslash\Omega$, it follows that $u \in \mathcal{S}$. Let $s = s^* + au$, where

$$a = \inf\{f(x) - s^*(x) : x \in \overline{B(y,2r)} \cap \overline{\Omega}\}.$$

Then

$$s(x) \leq f(x) + a\{u(x) - 1\} \leq f(x) \qquad (x \in \overline{B(y, 2r)} \cap \overline{\Omega})$$

and

$$s(x) = s^*(x) \leq f(x) \qquad (x \in \overline{\Omega} \backslash \overline{B(y, 2r)}),$$

so $s \in \mathcal{U}(f)$. Clearly $s \geq s^*$ on Ω. Further, since $\overline{B(y, 2r)} \cap Z = \emptyset$, it follows that $a > 0$, so $s > s^*$ on the non–empty open set $B(y, 2r) \cap \Omega$. This contradicts the assumption that s^* is a best one–sided L^1–approximant to f from $\mathcal{S}$. We thus conclude that $\partial(\check{\Omega}) \subseteq Z$.

Next suppose, again for the sake of contradiction, that s^* fails to be harmonic on the open set $\Omega \backslash Z$. Then there is a point z in $\Omega \backslash Z$ such that s^* is not harmonic on any neighbourhood of z. Let $k_0 \in \mathbb{N}$ be sufficiently large so that $\overline{B(z, 1/k_0)} \subseteq \Omega \backslash Z$. For each $k \geq k_0$ let s_k be the function defined to be equal to s^* on $\overline{\Omega} \backslash B(z, 1/k)$ and extended to all of $\overline{\Omega}$ by solving the Dirichlet problem on $B(z, 1/k)$ with boundary data s^*. Then $s_k \in \mathcal{S}$ and $s_k \downarrow s^*$ on $\overline{\Omega}$. Let

$$b = \inf\{f(x) - s^*(x) : x \in \overline{B(z, 1/k_0)}\}.$$

Then $b > 0$ so, by Dini's theorem, there exists $k_1 \geq k_0$ such that

$$s_{k_1}(x) - s^*(x) < b \qquad (x \in \overline{B(z, 1/k_0)}).$$

It follows that $s_{k_1} < f$ on $B(z, 1/k_1)$ and $s_{k_1} = s^* \leq f$ on $\overline{\Omega} \backslash B(z, 1/k_1)$, so $s_{k_1} \in \mathcal{U}(f)$. Since s^* is not harmonic on $B(z, 1/k_1)$, we see that $s_{k_1} > s^*$ on this ball. This contradicts the assumption that s^* is a best one–sided L^1–approximant to f from $\mathcal{S}$. Hence s^* must be harmonic on $\Omega \backslash Z$.

This completes the proof of part (a) of Theorem 2.

3.3 To prove part (b), let s^* be a best one–sided L^1–approximant to f from $\mathcal{S}$, and let $s \in \mathcal{U}(f)$. Then $\max\{s^*, s\} \in \mathcal{U}(f)$ and so $s \leq s^*$. Hence $s^* = \sup\{s : s \in \mathcal{U}(f)\}$ and so s^* is uniquely determined.

3.4 To prove part (c), suppose first that (ii) holds and let $g \in C(\overline{\Omega})$. Let

$$\mathcal{U}^{\check{}}(g) = \{u \in C(\overline{\Omega}) : u \text{ is subharmonic on } \check{\Omega} \text{ and } u \leq g \text{ on } \overline{\Omega}\}.$$

By the subharmonic extension result used earlier, $\mathcal{U}(g) = \mathcal{U}^{\smile}(g)$. Further, the regularity of $\check{\Omega}$ ensures that the function

$$x \mapsto \sup\{s(x) : s \in \mathcal{U}^{\smile}(g)\}$$

is continuous on the closure of $\check{\Omega}$, that is, on $\overline{\Omega}$ (see [13, Theorems 3, 4]). It follows easily that this supremum is subharmonic on Ω and so is the best one–sided L^1–approximant to g from $\mathcal{S}$. Hence (i) holds.

Conversely, suppose that (i) holds and define

$$g(x) = - \operatorname{dist}(x, \mathrm{I\!R}^n \backslash \check{\Omega}) \qquad (x \in \overline{\Omega}).$$

If s^* is the best one–sided L^1–approximant to g from $\mathcal{S}$, then s^* is subharmonic on $\check{\Omega}$ and $s^* < 0$ on $\check{\Omega}$. By (a), $s^* = g = 0$ on $\partial(\check{\Omega})$. Therefore $-s^*$ is a barrier for every boundary point of $\check{\Omega}$, and so $\check{\Omega}$ is regular for the Dirichlet problem (see [9, Theorem 8.22]).

4 The Question of Uniqueness in Theorem 1

4.1 In Theorem 1, a best one–sided L^1–approximant to f from $\mathcal{H}$ may be very far from unique, as the following example shows.

Example 2. If $\Omega = B(0,1) \subseteq \mathrm{I\!R}^2$ and $f(x) = (4/\pi)\tan^{-1}|x|$ on $\overline{\Omega}$, then every $h \in \mathcal{H}$ with $h(0) = 0$ and $|h| \leq 1$ on $\partial\Omega$ is a best one–sided L^1–approximant to f from $\mathcal{H}$.

To verify this, we note from [3, Theorems 6.16 and 6.19] that $h(x) \leq f(x)$ on $\overline{\Omega}$, and clearly $h(0) = f(0)$. Thus, by Theorem 1 (with ν a point measure at 0), h is a best one–sided L^1–approximant to f from $\mathcal{H}$.

4.2 Suppose that, in the notation of Theorem 1,

$$K_\nu \subseteq \hat{Z} \quad \text{and} \quad K_\nu \cap (\hat{Z})^\circ \neq \emptyset. \tag{5}$$

Then we claim that h^* is the *unique* best one–sided L^1–approximant to f from $\mathcal{H}$. To see this, suppose that h is another best one–sided L^1–approximant to f from $\mathcal{H}$ and choose x in $K_\nu \cap (\hat{Z})^\circ$. Then $h \leq f = h^*$ on Z and hence, by the maximum principle, $h \leq h^*$ on $\hat{Z}$ and therefore on K_ν. If $h < h^*$ at some point of K_ν, and hence on some set of positive ν–measure, then

$$\|f - h\|_1 - \|f - h^*\|_1 = \int_\Omega (h^* - h)\, d\lambda = \int_{K_\nu} (h^* - h)\, d\nu > 0,$$

contrary to our assumption that h is a best one–sided L^1–approximant to f from $\mathcal{H}$. Hence $h = h^*$ on K_ν, and in particular at x, so $h = h^*$ on the component of $(\hat{Z})^\circ$ which contains x and thus on all of Ω.

4.3 In fact, the second hypothesis in (5) can be relaxed: the argument in [1, § 2.3] shows that $(\hat{Z})^\circ$ can be replaced here by the fine interior of $\hat{Z}$ (a larger set). When $n = 2$ a further relaxation is possible, as we now indicate. For a Borel subset E of $\mathbb{R}^n$ and a point x we define

$$D(E, x, r) = \frac{\lambda(E \cap B(x, r))}{\lambda(B(x, r))} \qquad (r > 0).$$

Now suppose that $n = 2$ and, using the notation of Theorem 1, that $K_\nu \subset \hat{Z}$ and

$$\limsup_{r \to 0+} D(\hat{Z}, x, r) > \frac{1}{2} \tag{6}$$

for some x in K_ν. Then we claim that h^* is the unique best one–sided L^1–approximant to f from $\mathcal{H}$. The proof depends on the following simple lemma.

Lemma 1. *Suppose that H is harmonic on a domain in $\mathbb{R}^2$ containing a point x and let $E_+(H) = \{y : H(y) > 0\}$. If $H(x) = 0$ but $H \not\equiv 0$, then $D(E_+(H), x, r) \to 1/2$ as $r \to 0+$.*

In proving the lemma we may suppose that $x = 0$. In some neighbourhood of 0, we can write H as a sum $\sum_{j=m}^\infty H_j$, where H_j is a homogeneous harmonic polynomial of degree $j \geq 1$ and $H_m \not\equiv 0$. Then $H(y) = H_m(y) + O(|y|^{m+1})$ as $y \to 0$, and it follows that $D(E_+(H), 0, r)/D(E_+(H_m), 0, r) \to 1$ as $r \to 0+$. In polar co–ordinates, $H_m(\rho, \theta) = a\rho^m \cos(m\theta + \alpha)$ for some constants a and α, where $a \neq 0$, and so $D(E_+(H_m), 0, r) = 1/2$ for all $r > 0$. The lemma now follows.

To complete the proof of the above claim, suppose that (6) holds for some x in K_ν, let h be another best one–sided L^1–approximant to f from $\mathcal{H}$, and define $H = h - h^*$. As in § 4.2, we see that $h \leq h^*$ on $\hat{Z}$ with equality on K_ν. Thus $H(x) = 0$ and $E_+(H) \subseteq \mathbb{R}^2 \setminus \hat{Z}$. Hence

$$\liminf_{r\to 0+} D(E_+(H), x, r) < \frac{1}{2},$$

by (6), and it follows from Lemma 1 that $H \equiv 0$; that is, $h = h^*$ as required.

The above observation is sharp in the sense that it fails if equality is permitted in (6). To see this, consider the function f defined on the closed unit disc by $f(x_1, x_2) = |x_1|$ and note that both of the functions $(x_1, x_2) \mapsto \pm x_1$ are best one–sided L^1–approximants to f from $\mathcal{H}$.

We remark that Lemma 1 is false in higher dimensions. For example, if H is the harmonic polynomial defined on $\mathbb{R}^3$ by $H(x_1, x_2, x_3) = 2x_1^2 - x_2^2 - x_3^2$, then a short calculation shows that $D(E_+(H), 0, r) = 1 - 3^{-1/2}$ for all $r > 0$.

5   Concluding Remarks

5.1 It is natural to ask whether (2) implies (1). When $n = 2$ this is true because of the polygonal connectedness of fine domains (see [6, § 5]), and so (1) can be omitted in this case.

When $n \geq 3$ this implication does not hold. To see this, we recall that a set E is thin at a limit point y if and only if there is a subharmonic function s on $\mathbb{R}^n$ such that

$$\limsup_{x\to y, x\in E} s(x) < s(y).$$

Now let L be a compact set in $\mathbb{R}^2$ such that $L^{\circ} = \emptyset$ but $\mathbb{R}^2\backslash L$ is thin at 0 (see [7, Example 1.2] for an example of such a set). Thus there is a subharmonic function s on $\mathbb{R}^2$ such that

$$\limsup_{x\to 0, x\notin L} s(x) < s(0).$$

If we define $K_1 = L \times [-1, 1]$ and

$$s_1(x_1, x_2, x_3) = s(x_1, x_2) \qquad ((x_1, x_2, x_3) \in \mathbb{R}^3),$$

then

$$\limsup_{x\to 0, x\notin K_1} s_1(x) < s_1(0),$$

so $\mathbb{R}^3\backslash K_1$ is thin at 0. Similarly, if we define $K_2 = [-1, 1] \times L$ and

$$K_3 = \{(x_1, x_2, x_3) : (x_1, x_3) \in L \text{ and } x_2 \in [-1, 1]\},$$

then $\mathbb{R}^3 \backslash K_i$ is thin at 0 ($i = 2, 3$). Let $K = K_1 \cap K_2 \cap K_3$. Then $\mathbb{R}^3 \backslash K$ is thin at 0, so (1) fails (with $K_\nu = K$). However, it is easy to see that (2) holds. A similar counterexample can be given when $n \geq 4$.

It is, of course, clear that (1) does not imply (2); for example, if $n \geq 2$ and $K = \{(x_1, \ldots, x_n) \in \overline{B(0,1)} : x_n = 0\}$, then (1) holds but (2) fails.

5.2 To see more clearly why condition (1) is important, let $K \subset \mathbb{R}^3$ be the compact set constructed in § 5.1 and let ω be the fine component of the fine interior of K which contains 0. Further, let B be an open ball centred at 0 such that $K \subset B$, and let ν be the sweeping of the Dirac measure δ_0 at 0 onto $B \backslash \omega$; that is, ν is the measure which satisfies

$$\int_{K_\nu} G_B(x, y) \, d\nu(y) = \hat{R}^{B \backslash \omega}_{G_B(0, \cdot)}(x) \qquad (x \in B),$$

where $G_B(\cdot, \cdot)$ is the Green function for B and the regularized reduction on the right hand side of the equation is relative to superharmonic functions on B. Then $K_\nu \subseteq K$ and $\nu(\{0\}) = 0$ (see [5, 1.XI.3]) and

$$\int_{K_\nu} h \, d\nu = h(0) \quad (h \in \mathcal{H})$$

(see [5, 1.VI.3(h) and 1.X.2]), where $\mathcal{H}$ is defined relative to $\Omega = B$. Thus B is a quadrature domain for the measure $\lambda(B)\nu$.

Now let $f(x) = |x|$ and $h^* = 0$ on $\overline{B}$, and let $Z = Z(f - h^*)$. Since B is a quadrature domain for $\lambda(B)\delta_0$ and $Z = \{0\}$, it follows from Theorem 1 that the constant function 0 is a best one–sided L^1–approximant to f from $\mathcal{H}$. However, K_ν is not contained in $\hat{Z}$, so the characterization given in Theorem 1 is not valid for this choice of ν. Our construction of K ensured that (2) holds. Thus the reason for the failure of the characterization is that (1) is not satisfied.

The situation regarding (2) is less clear. We do not know whether it can be relaxed to the weaker condition that $\mathbb{R}^n \backslash K_\nu$ is connected. To see that the latter condition is indispensable consider a bounded domain V, a point $x \in V$ and a ball $B(x, r)$ containing $\overline{V}$. Then $B(x, r)$ is a quadrature domain for $\lambda(B(x,r))\mu$, where μ denotes harmonic measure for V and x, but the characterization of best approximants in Theorem 1 is clearly not valid in this case.

5.3 A discussion of the possible non–existence of best harmonic one–sided L^1–approximants may be found in [1, § 4.1].

Added in proof. Let $n = 2$ and let Ω be the ellipse of Example 1. Harold Shapiro has kindly communicated to us an argument, due to Björn Gustafsson, which shows the existence of a measure μ such that Ω is a quadrature domain associated with μ but K_μ does not contain the interfocal line segment $\{0\} \times [-1, 1]$. This implies a negative answer to the question left open at the end of Example 1. It also shows that, in Theorem 1, (2) cannot be relaxed to the weaker condition that $\mathbb{R} \setminus K_\nu$ is connected (see the last paragraph of § 5.2).

References

[1]   D. H. Armitage, S. J. Gardiner, W. Haussmann, L. Rogge: *Best one–sided L^1–approximation by harmonic functions*, Manuscripta Math. **96** (1998), 181–194.

[2]   D. H. Armitage, W. Haussmann, K. Zeller: *Best harmonic and polyharmonic approximation*, in: *Approximation and Optimization*, Vol. I, D. D. Stancu et al. (eds.), 17–34, Transilvania Press, Cluj–Napoca, 1997.

[3]   S. Axler, P. Bourdon, W. Ramey: *Harmonic Function Theory*, Springer, New York, 1992.

[4]   R. Courant, D. Hilbert: *Methods of Mathematical Physics*, Vol. II, Wiley, New York, 1962.

[5]   J. L. Doob: *Classical Potential Theory and its Probabilistic Counterpart*, Springer, New York, 1983.

[6]   B. Fuglede: *Asymptotic paths for subharmonic functions*, Math. Ann. **213** (1975), 261–274.

[7]   S. J. Gardiner: *Harmonic Approximation*, London Math. Soc. Lect. Notes **221**, Cambridge Univ. Press, Cambridge, 1995.

[8]   B. Gustafsson, M. Sakai, H. S. Shapiro: *On domains in which harmonic functions satisfy mean value properties*, Potential Anal. **7** (1997), 467–484.

[9]   L. L. Helms: *Introduction to Potential Theory*, Krieger, New York, 1975.

[10]   L. Karp, A. S. Margulis: *Newtonian potential theory for unbounded sources and applications to free boundary problems*, J. Analyse Math. **70** (1996), 1–63.

[11]   A. Pinkus: *On L^1-Approximation*, Cambridge Univ. Press, Cambridge, 1989.

[12]   M. Sakai: *Sharp estimates of the distance from a fixed point to the frontier of a Hele–Shaw flow*, Potential Anal. **8** (1998), 277–302.

[13]   D. Zwick: *The obstacle problem and best superharmonic approximation*, in: *Multivariate Approximation and Interpolation*, W. Haussmann et al. (eds.), Internat. Ser. Numer. Math. **94**, Birkhäuser, Basel, 1990.

Addresses:

David H. Armitage
Department of Pure Mathematics
Queen's University
Belfast BT7 1NN
Northern Ireland

Stephen J. Gardiner
Department of Mathematics
University College Dublin
Dublin 4
Ireland

Numerical Stability of Fast Fourier Transforms

Günter Baszenski, Rao Xuefang and Manfred Tasche

Dedicated to Professor M. Reimer on the occasion of his 65th birthday

Abstract

In this paper we consider the numerical stability of fast Fourier transforms (FFT). We give a new approach to the results in [5], [1] and [7]. Furthermore, we extend these results to the consideration of precomputed roots of unity and to two–dimensional discrete Fourier transforms. We show that for different precomputation schemes, the roundoff error of precomputed roots of unity has a very strong influence on the numerical stability of fast Fourier transforms. We close with numerical tests which illustrate our theoretical results.

1 Introduction

Discrete Fourier transforms and related discrete trigonometric transforms (such as discrete cosine transforms, discrete sine transforms, and discrete Hartley transforms) have found wide applications in numerical analysis and signal processing (see [6], [2]). Repeated use of discrete Fourier transforms occurs in some applications. If the transform length is large, then it is important to have fast and numerically stable realizations of discrete Fourier transforms.

In this paper we consider the numerical stability of the FFT. In a worst case study, we show that various precomputation schemes of roots of unity lead to a different behaviour of numerical stability of the FFT. The method of direct call of the trigonometric values needed leads to the best numerical stability. On the other hand, the method of repeated multiplication leads to a bad stability of the FFT.

We use the following concept of numerical stability. Let $N \in \mathbb{N}$ with $N \geq 2$ be given. Let $\mathbf{x} \in \mathbb{C}^N$ be an arbitrary input vector, $\mathbf{F}_N \in \mathbb{C}^{N \times N}$ be the unitary Fourier matrix

$$\mathbf{F}_N := \frac{1}{\sqrt{N}} \left(w_N^{jk} \right)_{j,k=0}^{N-1}, \quad w_N := \mathrm{e}^{-2\pi\mathrm{i}/N} \tag{1.1}$$

Advances in Multivariate Approximation; W. Haußmann, K. Jetter and M. Reimer (eds.)
Mathematical Research, Vol. 107, pp. 57–72, ISBN 3–527–40236–5
© WILEY–VCH, Berlin 1999

and $\mathbf{y} := \mathbf{F}_N \mathbf{x}$ be the exact output vector. Let $\hat{\mathbf{y}} \in \mathbb{C}^N$ be the computed vector using floating point arithmetic with roundoff unit u. Then $\hat{\mathbf{y}}$ can be represented in the form

$$\hat{\mathbf{y}} = \mathbf{F}_N \left(\mathbf{x} + \Delta\mathbf{x} \right) \quad (\Delta\mathbf{x} \in \mathbb{C}^N).$$

An algorithm for computing $\mathbf{y} = \mathbf{F}_N \mathbf{x}$ is called *normwise backward stable* (see [4], p. 142), if for all $\mathbf{x} \in \mathbb{C}^N$ we have

$$\|\Delta\mathbf{x}\|_2 \le (k_N u + \mathcal{O}(u^2))\|\mathbf{x}\|_2, \tag{1.2}$$

where k_N is a positive constant with $k_N u \ll 1$. The size of k_N is a measure of numerical stability.

An important example of a fast *and* numerically stable algorithm is the FFT. For the Cooley–Tukey and Gentleman–Sande algorithms with $N = 2^n$ (see [6], pp. 17–22 and 64–75), G. U. Ramos [5], M. Arioli et al. [1] and P. Y. Yalamov [7] have shown that $k_N = \mathcal{O}(\log_2 N)$ under the assumption that all roots of unity are exactly known or precomputed by direct calls (see also [3], pp. 9–14). These results can be proved by a clever factorization of the Fourier matrix into a product of sparse unitary matrices (see [6], pp. 20–22).

Our paper is organized as follows: In Section 2, we consider three methods for precomputation of roots of unity. Further methods for precomputation of roots of unity can be found in [6], pp. 25–28. All these methods compute the roots of unity with roundoff errors (see (2.3)–(2.5)). Our main result is Theorem 3.2 in Section 3. We show that the Cooley–Tukey algorithm has the "best" numerical stability (1.2) with a constant $k_N = \mathcal{O}(\log_2 N)$, if all roots of unity are precomputed accurately by direct calls. In Section 4, we present numerical tests which yield the same results.

2   Precomputation of Roots of Unity

In the following, we use the standard model of real floating point arithmetic (see [4], p. 44): For arbitrary $\xi, \eta \in \mathbb{R}$ and any operation $\circ \in \{+, -, \times, /\}$, the exact value $\xi \circ \eta$ and the computed value $\mathrm{fl}\,(\xi \circ \eta)$ are related by

$$\mathrm{fl}\,(\xi \circ \eta) = (\xi \circ \eta)(1 + \delta) \quad (|\delta| \le u),$$

where u denotes the *roundoff unit* (or *machine precision*). In the IEEE arithmetic of single precision (24 bits for the mantissa (with 1 sign bit), 8 bits for the exponent), we have

$$u = 2^{-24} \approx 5.96 \times 10^{-8}$$

and for double precision (53 bits for the mantissa (with 1 sign bit), 11 bits for the exponent)

$$u = 2^{-53} \approx 1.11 \times 10^{-16}$$

(see [4], p. 45). Since complex arithmetic is implemented using real operations, the complex floating point error model is a consequence of the corresponding real arithmetic model (see [4], pp. 78–80): For arbitrary $\xi, \eta \in \mathbb{C}$, we have

$$\mathrm{fl}\,(\xi + \eta) = (\xi + \eta)(1 + \delta) \quad (|\delta| \le u),$$
$$\mathrm{fl}\,(\xi\eta) = \xi\eta\,(1 + \delta) \quad (|\delta| \le \frac{2\sqrt{2}u}{1 - 2u}). \tag{2.1}$$

In particular, if $\xi \in \mathbb{R} \cup i\mathbb{R}$ and $\eta \in \mathbb{C}$, then

$$\mathrm{fl}\,(\xi\eta) = \xi\eta\,(1 + \delta) \quad (\delta| \le u). \tag{2.2}$$

We compare three different methods for precomputing the roots of unity w_N^k ($k = 1, \ldots, N_1 - 1$) with $N = 2^n$ and $N_1 := N/2$ (see [6], pp. 23–28). The most obvious method is to call repeatedly library routines for cosine and sine functions.

Algorithm 2.1. (Direct call)

Input: $N := 2^n$ ($n \in \mathbb{N}, n \ge 2$), $N_1 := N/2$, $\varphi := 2\pi/N$.

For $k = 1\,(1)\,N_1 - 1$ *form*

$$\hat{w}_N^k := \mathrm{fl}\,(\cos(k\varphi)) - i\,\mathrm{fl}\,(\sin(k\varphi)).$$

Output: $\hat{w}_N^k$ *precomputed value of* w_N^k ($k = 1, \ldots, N_1 - 1$).

This algorithm involves almost N trigonometric function calls. Using symmetries of real and imaginary parts of the roots of unity w_N^k we only need $N/4$ calls of $\cos(2\pi k/N)$ ($k = 1, \ldots, N/4$). If the cosine and sine functions are quality library routines, then very accurate roots of unity are precomputed such that with some $c > 0$

$$|\hat{w}_N^k - w_N^k| \le c\,u \quad (k = 1, \ldots, N_1 - 1). \tag{2.3}$$

If

$$| \mathrm{fl}\left(\cos\left(k\varphi\right)\right) - \cos\left(k\varphi\right)| \leq \frac{1}{2}u, \quad | \mathrm{fl}\left(\sin\left(k\varphi\right)\right) - \sin\left(k\varphi\right)| \leq \frac{1}{2}u,$$

then we can choose $c = \frac{1}{2}\sqrt{2}$ in (2.3).

The next algorithm uses only two trigonometric function calls and is based on repeated multiplication $w_N^k = w_N\, w_N^{k-1}$.

Algorithm 2.2. (Repeated multiplication)

Input: $N := 2^n$, $(n \in \mathbb{N}, n \geq 2)$, $N_1 := N/2$, $\varphi := 2\pi/N$.

1. *Form by direct call*

$$\hat{w}_N := \mathrm{fl}\left(\cos\,\varphi\right) - \mathrm{i}\,\mathrm{fl}\left(\sin\,\varphi\right).$$

2. For $k = 2\,(1)\,N_1 - 1$ *form*

$$\hat{w}_N^k := \mathrm{fl}\left(\hat{w}_N\, \hat{w}_N^{k-1}\right).$$

Output: $\hat{w}_N^k$ *precomputed value of* w_N^k $(k = 1, \ldots, N_1 - 1)$.

Since one complex multiplication requires 4 real multiplications and 2 real additions (i.e. 6 flops), a total of $6\,(N_1 - 1) = 3N - 6$ flops are involved. Since $\hat{w}_N^k$ $(k = 2, \ldots, N_1 - 1)$ is the result of $6\,(k - 1)$ flops, we obtain by (2.1) the estimate

$$|\hat{w}_N^k - w_N^k| \leq c\,k\,u \leq c\,(N_1 - 1)\,u \quad (k = 1, \ldots, N_1 - 1) \qquad (2.4)$$

with some $c > 0$.

The third method combines the accuracy of Algorithm 2.1 with the arithmetical simplicity of Algorithm 2.2 by using the fact that

$$\left(w_N^k\right)_{k=2^j+1}^{2^{j+1}-1} = w_N^{2^j}\left(w_N^k\right)_{k=1}^{2^j-1} \quad (j = 1, \ldots, n - 2).$$

Algorithm 2.3. (Repeated subvector scaling)

Input: $N := 2^n$ $(n \in \mathbb{N}, n \geq 3)$, $N_j := N/2^j$ $(j = 0, \ldots, n - 2)$.

1. For $j = 0\,(1)\,n - 3$ *form*

$$\hat{w}_{N_j} = \hat{w}_N^{2^j} := \mathrm{fl}\left(\cos\,\varphi_j\right) - \mathrm{i}\,\mathrm{fl}\left(\sin\,\varphi_j\right)$$

with $\varphi_j := 2\pi/N_j$ by direct call.

2. For $j = 1\,(1)\,n - 3$ *multiply*

$$\left(\hat{w}_N^k\right)_{k=2^j+1}^{2^{j+1}-1} := \mathrm{fl}\left(\hat{w}_{N_j}\left(\hat{w}_N^k\right)_{k=1}^{2^j-1}\right).$$

3. *Form*

$$\left(\hat{w}_N^k\right)_{k=N_2+1}^{N_1-1} := -\mathrm{i}\,\left(\hat{w}_N^k\right)_{k=1}^{N_2-1}.$$

Output: $\hat{w}_N^k$ *precomputed value of* w_N^k $(k = 1, \ldots, N_1 - 1)$.

Note that w_N^k $(k = 1, \ldots, N_2 - 1)$ is the result of at most $\lfloor \log_2 k \rfloor + 1$ complex multiplications, where all factors are obtained by direct calls. Hence by (2.1), it follows that for $k = 1, \ldots, N_2 - 1$

$$|\hat{w}_N^k - w_N^k| \le c\,u\,(\lfloor \log_2 k \rfloor + 1) \le c\,u\,(n - 2).$$

By step 3 of Algorithm 2.3, we have for $k = 1, \ldots, N_2 - 1$

$$|\hat{w}_N^{N_2+k} - w_N^{N_2+k}| = |(-\mathrm{i})(\hat{w}_N^k - w_N^k)| = |\hat{w}_N^k - w_N^k|$$

such that

$$|\hat{w}_N^k - w_N^k| \le c\,u\,(n - 2) \quad (k = 1, \ldots, N_1 - 1). \tag{2.5}$$

3 Stability of the Cooley–Tukey Algorithm

Let $\mathbf{x} \in \mathbb{C}^N$ be an arbitrary input vector and $\mathbf{y} := \mathbf{F}_N\mathbf{x}$ be the exact output vector. Further let $\hat{\mathbf{y}}$ be the computed output vector using floating point arithmetic with roundoff unit u. Since $\mathbf{F}_N$ is regular, $\hat{\mathbf{y}}$ can be represented in the form

$$\hat{\mathbf{y}} = \mathbf{F}_N(\mathbf{x} + \Delta\mathbf{x})$$

with $\Delta\mathbf{x} \in \mathbb{C}^N$.

An algorithm used for computing $\mathbf{F}_N\mathbf{x}$ is called *normwise backward stable* (see [4], p. 142), if for all vectors $\mathbf{x} \in \mathbb{C}^N$ there exists a positive constant k_N such that

$$\|\Delta\mathbf{x}\|_2 \le (k_N u + \mathcal{O}(u^2))\|\mathbf{x}\|_2 \tag{3.1}$$

and $k_N u \ll 1$. Since $\mathbf{F}_N$ is unitary, we conclude that

$$\|\Delta\mathbf{x}\|_2 = \|\mathbf{F}_N(\Delta\mathbf{x})\|_2 = \|\hat{\mathbf{y}} - \mathbf{y}\|_2, \quad \|\mathbf{x}\|_2 = \|\mathbf{F}_N\mathbf{x}\|_2 = \|\mathbf{y}\|_2.$$

Here we have also *normwise forward stability* by

$$\|\hat{\mathbf{y}} - \mathbf{y}\|_2 \le (k_N u + \mathcal{O}(u^2))\|\mathbf{y}\|_2.$$

As known (see [6], pp. 17–18), the Cooley–Tukey algorithm for computing $\mathbf{F}_N\mathbf{x}$ corresponds to a factorization of $\mathbf{F}_N$ into a product of sparse unitary matrices, namely

$$\mathbf{F}_N = 2^{-n/2}\,\mathbf{M}_N^{(n)}\,\mathbf{M}_N^{(n-1)}\ldots\mathbf{M}_N^{(1)}\,\mathbf{B}_N \quad (N = 2^n), \tag{3.2}$$

where $\mathbf{B}_N$ is the bit–reversal permutation matrix and $\mathbf{M}_N^{(j)}$ is the Kronecker product

$$\mathbf{M}_N^{(j)} := \mathbf{I}_{N_j} \otimes \mathbf{A}_{2^j} \quad (j = 1,\ldots,n;\ N_j := N/2^j) \tag{3.3}$$

with

$$\mathbf{A}_{2^j} := \begin{pmatrix} \mathbf{I}_{2^{j-1}} & \mathbf{W}_{2^{j-1}} \\ \mathbf{I}_{2^{j-1}} & -\mathbf{W}_{2^{j-1}} \end{pmatrix}, \quad \mathbf{W}_{2^{j-1}} := \operatorname{diag}\left(w_{2^j}^k\right)_{k=0}^{2^{j-1}-1}. \tag{3.4}$$

Clearly, $\mathbf{M}_N^{(j)}$ contains only 2 nonzero entries in each row and column, respectively. Furthermore, $2^{-1/2}\,\mathbf{M}_N^{(j)}$ is unitary, since

$$\mathbf{A}_{2^j}(\bar{\mathbf{A}}_{2^j})^{\mathrm{T}} = 2\,\mathbf{I}_{2^j}$$

and hence

$$\frac{1}{2}\,\mathbf{M}_N^{(j)}(\bar{\mathbf{M}}_N^{(j)})^{\mathrm{T}} = \frac{1}{2}\,(\mathbf{I}_{N_j} \otimes \mathbf{A}_{2^j}\,\bar{\mathbf{A}}_{2^j}^{\mathrm{T}}) = \mathbf{I}_{N_j} \otimes \mathbf{I}_{2^j} = \mathbf{I}_N.$$

Consequently, for $0 \le j < k \le n$ we see that the matrix product

$$(2^{-1/2}\,\mathbf{M}_N^{(k)}) \ldots (2^{-1/2}\,\mathbf{M}_N^{(j+1)})$$

is also unitary and has the spectral norm 1, i.e.

$$\|\mathbf{M}_N^{(k)}\ldots\mathbf{M}_N^{(j+1)}\|_2 = 2^{(k-j)/2}. \tag{3.5}$$

We summarize:

Algorithm 3.1. (Cooley–Tukey algorithm)

Input: $N := 2^n$ $(n \in \mathbb{N}, n \geq 3), \mathbf{x} \in \mathbb{C}^N$.

0. *Precompute* w_N^k $(k = 1, \ldots, N_1 - 1)$.

1. *Permute*

$$\mathbf{x}_0 := \mathbf{B}_N \mathbf{x}.$$

2. *For* $j = 1\,(1)\,n$ *form*

$$\mathbf{x}_j := \mathbf{M}_N^{(j)} \mathbf{x}_{j-1}.$$

3. *Multiply*

$$\mathbf{y} := 2^{-n/2} \mathbf{x}_n.$$

Output: $\mathbf{y} = \mathbf{F}_N \mathbf{x}$.

Since the roots of unity w_N^k $(k = 1, \ldots, N_1 - 1)$ are precomputed by one of the Algorithms 2.1–2.3, we use $\mathbf{M}_N^{(j)}$ with precomputed entries in step 2. This matrix is denoted by $\hat{\mathbf{M}}_N^{(j)}$.

Theorem 3.2. *Let* $N = 2^n$ $(n \in \mathbb{N}, n \geq 3)$ *and* $\mathbf{x} \in \mathbb{C}^N$. *Assume that* w_N^k $(k = 1, \ldots, N_1 - 1)$ *are precomputed by direct call, repeated multiplication, and repeated subvector scaling, respectively. Let* $c > 0$ *be the corresponding constant in* (2.3)–(2.5). *Then the Cooley–Tukey Algorithm 3.1 is normwise backward stable with the constant*

$$k_N = 3 + \sqrt{2}\,k_N'(n - 2),$$

where

$$k_N' := \begin{cases} 2\sqrt{2} + 1 + c & \textit{for direct call,} \\ 2\sqrt{2} + 1 + c\,(N_1 - 1) & \textit{for repeated multiplication,} \\ 2\sqrt{2} + 1 + c\,(n - 2) & \textit{for repeated subvector scaling.} \end{cases} \qquad (3.6)$$

Remark. We observe that the precomputation error has a strong influence on the numerical stability of the FFT. Theorem 3.2 implies that

$$k_N = \begin{cases} \mathcal{O}(\log_2 N) & \text{for direct call,} \\ \mathcal{O}(N \log_2 N) & \text{for repeated multiplication,} \\ \mathcal{O}((\log_2 N)^2) & \text{for repeated subvector scaling.} \end{cases} \qquad (3.7)$$

The best asymptotic behaviour of k_N is obtained for direct call. For the other precomputation methods, the size of k_N is mainly determined by the precomputation error.

Proof. Let $\hat{\mathbf{x}}_0 = \mathbf{x}_0 := \mathbf{B}_N \mathbf{x}$. By $\hat{\mathbf{x}}_j$, we denote the computed vector $\hat{\mathbf{M}}_N^{(j)} \hat{\mathbf{x}}_{j-1}$ $(j = 1, \dots, n)$, i.e.

$$\hat{\mathbf{x}}_j = \mathrm{fl}\left(\hat{\mathbf{M}}_N^{(j)} \hat{\mathbf{x}}_{j-1}\right).$$

Then we introduce the error vector $\mathbf{e}_j \in \mathbb{C}^N$ of step j $(j = 1, \dots, n)$ by

$$\hat{\mathbf{x}}_j = \hat{\mathbf{M}}_N^{(j)} \hat{\mathbf{x}}_{j-1} + \mathbf{e}_j, \tag{3.8}$$

i.e., $\mathbf{e}_j$ contains the error of floating point arithmetic for computing $\hat{\mathbf{M}}_N^{(j)} \hat{\mathbf{x}}_{j-1}$ and the precomputation errors of $w_{2^j}^k$ $(k = 1, \dots, 2^{j-1} - 1)$.

1) Step $j = 1$: Note that

$$\hat{\mathbf{M}}_N^{(1)} = \mathbf{M}_N^{(1)} = \mathbf{I}_{2^{n-1}} \otimes \mathbf{A}_2, \quad \mathbf{A}_2 = \begin{pmatrix} 1 & 1 \\ 1 & -1 \end{pmatrix}.$$

Hence the first step consists of $N/2$ butterfly operations

$$\begin{aligned} \eta_1 &:= \xi_1 + \xi_2, \\ \eta_2 &:= \xi_1 - \xi_2 \end{aligned} \quad (\xi_1, \xi_2 \in \mathbb{C}). \tag{3.9}$$

By (2.1) we obtain for the computed values

$$\begin{aligned} \hat{\eta}_1 &:= (\xi_1 + \xi_2)(1 + \delta_1) = \eta_1 + (\xi_1 + \xi_2)\delta_1, \\ \hat{\eta}_2 &:= (\xi_1 - \xi_2)(1 + \delta_2) = \eta_2 + (\xi_1 + \xi_2)\delta_2 \end{aligned} \quad (|\delta_1|, |\delta_2| \le u)$$

and hence

$$|\hat{\eta}_1 - \eta_1|^2 + |\hat{\eta}_2 - \eta_2|^2 \le u^2 \left(|\xi_1 + \xi_2|^2 + |\xi_1 - \xi_2|^2\right) = 2u^2 \left(|\xi_1|^2 + |\xi_2|^2\right).$$

Consequently, we have

$$\|\mathbf{e}_1\|_2 \le \sqrt{2}u\,\|\mathbf{x}_0\|_2 = \sqrt{2}u\,\|\mathbf{x}\|_2. \tag{3.10}$$

2) Step $j = 2$: Note that

$$\hat{\mathbf{M}}_N^{(2)} = \mathbf{M}_N^{(2)} = \mathbf{I}_{2^{n-2}} \otimes \mathbf{A}_4$$

with

$$\mathbf{A}_4 = \begin{pmatrix} \mathbf{I}_2 & \mathbf{W}_2 \\ \mathbf{I}_2 & -\mathbf{W}_2 \end{pmatrix}, \quad \mathbf{W}_2 = \begin{pmatrix} 1 & 0 \\ 0 & -i \end{pmatrix}.$$

Then the second step consists of $N/4$ butterfly operations (3.9) and of $N/4$ butterfly operations

$$\begin{aligned} \eta_1 &:= \xi_1 - i\,\xi_2, \\ \eta_2 &:= \xi_1 + i\,\xi_2 \end{aligned} \quad (\xi_1, \xi_2 \in \mathbb{C}).$$

Analogously to the first step, we get the estimate

$$\|\mathbf{e}_2\|_2 \le \sqrt{2}u \, \|\hat{\mathbf{x}}_1\|_2.$$

By (3.8) and (3.10), we see that

$$\|\hat{\mathbf{x}}_1\|_2 \le \|\mathbf{M}_N^{(1)}\| \, \|\hat{\mathbf{x}}_0\|_2 + \|\mathbf{e}_1\|_2 \le \sqrt{2}\,(1+u)\,\|\mathbf{x}\|_2$$

and therefore

$$\|\mathbf{e}_2\|_2 \le (2u + \mathcal{O}(u^2))\,\|\mathbf{x}\|_2. \tag{3.11}$$

3) Step j ($j \in \{3, \dots, n\}$): Using the structure of $\mathbf{M}_N^{(j)}$ (see (3.3)–(3.4)), it follows that step j consists of 2^{n-j} butterfly operations of the form

$$\begin{aligned} \eta_1 &:= \xi_1 + w_{2^j}^k\,\xi_2, \\ \eta_2 &:= \xi_1 - w_{2^j}^k\,\xi_2 \end{aligned} \quad (\xi_1, \xi_2 \in \mathbb{C};\, k = 0, \dots, 2^{j-1} - 1). \tag{3.12}$$

Now $w_{2^j}^k$ is precomputed by

$$\hat{w}_{2^j} = w_{2^j}^k + \varepsilon_{2^j}^k \quad (\varepsilon_{2^j}^0 := 0). \tag{3.13}$$

Note that $w_{2^j}^k = w_N^{N_j k}$. Thus by (2.3)–(2.5), we can estimate

$$\left|\varepsilon_{2^j}^k\right| \le \begin{cases} c\,u & \text{by direct call,} \\ c\,(N_1 - 1)\,u & \text{by repeated multiplication,} \\ c\,(n - 2)\,u & \text{by repeated subvector scaling.} \end{cases} \tag{3.14}$$

Using complex floating point arithmetic (see (2.1)), we obtain the computed values

$$\begin{aligned} \hat{\eta}_1 &= (\xi_1 + \hat{w}_{2^j}^k\,\xi_2\,(1+\delta_1))(1+\delta_2), \\ \hat{\eta}_2 &= (\xi_1 - \hat{w}_{2^j}^k\,\xi_2\,(1+\delta_1))(1+\delta_3) \end{aligned} \tag{3.15}$$

with

$$|\delta_1| \le \frac{2\sqrt{2}\,u}{1 - 2u}, \quad |\delta_2|,\, |\delta_3| \le u. \tag{3.16}$$

By (3.13) and (3.15), it follows that

$$\hat{\eta}_1 - \eta_1 = \xi_1 \delta_2 + \hat{w}_{2^j}^k \, \xi_2 (\delta_1 + \delta_2 + \delta_1 \delta_2) + \varepsilon_{2^j}^k \, \xi_2.$$

Using (3.6), (3.14) and (3.16), we obtain that

$$|\hat{\eta}_1 - \eta_1| \le |\xi_1|\, u + |\xi_2|\, (k_N' u + \mathcal{O}(u^2))$$

and hence

$$|\hat{\eta}_1 - \eta_1|^2 \le 2\,(k_N' u + \mathcal{O}(u^2))^2 (|\xi_1|^2 + |\xi_2|^2).$$

The same estimate is true for $|\hat{\eta}_2 - \eta_2|^2$ such that

$$|\hat{\eta}_1 - \eta_1|^2 + |\hat{\eta}_2 - \eta_2|^2 \le 4\,(k_N' u + \mathcal{O}(u^2))^2 (|\xi_1|^2 + |\xi_2|^2).$$

Thus for the error vector $\mathbf{e}_j$, we obtain the estimate

$$\|\mathbf{e}_j\|_2 \le 2\,(k_N' u + \mathcal{O}(u^2))\, \|\hat{\mathbf{x}}_{j-1}\|_2.$$

From (3.5) and (3.8), it follows that

$$\|\hat{\mathbf{x}}_{j-1}\|_2 \le (2^{(j-1)/2} + \mathcal{O}(u))\, \|\mathbf{x}\|_2$$

and hence

$$\|\mathbf{e}_j\|_2 \le 2^{(j+1)/2}\,(k_N' u + \mathcal{O}(u^2))\, \|\mathbf{x}\|_2. \tag{3.17}$$

4) Now we estimate the roundoff error $\|\hat{\mathbf{x}}_n - \mathbf{x}_n\|_2$. Applying repeatedly (3.8), we obtain

$$\begin{aligned}
\hat{\mathbf{x}}_n &= \mathbf{M}_N^{(n)} \hat{\mathbf{x}}_{n-1} + \mathbf{e}_n \\
&= \mathbf{M}_N^{(n)} \mathbf{M}_N^{(n-1)} \hat{\mathbf{x}}_{n-2} + \mathbf{M}_N^{(n)} \mathbf{e}_{n-1} + \mathbf{e}_n \\
&\;\;\vdots \\
&= \mathbf{M}_N^{(n)} \dots \mathbf{M}_N^{(1)} \mathbf{B}_N \mathbf{x} + \mathbf{M}_N^{(n)} \dots \mathbf{M}_N^{(2)} \mathbf{e}_1 + \dots + \mathbf{M}_N^{(n)} \mathbf{e}_{n-1} + \mathbf{e}_n
\end{aligned}$$

such that by

$$\mathbf{x}_n = \mathbf{M}_N^{(n)} \dots \mathbf{M}_N^{(1)} \mathbf{B}_N \mathbf{x}$$

(see Algorithm 3.1) and by (3.5), (3.10), (3.11) and (3.17)

$$\|\hat{\mathbf{x}}_n - \mathbf{x}_n\|_2 \ \leq \ \|\mathbf{M}_N^{(n)}\ldots\mathbf{M}_N^{(2)}\|_2\|\mathbf{e}_1\|_2 + \ldots + \|\mathbf{M}_N^{(n)}\|_2\|\mathbf{e}_{n-1}\|_2 + \|\mathbf{e}_n\|_2$$
$$\leq \ 2^{n/2}(2u + (n-2)(\sqrt{2}k'_N u + \mathcal{O}(u^2)))\|\mathbf{x}\|_2.$$

5) The final step of Algorithm 3.1 is the scaling $\mathbf{y} = 2^{-n/2}\mathbf{x}_n$.

Let $\hat{\mathbf{y}} := \mathrm{fl}(2^{-n/2}\hat{\mathbf{x}}_n)$. Using (2.2), (3.5) and (3.8), we obtain that

$$\|\hat{\mathbf{y}} - 2^{-n/2}\hat{\mathbf{x}}_n\|_2 \leq 2^{-n/2}u\|\hat{\mathbf{x}}_n\|_2 \leq (u + \mathcal{O}(u^2))\|\mathbf{x}\|_2$$

and hence

$$\|\hat{\mathbf{y}} - \mathbf{y}\|_2 \ \leq \ \|\hat{\mathbf{y}} - 2^{-n/2}\hat{\mathbf{x}}_n\|_2 + 2^{-n/2}\|\hat{\mathbf{x}}_n - \mathbf{x}_n\|_2$$
$$\leq \ (3u + (n-2)(\sqrt{2}k'_N u + \mathcal{O}(u^2)))\|\mathbf{x}\|_2.$$

This completes the proof. $\square$

Now we consider the numerical stability of the FFT with respect to other norms. Note that $\|\mathbf{F}_N\|_2 = 1$, but

$$\|\mathbf{F}_N\|_1 = \|\mathbf{F}_N\|_\infty = \sqrt{N}.$$

Corollary 3.3. *Let $p \in \{1, \infty\}$. Under the assumptions of Theorem 3.2, we have the estimate*

$$\|\Delta\mathbf{x}\|_p \leq (\sqrt{N}\,k_N\,u + \mathcal{O}(u^2))\,\|\mathbf{x}\|_p$$

with the constant k_N defined in (3.6).

Proof. Note that for $\mathbf{x} \in \mathbb{C}^N$ we have

$$\|\mathbf{x}\|_1 \leq \sqrt{N}\,\|\mathbf{x}\|_2 \leq \sqrt{N}\,\|\mathbf{x}\|_1, \quad \|\mathbf{x}\|_\infty \leq \|\mathbf{x}\|_2 \leq \sqrt{N}\,\|\mathbf{x}\|_\infty.$$

Hence by Theorem 3.2 we obtain for $p = 1$ that

$$\|\Delta\mathbf{x}\|_1 \ \leq \ \sqrt{N}\,\|\Delta\mathbf{x}\|_2 \leq \sqrt{N}\,(k_N u + \mathcal{O}(u^2))\,\|\mathbf{x}\|_2$$
$$\leq \ (\sqrt{N}\,k_N u + \mathcal{O}(u^2))\,\|\mathbf{x}\|_1.$$

The case $p = \infty$ can be handled analogously. $\square$

Finally, we extend Theorem 3.2 to the two–dimensional FFT. Let $\mathbf{X} \in \mathbb{C}^{N \times N}$. The Fourier–transformed matrix of $\mathbf{X}$ is defined by

$$\mathbf{Y} := \mathbf{F}_N\,\mathbf{X}\,\mathbf{F}_N.$$

We can realize this transform by the simple row–column method, which is based on the following observation

$$\mathbf{Y} = (\mathbf{F}_N \mathbf{X})\mathbf{F}_N = \mathbf{Z}\,\mathbf{F}_N = (\mathbf{F}_N \mathbf{Z}^{\mathrm{T}})^{\mathrm{T}} \quad (\mathbf{Z} := \mathbf{F}_N \mathbf{X}).$$

Algorithm 3.4. (Row–column method for two–dimensional FFT)

Input: $N := 2^n$ ($n \in \mathbb{N}, n \geq 3$), $\mathbf{X} = (\mathbf{x}_0, \dots, \mathbf{x}_{N-1}) \in \mathbb{C}^{N \times N}$.

1. For $k = 0\,(1)\,N - 1$ *compute by Algorithm 3.1*

$$\mathbf{z}_k := \mathbf{F}_N\,\mathbf{x}_k.$$

2. Set $\mathbf{Z} := (\mathbf{z}_0, \dots, \mathbf{z}_{N-1})$ *and* $(\mathbf{u}_0, \dots, \mathbf{u}_{N-1}) := \mathbf{Z}^{\mathrm{T}}.$

3. For $k = 0\,(1)\,N - 1$ *compute by Algorithm 3.1*

$$\mathbf{v}_k := \mathbf{F}_N\,\mathbf{u}_k.$$

4. *Form* $\mathbf{Y} := (\mathbf{v}_0, \dots, \mathbf{v}_{N-1})^{\mathrm{T}}.$

Output: $\mathbf{Y} = \mathbf{F}_N \mathbf{X}\,\mathbf{F}_N.$

Let $\mathbf{X} \in \mathbb{C}^{N \times N}$ be given. By $\mathbf{Y} = \mathbf{F}_N \mathbf{X}\,\mathbf{F}_N$ we denote the exact Fourier–transformed matrix of $\mathbf{X}$. Further let $\hat{\mathbf{Y}}$ be the transformed matrix using floating point arithmetic with roundoff unit u. Then $\hat{\mathbf{Y}}$ can be represented in the form

$$\hat{\mathbf{Y}} = \mathbf{F}_N(\mathbf{X} + \Delta\mathbf{X})\,\mathbf{F}_N$$

with $\Delta\mathbf{X} \in \mathbb{C}^{N \times N}$.

Theorem 3.5. *Let* $N = 2^n$ ($n \in \mathbb{N}, n \geq 3$) *and* $\mathbf{X} \in \mathbb{C}^{N \times N}$. *Assume that* w_N^k ($k = 1, \dots, N_1 - 1$) *are precomputed by direct call, repeated multiplication, and repeated subvector scaling, respectively. Let* $c > 0$ *be the corresponding constant in* (2.3)–(2.5). *Then Algorithm 3.4 is normwise backward stable with respect to the Frobenius norm, i.e.*

$$\|\Delta\mathbf{X}\|_F \leq (2\sqrt{N}\,k_N u + \mathcal{O}(u^2))\|\mathbf{X}\|_F$$

with the constant k_N *defined in* (3.6).

Proof. Since the Frobenius norm is unitary invariant, we have

$$\|\Delta\mathbf{X}\|_F = \|\mathbf{Y} - \hat{\mathbf{Y}}\|_F, \quad \|\mathbf{X}\|_F = \|\mathbf{Y}\|_F. \tag{3.18}$$

Note that $\|\mathbf{F}_N\|_F = \sqrt{N}$. By Theorem 3.2, we know that the computed vectors $\hat{\mathbf{z}}_k$ $(k = 0, \ldots, N-1)$ in step 1 of Algorithm 3.4 fulfill the estimate

$$\|\hat{\mathbf{z}} - \mathbf{z}\|_2^2 \leq (k_N u + \mathcal{O}(u^2))^2 \|\mathbf{x}_k\|_2^2 \quad (k = 0, \ldots, N-1).$$

Summing up, this yields

$$\|\hat{\mathbf{Z}} - \mathbf{Z}\|_F \leq (k_N u + \mathcal{O}(u^2)) \|\mathbf{X}\|_F \tag{3.19}$$

with $\hat{\mathbf{Z}} := (\hat{\mathbf{z}}_0, \ldots, \hat{\mathbf{z}}_{N-1})$. Set $(\hat{\mathbf{u}}_0, \ldots, \hat{\mathbf{u}}_{N-1}) := \hat{\mathbf{Z}}^{\mathrm{T}}$. Let $\tilde{\mathbf{v}}_k := \mathbf{F}_N \hat{\mathbf{u}}_k$ and $\hat{\mathbf{v}}_k$ be the corresponding computed vector. Applying again Theorem 3.2, we get

$$\|\hat{\mathbf{v}}_k - \tilde{\mathbf{v}}_k\|_2^2 \leq (k_n u + \mathcal{O}(u^2))^2 \|\hat{\mathbf{u}}_k\|_2^2$$

and hence

$$\|\hat{\mathbf{Y}} - \tilde{\mathbf{Y}}\|_F \leq (k_N u + \mathcal{O}(u^2))\|\hat{\mathbf{Z}}\|_F \tag{3.20}$$

with $\hat{\mathbf{Y}} := (\hat{\mathbf{v}}_0, \ldots, \hat{\mathbf{v}}_{N-1})^{\mathrm{T}}$ and $\tilde{\mathbf{Y}} := (\tilde{\mathbf{v}}_0, \ldots, \tilde{\mathbf{v}}_{N-1})^{\mathrm{T}} = \hat{\mathbf{Z}}\,\mathbf{F}_N$. By (3.18), we can estimate

$$\|\Delta\mathbf{X}\|_F = \|\mathbf{Y} - \hat{\mathbf{Y}}\|_F \leq \|\mathbf{Y} - \tilde{\mathbf{Y}}\|_F + \|\tilde{\mathbf{Y}} - \hat{\mathbf{Y}}\|_F. \tag{3.21}$$

Since the Frobenius norm is consistent, it follows by (3.19) that

$$\begin{aligned}
\|\mathbf{Y} - \tilde{\mathbf{Y}}\|_F &\leq \|\mathbf{Z} - \hat{\mathbf{Z}}\|_F \|\mathbf{F}_N\|_F \\
&= \sqrt{N}\|\mathbf{Z} - \hat{\mathbf{Z}}\|_F \\
&\leq (\sqrt{N}\,k_N u + \mathcal{O}(u^2))\|\mathbf{X}\|_F.
\end{aligned} \tag{3.22}$$

Further by (3.19), we obtain

$$\begin{aligned}
\|\hat{\mathbf{Z}}\|_F &\leq \|\mathbf{Z}\|_F + \|\hat{\mathbf{Z}} - \mathbf{Z}\|_F \\
&\leq \|\mathbf{F}_N\|_F \|\mathbf{X}\|_F + \mathcal{O}(u)\|\mathbf{X}\|_F \\
&= (\sqrt{N} + \mathcal{O}(u))\|\mathbf{X}\|_F,
\end{aligned}$$

such that by (3.20)

$$\|\hat{\mathbf{Y}} - \tilde{\mathbf{Y}}\|_F \leq (\sqrt{N}\,k_N u + \mathcal{O}(u^2))\,\|\mathbf{X}\|_F. \tag{3.23}$$

From (3.21)–(3.23) the result follows. $\qquad\square$

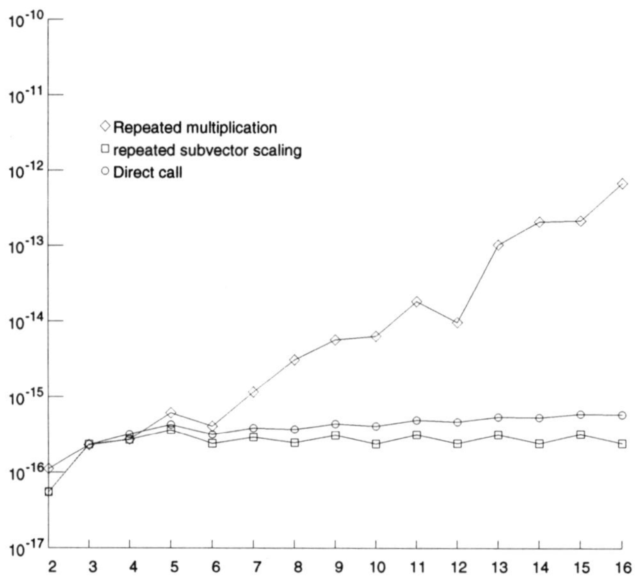

Figure 1: Relative roundoff errors $\| \hat{\mathbf{y}} - \mathbf{y} \|_2 / \| \mathbf{y} \|_2$ of the computed example with $N = 2^n$ $(n = 2, \ldots, 16)$

4   Numerical Experiments

As an example we consider the computation of $\mathbf{y} = \mathbf{F}_N \mathbf{x}$ with $\mathbf{x} = (2^{-k})_{k=0}^{N-1}$ and $N = 2^n$ $(n = 2, \ldots, 16)$. In order to evaluate this Fourier transform algebraically, we use the geometric sum formula

$$\sum_{k=0}^{N-1} 2^{-k} w^k = \frac{\left(\frac{w}{2}\right)^N - 1}{\frac{w}{2} - 1} \quad (w \in \mathbb{C} \setminus \{2\}). \tag{4.1}$$

For $w := w_N^j = \mathrm{e}^{-2\pi \mathrm{i} j / N}$ we obtain from (4.1) the values

$$y_j = \frac{1}{\sqrt{N}} \frac{2^{1-N} - 2}{w_N^j - 2} \quad (j = 0, \ldots, N - 1).$$

By $\hat{\mathbf{y}}$ we denote the result of numerical computation in double precision. Figure

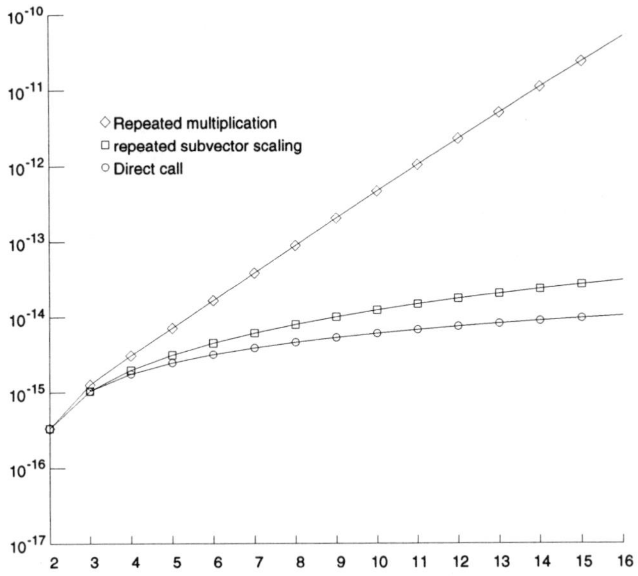

Figure 2: Theoretical bounds $k_N u$ of the relative roundoff errors

1 shows the corresponding relative roundoff errors $\|\hat{\mathbf{y}} - \mathbf{y}\|_2 / \|\mathbf{y}\|_2$. Note that

$$\frac{\|\hat{\mathbf{y}} - \mathbf{y}\|_2}{\|\mathbf{y}\|_2} = \frac{\|\Delta\mathbf{x}\|_2}{\|\mathbf{x}\|_2}$$

with $\hat{\mathbf{y}} = \mathbf{F}_N(\mathbf{x} + \Delta\mathbf{x})$. For comparison, Figure 2 shows the theoretical bounds $k_N u$ of Theorem 3.2.

Acknowledgment. The work of the second author (during a research stay at the University of Rostock) was supported by the German Academic Exchange Program (DAAD).

References

[1]   M. Arioli, H. Munthe–Kaas, L. Valdettaro: *Componentwise error analysis for FFTs with applications to fast Helmholtz solvers,* Numer. Algorithms **12** (1996), 65–88.

[2]   W. L. Briggs, V. E. Henson: *The DFT*, Soc. Ind. Appl. Math., Philadelphia 1995.

[3]   P. Henrici: *Applied and Computational Complex Analysis, Vol. 3*, Wiley, New York 1993.

[4]   N. J. Higham: *Accuracy and Stability of Numerical Algorithms*, Soc. Ind. Appl. Math., Philadephia 1996.

[5]   G. U. Ramos: *Roundoff error analysis of the fast Fourier transform*, Math. Comp. **25** (1971), 757–768.

[6]   C. F. Van Loan: *Computational Framework for the Fast Fourier Transform*, Soc. Ind. Appl. Math., Philadelphia 1992.

[7]   P. Y. Yalamov: *Improvements of some bounds on the stability of fast Helmholtz solvers*, preprint, Univ. of Rousse 1998.

Addresses:

GÜNTER BASZENSKI
Fachbereich Nachrichtentechnik
Fachhochschule Dortmund
Sonnenstr. 96–100
D–44139 Dortmund
Germany

RAO XUEFANG
Beijing Institute of Tracking
and Telecommunications Technology
Beijing PO Box 5131
Beijing 100094
P. R. China

MANFRED TASCHE
Institut für Mathematik
Medizinische Universität zu Lübeck
Wallstr. 40
D–23560 Lübeck
Germany

The Saturation Theorem for Box Spline Orthogonal Projection

Marek Beśka and Karol Dziedziul

To Prof. Manfred Reimer on his 65th birthday

Abstract

In this paper we prove the saturation theorem for box spline orthogonal projection. This is a continuation of our research in [2], where we dealt with the box spline interpolation and the Bernstein–Schnabl operator.

1 Introduction

We start by recalling some basic facts and terminology. The best general reference here is the book [3].

Let $W_2^k(\mathbb{R}^d)$ denote the Sobolev spaces with the norm

$$\|f\|_{k,2} = \sum_{|\beta| \leq k} \|D^\beta f\|_2,$$

where

$$D^\beta f = \frac{\partial^{|\beta|} f}{\partial x_1^{\beta_1} \cdots \partial x_d^{\beta_d}}, \quad \beta = (\beta_1, \cdots, \beta_d), \quad |\beta| = \beta_1 + \cdots + \beta_d,$$

and

$$\|f\|_2 = \left(\int_{\mathbb{R}^d} |f|^2 \right)^{1/2}.$$

We put

$$|f|_{k,2} = \sum_{|\beta|=k} \|D^\beta f\|_2.$$

We use the standard notation for convolution

$$f * g(x) = \int_{\mathbb{R}^d} f(y) g(x - y) \, dy.$$

Advances in Multivariate Approximation; W. Haußmann, K. Jetter and M. Reimer (eds.)
Mathematical Research, Vol. 107, pp. 73–83, ISBN 3-527-40236-5
© WILEY-VCH, Berlin 1999

Let $V = \{v_1, v_2, \cdots, v_n\}$ denote a set of not necessarily distinct, non–zero vectors in $\mathbb{Z}^d$, such that

$$\mathrm{span}\{V\} = \mathbb{R}^d.$$

We call such a set admissible. The box spline corresponding to V, denoted by $B_V(\cdot)$, is defined by requiring that

$$\int_{\mathbb{R}^d} f(x) B_V(x)\, dx = \int_{[0,1]^n} f(Vu)\, du \tag{1.1}$$

holds for any continuous function f on $\mathbb{R}^d$.

The Fourier transform is given by

$$\widehat{f}(\xi) = \int_{\mathbb{R}^d} f(t) e^{-2\pi i \xi \cdot t}\, dt.$$

Here and subsequently "$\cdot$" denotes the scalar product in $\mathbb{R}^d$. From (1.1) we get by simple calculation that

$$\widehat{B}_V(\xi) = \prod_{v \in V} \frac{1 - e^{-2\pi i \xi \cdot v}}{2\pi i \xi \cdot v}. \tag{1.2}$$

For an admissible set V let

$$\varrho_V = \max\{\, r : \forall\, W \subset V \text{ such that } |W| = r,\ \mathrm{span}\{V \setminus W\} = \mathbb{R}^d \,\}.$$

This parameter determines the smoothness of a box spline:

$$B_V(\cdot) \in C^{\varrho_V - 1}(\mathbb{R}^d) \setminus C^{\varrho_V}(\mathbb{R}^d).$$

Moreover, if Π_k denotes the space of all d–variable polynomials of total degree at most k, then

$$\Pi_{\varrho_V} \subset \mathrm{span}\{ B_V(\cdot - \alpha) : \alpha \in \mathbb{Z}^d \}.$$

Let us define

$$S_{L^2}(hV) = \overline{\mathrm{span}}\{ B_{hV}(\cdot - \alpha) : \alpha \in h\mathbb{Z}^d \},$$

where $h > 0$ and the closure is taken in the $L^2(\mathbb{R}^d)$ topology. The orthogonal projection from $L^2(\mathbb{R}^d)$ onto $S_{L^2}(hV)$ is denoted by P_h.

By $-V$ we denote the family of vectors such that

$$-v \in V \quad \text{if and only if} \quad v \in -V.$$

We denote by $|V|$ the cardinality of the set V. A family V is unimodular if $|\det W| \leq 1$ for all $W \subset V$ such that $|W| = d$. Put

$$X = V \cup -V.$$

The following conditions are equivalent:

i) V is unimodular.

ii) For all $x \in \mathbb{R}^d$

$$w(x) = \sum_{\alpha \in \mathbf{Z}^d} B_X(\alpha) e^{2\pi i \alpha \cdot x} \neq 0. \tag{1.3}$$

Moreover, in this case, the sequence $\{b(\alpha)\}$ of Fourier coefficients of the function $1/w(x)$ is symmetric, i.e.

$$b(\alpha) = b(-\alpha), \tag{1.4}$$

and decays exponentially, i.e. there are constants $C > 0$ and $0 < q < 1$ such that for all $\alpha \in \mathbf{Z}^d$

$$|b(\alpha)| \leq Cq^{\|\alpha\|}.$$

The function

$$B_V^*(x) = \sum_{\alpha \in \mathbf{Z}^d} b(\alpha) B_V(x - \alpha) \tag{1.5}$$

is biorthogonal, i.e.

$$\int_{\mathbb{R}^d} B_V^*(t) B_V(t - \alpha)\, dt = \begin{cases} 0 & \text{for } \alpha \neq 0, \\ 1 & \text{for } \alpha = 0. \end{cases}$$

Denoting the inner product in $L^2(\mathbb{R}^d)$ by $(\cdot, \cdot)$, the orthogonal projection onto $S_{L^2}(hV)$ can be written as

$$P_h = \sigma_h \circ P \circ \sigma_{1/h},$$

where

$$\sigma_h f(x) = f(x/h),$$

and

$$Pf(x) = \sum_{\alpha \in \mathbf{Z}^d} (f, B_V^*(\cdot - \alpha)) B_V(x - \alpha)$$

is the orthogonal projection onto $S_{L^2}(V)$.

2   The Saturation Theorem

The rate of convergence of P_h is well known in the theory of wavelets.

Theorem 2.1. *Let V be admissible. There is a constant C_V such that for all $f \in W_2^{\varrho_V + 1}(\mathbb{R}^d)$*

$$\|P_h f - f\|_2 \le C_V h^{\varrho_V + 1} |f|_{\varrho_V + 1, 2}. \tag{2.1}$$

The proof is based on the fact that the norm of the operators P_h are bounded by the same constant. When the set V is unimodular, we can prove more, namely

Theorem 2.2. *Let V be unimodular and $f \in W_2^{\varrho_V + 1}(\mathbb{R}^d)$. If $h \to 0$ then*

$$\frac{P_h f - f}{h^{\varrho_V + 1}} \longrightarrow 0 \tag{2.2}$$

in the weak topology in $L^2(\mathbb{R}^d)$ and

$$\left\| \frac{P_h f - f}{h^{\varrho_V + 1}} \right\|_2^2 \longrightarrow \left(\frac{1}{4\pi^2} \right)^{\varrho_V + 1} \sum_{U \in \Lambda} \|D_U f\|_2^2 \sum_{\alpha \perp (V \setminus U), \alpha \neq 0} \prod_{v \in U} \frac{1}{(\alpha \cdot v)^2}, \tag{2.3}$$

where

$$\Lambda = \{ U \subset V : |U| = \varrho_V + 1, \quad \mathrm{span}\{V \setminus U\} \neq \mathbb{R}^d \}$$

and

$$D_U = \prod_{u \in U} D_u;$$

here D_u is the directional derivative in the direction u.

We have put $\beta \perp (V \setminus W) \quad \Longleftrightarrow \quad \forall_{v \in (V \setminus W)} \quad v \cdot \beta = 0.$

An easy computation shows that it suffices to prove Theorem 2.2 when $\hat{f}$ has compact support. Let us first prove two lemmas.

Lemma 2.3. *For $f \in W_2^{\varrho_V + 1}(\mathbb{R}^d)$ as above we have*

$$\widehat{P_h f}(x) = \frac{\widehat{B_V}(hx)}{w(-hx)} \sum_{\alpha \in \mathbf{Z}^d} \widehat{B_{-V}}(hx + \alpha) \hat{f}(x + \alpha/h), \tag{2.4}$$

where $w(\cdot)$ is given by (1.3).

Proof. By definition (1.1) and (1.5) it follows that

$$B_{-V}(t) = B_V(-t)$$

and

$$B^*_{-V}(t) = B^*_V(-t).$$

Thus from the properties of the Fourier transform we get

$$\widehat{P_h f}(x) = h^d \widehat{B_V}(hx) \sum_{\alpha \in \mathbf{Z}^d} (\sigma_{1/h} f, B^*_V(\cdot - \alpha)) e^{-2\pi i h \alpha \cdot x} \tag{2.5}$$

$$= h^d \widehat{B_V}(hx) \sum_{\alpha \in \mathbf{Z}^d} (\sigma_{1/h} f, B^*_{-V}(\alpha - \cdot)) e^{-2\pi i h \alpha \cdot x}$$

$$= h^d \widehat{B_V}(hx) \sum_{\alpha \in \mathbf{Z}^d} (\sigma_{1/h} f * B^*_{-V})(\alpha) e^{-2\pi i h \alpha \cdot x}.$$

We use the Poisson summation formula [7]

$$\sum_{\alpha \in \mathbf{Z}^d} \widehat{F}(\alpha) e^{2\pi i \alpha \cdot x} = \sum_{\alpha \in \mathbf{Z}^d} F(x - \alpha) \tag{2.6}$$

for the function

$$F(-t) = \frac{1}{h^d} \widehat{B^*_{-V}}(t) \widehat{f}(t/h) \tag{2.7}$$

and for its Fourier transform

$$\widehat{F}(x) = (\sigma_{1/h} f * B^*_{-V})(x). \tag{2.8}$$

Consequently (2.6), (2.7) and (2.8) imply that

$$\sum_{\alpha \in \mathbf{Z}^d} \widehat{F}(\alpha) e^{2\pi i \alpha \cdot (-hx)} = \sum_{\alpha \in \mathbf{Z}^d} F(-hx - \alpha) = \frac{1}{h^d} \sum_{\alpha \in \mathbf{Z}^d} \widehat{B^*_{-V}}(hx + \alpha) \widehat{f}(\frac{hx + \alpha}{h}).$$

Thus (2.5) gives

$$\widehat{P_h f}(x) = \widehat{B_V}(hx) \sum_{\alpha \in \mathbf{Z}^d} \widehat{B^*_{-V}}(hx + \alpha) \widehat{f}(x + \alpha/h). \tag{2.9}$$

(1.4) and (1.5) imply that

$$\widehat{B^*_{-V}}(x) = \widehat{B_{-V}}(x) \sum_{\gamma \in \mathbf{Z}^d} b(\gamma) e^{-2\pi i \gamma \cdot x}. \tag{2.10}$$

From (2.9), (2.10) and (1.3) we get (2.4). $\square$

Lemma 2.4. *Let* $f \in W_2^{\varrho v + 1}(\mathbb{R}^d)$ *and* $\operatorname{supp} \widehat{f} \subset [-N, N]^d = C \subset \mathbb{R}^d$ *for some* $N \in \mathbb{R}$. *Then*

$$\int_C \left| \frac{\widehat{P_h f} - \widehat{f}}{h^{\varrho v + 1}} \right|^2 \longrightarrow 0 \tag{2.11}$$

as $h \to 0$.

Proof. From (2.4) for sufficiently small $h > 0$ we get

$$\int_C \left| \frac{\widehat{P_h f} - \widehat{f}}{h^{\varrho v + 1}} \right|^2 = \int_C \left| \frac{\widehat{f}(x)\left(\widehat{B_V}(hx)\widehat{B_{-V}}(hx) - w(-hx)\right)}{w(-hx)\, h^{\varrho v + 1}} \right|^2 dx.$$

From the convolution properties of box spline we have

$$\widehat{B_V}\widehat{B_{-V}} = \widehat{B_V * B_{-V}} = \widehat{B_X}. \tag{2.12}$$

Now from (1.3) and (2.12) it follows that

$$\int_C \left| \frac{\widehat{P_h f} - \widehat{f}}{h^{\varrho v + 1}} \right|^2 = \int_C \left| \frac{\widehat{f}(x)\left(\widehat{B_X}(hx) - \sum_{\alpha \in \mathbf{Z}^d} B_X(\alpha)e^{2\pi i \alpha \cdot (-hx)}\right)}{w(-hx)\, h^{\varrho v + 1}} \right|^2 dx.$$

Note that

$$\sum_{\alpha \in \mathbf{Z}^d} B_X(\alpha)e^{2\pi i \alpha \cdot (-hx)} = \sum_{n=0}^{\infty} \frac{(-2\pi i)^n h^n}{n!} \sum_{|\beta|=n} \frac{n!}{\beta!} x^\beta \sum_{\alpha \in \mathbf{Z}^d} \alpha^\beta B_X(\alpha), \tag{2.13}$$

where

$$\beta = (\beta_1, \cdots, \beta_d), \quad \beta! = \beta_1! \ldots \beta_d!.$$

From Poisson's summation formula we have

$$\sum_{\alpha \in \mathbf{Z}^d} \alpha^\beta B_X(\alpha) = \sum_{\alpha \in \mathbf{Z}^d} \frac{D^\beta \widehat{B_X}(\alpha)}{(-2\pi i)^\alpha}. \tag{2.14}$$

By (1.2) we have that

$$D^\beta \widehat{B_X}(\alpha) = 0 \tag{2.15}$$

for $\alpha \neq 0$ and $|\beta| \leq \varrho_X$. Thus from (2.13), (2.14) and (2.15) we get

$$\sum_{\alpha \in \mathbf{Z}^d} B_X(\alpha)e^{-2\pi i \alpha \cdot hx} = \sum_{n=0}^{\varrho_X} h^n \sum_{|\beta|=n} \frac{x^\beta}{\beta!} D^\beta \widehat{B_X}(0) + \tag{2.16}$$

$$+ \sum_{n=\varrho_X+1}^{\infty} h^n (-2\pi i)^n \sum_{|\beta|=n} \frac{x^\beta}{\beta!} \sum_{\alpha \in \mathbf{Z}^d} \alpha^\beta B_X(\alpha).$$

On the other hand,

$$\widehat{B_X}(hx) = \sum_{m=0}^{\infty} h^m \sum_{|\beta|=m} \frac{D^\beta \widehat{B_X}(0)}{\beta!} x^\beta. \tag{2.17}$$

Since

$$\frac{1}{w(-hx)} \longrightarrow \sum_{\alpha \in \mathbf{Z}^d} B_X(\alpha) = 1, \qquad h \to 0,$$

(2.16) and (2.17) show that

$$\frac{\widehat{B_X}(hx) - \sum_{\alpha \in \mathbf{Z}^d} B_X(\alpha) e^{-2\pi i \alpha \cdot hx}}{w(-hx) \, h^{\varrho v + 1}} \longrightarrow 0, \qquad h \to 0$$

uniformly for $x \in C$, which proves Lemma 2.4. $\square$

Now we can prove (2.2). By Theorem 2.1 we shall have established (2.2) if we prove the following:

$$\int_{\mathbb{R}^d} \frac{P_h f - f}{h^{\varrho v + 1}} g \longrightarrow 0, \qquad h \to 0, \tag{2.18}$$

for $\widehat{g}$ having a compact support ($g \in L^2(\mathbb{R}^d)$). Because $\widehat{f}$ also has compact support there exists $N \in \mathbb{R}$ such that

$$\operatorname{supp} \widehat{f}, \ \operatorname{supp} \widehat{g} \subset [-N, N]^d = C.$$

Hence (2.2) follows from

$$\int_{\mathbb{R}^d} \frac{P_h f - f}{h^{\varrho v + 1}} g = \int_{\mathbb{R}^d} \frac{\widehat{P_h f} - \widehat{f}}{h^{\varrho v + 1}} \widehat{g} = \int_C \frac{\widehat{P_h f} - \widehat{f}}{h^{\varrho v + 1}} \widehat{g} \longrightarrow 0, \qquad h \to 0$$

(by Lemma 2.4).

Now we turn to the proof of (2.3). From Plancherel's formula and Lemma 2.4, it suffices to show that

$$\int_{\mathbb{R}^d \setminus C} \left| \frac{\widehat{P_h f}}{h^{\varrho v + 1}} \right|^2 \to \left(\frac{1}{4\pi^2} \right)^{\varrho v + 1} \sum_{W \in \Lambda} \| D_W f \|_2^2 \sum_{\alpha \perp (V \setminus W), \alpha \neq 0} \prod_{v \in W} \frac{1}{(\beta \cdot v)^2}, \tag{2.19}$$

when $h \to 0$. From (2.4) we have

$$\int_{\mathbb{R}^d \setminus C} \left| \frac{\widehat{P_h f}}{h^{\varrho v + 1}} \right|^2 = \int_{\mathbb{R}^d \setminus C} \left| \frac{\widehat{B_V}(hx) \sum_{\alpha \in \mathbf{Z}^d, \alpha \neq 0} \widehat{B_{-V}}(hx + \alpha) \hat{f}(x + \alpha/h)}{w(-hx)\, h^{\varrho v + 1}} \right|^2 \, dx.$$

Since $\hat{f}$ has compact support, we have for sufficiently small $h > 0$

$$\int_{\mathbb{R}^d \setminus C} \left| \frac{\widehat{P_h f}}{h^{\varrho v + 1}} \right|^2 = \tag{2.20}$$

$$= \sum_{\delta \in \mathbf{Z}^d, \delta \neq 0} \int_{C + \delta/h} \left| \frac{\widehat{B_V}(hx) \widehat{B_{-V}}(hx - \delta) \hat{f}(x - \delta/h)}{w(-hx)\, h^{\varrho v + 1}} \right|^2 \, dx$$

$$= \sum_{\delta \in \mathbf{Z}^d, \delta \neq 0} \int_C \left| \frac{\widehat{B_V}(hu + \delta) \widehat{B_{-V}}(hu) \hat{f}(u)}{w(-hu)\, h^{\varrho v + 1}} \right|^2 \, du.$$

If $h \longrightarrow 0$ then

$$w(-hu) \longrightarrow 1$$

and

$$\widehat{B_{-V}}(hu) \longrightarrow \widehat{B_{-V}}(0) = 1$$

uniformly on C. To prove (2.19), we need only consider the convergence of the expression

$$\sum_{\delta \in \mathbf{Z}^d, \delta \neq 0} \int_C \left| \frac{\widehat{B_V}(hu + \delta) \hat{f}(u)}{h^{\varrho v + 1}} \right|^2 \, du$$

as $h \to 0$. From (1.2) we have

$$\sum_{\delta \in \mathbf{Z}^d, \delta \neq 0} \left| \frac{\widehat{B_V}(hu + \delta)}{h^{\varrho v + 1}} \right|^2 = \sum_{\delta \in \mathbf{Z}^d, \delta \neq 0} \left| \frac{\prod_{v \in V} \dfrac{1 - e^{-2\pi i h u \cdot v}}{2\pi i (\delta \cdot v + h u \cdot v)}}{h^{\varrho v + 1}} \right|^2. \tag{2.21}$$

Define

$$S_0 = \{ \alpha \in \mathbb{Z}^d \setminus \{0\} : \text{there is a } v \in V \text{ with } \alpha \cdot v = 0 \},$$

$$S_1 = \{ \alpha \in \mathbb{Z}^d \setminus \{0\} : \alpha \cdot v \neq 0 \text{ for all } v \in V \}.$$

We divide the range of summation (2.20) into the sets S_0 and S_1. If $h \to 0$, then the sum for $\alpha \in S_1$ disappears. Consider

$$\sum_{\alpha \in S_0} \left| \frac{\displaystyle\prod_{v \in V} \frac{1 - e^{-2\pi i h u \cdot v}}{2\pi i (\delta \cdot v + h u \cdot v)}}{h^{\varrho v + 1}} \right|^2 \tag{2.22}$$

$$= \sum_{\alpha \in S_0} \left| \left(\prod_{v \in V, \alpha \cdot v = 0} \frac{1 - e^{-2\pi i h u \cdot v}}{2\pi i (h u \cdot v)} \right) \frac{\displaystyle\prod_{v \in V, \alpha \cdot v \neq 0} \frac{1 - e^{-2\pi i h u \cdot v}}{2\pi i (\alpha \cdot v + h u \cdot v)}}{h^{\varrho v + 1}} \right|^2 .$$

If $h \to 0$ then

$$\prod_{v \in V, \alpha \cdot v = 0} \frac{1 - e^{-2\pi i h u \cdot v}}{2\pi i (h u \cdot v)} \longrightarrow 1 \tag{2.23}$$

uniformly on C. For $\alpha \in S_0$ we define

$$J_\alpha = \{ v \in X : v \cdot \alpha \neq 0 \}.$$

Note that
$$|J_\alpha| \geq \varrho v + 1.$$

Moreover, there exists an $\alpha \in S_0$ such that

$$|J_\alpha| = \varrho v + 1.$$

Let
$$I = \{ \alpha \in S_0 : |J_\alpha| = \varrho v + 1 \}.$$

If $h \to 0$ and $\alpha \in S_0 \setminus I$ then

$$\frac{1}{h^{\varrho v + 1}} \prod_{v \in V, \alpha \cdot v \neq 0} \frac{1 - e^{-2\pi i h u \cdot v}}{2\pi i (\alpha \cdot v + h u \cdot v)} \longrightarrow 0. \tag{2.24}$$

It follows from (2.21), (2.22), (2.23) and (2.24) that, if $h \to 0$, then

$$\sum_{\delta \in \mathbf{Z}^d, \delta \neq 0} \left| \frac{\widehat{B_V}(h u + \delta)}{h^{\varrho v + 1}} \right|^2 \longrightarrow \sum_{\alpha \in I} \prod_{v \in J_\alpha} \left| \frac{2\pi i u \cdot v}{2\pi i \alpha \cdot v} \right|^2$$

uniformly for $u \in C$. Therefore, if $h \to 0$, then

$$\int_C \sum_{\delta \in \mathbf{Z}^d, \delta \neq 0} \left| \frac{\widehat{B_V}(hu + \delta)\widehat{f}(u)}{h^{\varrho v + 1}} \right|^2 du \longrightarrow \int_C \sum_{\alpha \in I} \prod_{v \in J_\alpha} \left| \frac{2\pi i u \cdot v}{2\pi i \alpha \cdot v} \widehat{f}(u) \right|^2 du$$

$$= \sum_{\alpha \in I} \prod_{v \in J_\alpha} \left| \frac{1}{\alpha \cdot v} \right|^2 \int_C \prod_{v \in J_\alpha} \left| u \cdot v \widehat{f}(u) \right|^2 du.$$

From Plancherel's formula it follows that

$$\int_C \sum_{\delta \in \mathbf{Z}^d, \delta \neq 0} \left| \frac{\widehat{B_V}(hu + \delta)\widehat{f}(u)}{h^{\varrho v + 1}} \right|^2 du \longrightarrow \left(\frac{1}{4\pi^2} \right)^{\varrho v + 1} \sum_{\alpha \in I} \prod_{v \in J_\alpha} \left| \frac{1}{\alpha \cdot v} \right|^2$$

$$\times \int_C |D_{J_\alpha} f(u)|^2 du,$$

where

$$D_{J_\alpha} = \prod_{v \in J_\alpha} D_v.$$

Note that if $\alpha \in I$, then

$$J_\alpha = W \quad \rightarrow \quad \alpha \perp (V \setminus W) \quad \text{and} \quad W \in \Lambda.$$

Moreover, for all $W \in \Lambda$ there is an $\alpha \in I$ such that

$$W = J_\alpha.$$

It follows that

$$\sum_{\alpha \in I} \prod_{v \in J_\alpha} \left| \frac{1}{\alpha \cdot v} \right|^2 \int_C |D_{J_\alpha} f(u)|^2 du = \sum_{W \in \Lambda} \|D_W f\|_2^2 \sum_{\alpha \perp (V \setminus W), \alpha \neq 0} \prod_{v \in W} \frac{1}{(\beta \cdot v)^2},$$

which completes the proof of Theorem 2.2. $\qquad\qquad\square$

References

[1]   M. Beśka, K. Dziedziul: *Multiresolution approximation and Hardy spaces*, J. Approx. Theory **88** (1997), 154–167.

[2]   M. Beśka, K. Dziedziul: *The saturation theorem for interpolation and Bernstein–Schnabl operator,* Math. Comp., in press (1999).

[3]   C. de Boor, K. Höllig, S. Riemenschneider: *Box Splines,* Springer, New York–Berlin–London 1993.

[4]   C. K. Chui, K. Jetter, J. D. Ward: *Cardinal interpolation by multivariate splines,* Math. Comp. **48** (1987), 711–724.

[5]   K. Dziedziul: *The saturation theorem for quasiprojections,* Studia Sci. Math. Hungarica **35** (1999), in press.

[6]   K. Dziedziul: *Box Splines* (Polish); Wydawnictwo Politechniki Gdańskiej (1997).

[7]   E. M. Stein, G. Weiss: *Introduction to Fourier Analysis on Euclidean Spaces,* Princeton Univ. Press, Princeton, New Jersey 1971.

Address:

MAREK BEŚKA; KAROL DZIEDZIUL
Department of Mathematics
Technical University of Gdańsk
Narutowicza 11/12
80–952 Gdańsk
Poland

Problem on Monotonic Behaviour of Bernstein Operators

Marek Beśka and Karol Dziedziul

The set of all algebraic polynomials of degree $\leq n$ on $\mathbb{R}^d$ is denoted by Π_n. The standard simplex in $\mathbb{R}^d$ will be denoted by

$$Q := \{ (x_1, ..., x_d) \in \mathbb{R}^d \; : \; x_1 + x_2 + ... + x_d \leq 1, \; x_1, x_2, ..., x_d \leq 1 \}.$$

Let $f \in C(Q)$ and $n \in \mathbb{N}$. Then the Bernstein operator is defined by

$$B_n(f)(x) := \sum_{|\alpha| \leq n} f(\frac{\alpha}{n}) \, N_{\alpha,n}(x), \qquad x \in Q,$$

where we have $N_{\alpha,n}(x) = \dfrac{n!}{\alpha_0! \alpha!} x_0^{\alpha_0} x^{\alpha}$ (the Bernstein basis for Π_n), $x = (x_1, ..., x_d)$, $\alpha = (\alpha_1, ..., \alpha_d)$, $|\alpha| = \alpha_1 + ... + \alpha_d$, $\alpha_0 = n - |\alpha|$, and $x_0 = 1 - (x_1 + ... + x_d)$ for $\alpha \in \mathbb{N}^d \cup \{0\}$, $|\alpha| \leq n$. Let us introduce for $f \in C^2(Q)$ the differential operator

$$T(f)(x) = \frac{1}{2} \left\{ \sum_{1 \leq i \leq d} x_i \, (1 - x_i) \frac{\partial^2 f(x)}{\partial x_i^2} - 2 \sum_{1 \leq i < j \leq d} x_i x_j \frac{\partial^2 f(x)}{\partial x_i \partial x_j} \right\},$$

where $x = (x_1, \ldots, x_d) \in Q$. It is well known that for every $f \in C^2(Q)$

$$\lim_{n \to \infty} n \Big(B_n(f)(x) - f(x) \Big) = T(f)(x), \qquad x = (x_1, \ldots, x_d) \in Q$$

holds uniformly on Q.

Conjecture. Let $f \in C^2(Q)$. Then the following are equivalent:

1) $B_{n+1}(f)(x) \leq B_n(f)(x), \; x \in Q, \quad n \in \mathbb{N},$

2) $f(x) \leq B_n(f)(x), \; x \in Q, \quad n \in \mathbb{N},$

3) $T(f)(x) \geq 0, \; x \in Q.$

It is easy to check that for $d = 1$ this conjecture is true.

Advances in Multivariate Approximation; W. Haußmann, K. Jetter and M. Reimer (eds.)
Mathematical Research, Vol. 107, p. 84, ISBN 3–527–40236–5
© WILEY–VCH, Berlin 1999

Best One–Sided L^1–Approximation by Blending Functions

Borislav D. Bojanov, Dimiter P. Dryanov, Werner Haußmann
and Geno P. Nikolov

Dedicated to Manfred Reimer on the occasion of his 65th birthday

Abstract

We consider best one–sided L^1–approximation by the infinite–dimensional space $B^{2m,2n} := \{g \in C^{2m,2n}([-1,1]^2) \ : \ \partial^{m+n}/\partial x^m \partial y^n g = 0\}$ of blending functions of order $(2m, 2n)$. By analogy to the univariate one–sided L^1–approximation by algebraic polynomials of degree $2m - 1$, one would expect that a best one–sided approximant might be given by osculatory Hermite interpolation with respect to an appropriate point set. We show that in the multivariate case this is not true in general, but that it holds if one approximates with respect to a smaller function space rather than $B^{2m,2n}$. We shall characterize such best approximants and we discuss the question of uniqueness.

1   Introduction and Statement of Results

Blending functions are used in various fields of mathematics such as numerical treatment of partial differential equations, cubature formulae and computer aided geometric design. They also prove to be a natural tool for solving multivariate approximation problems. Some classical results in approximation theory have their multivariate counterparts in approximation by blending functions. A typical example is the problem of finding the best L^1–approximant on $I^2 := [-1,1]^2$ to $x^m y^n$ from the space of *blending functions*

$$B^{m,n}(I^2) := \{ \, g \in C^{m,n}(I^2) \ : \ D^{m,n}g := \frac{\partial^{m+n}}{\partial x^m \partial y^n} \, g = 0 \, \},$$

where

$$C^{m,n}(I^2) := \{ \, f : I^2 \to \mathbb{R} \ : \ D^{k,\ell}f \text{ continuous}, \ 0 \le k \le m, \ 0 \le \ell \le n \, \}.$$

Advances in Multivariate Approximation; W. Haußmann, K. Jetter and M. Reimer (eds.)
Mathematical Research, Vol. 107, pp. 85–106, ISBN 3–527–40236–5
© WILEY–VCH, Berlin 1999

The minimal deviation is attained by the normalized product $2^{-m-n}U_m(x)U_n(y)$ of two Chebyshev polynomials of second kind (see Haußmann–Zeller [9, p. 551]). This result is a special case of the following

Theorem A. (see [9]) *Let $f \in C^{m,n}(I^2)$ with $D^{m,n}f \geq 0$. Then f possesses exactly one best L^1–approximant in $B^{m,n}(I^2)$. It is given by the blending function g which coincides with f on the point set $G := \{ (x,y) \in I^2 : U_m(x)U_n(y) = 0 \}$.*

Theorem A states that the best L^1–approximant by blending functions for $f \in C^{m,n}(I^2)$ with $D^{m,n}f \geq 0$ is given by blending interpolation as considered in [9]. There it was shown that for any $f \in C^{m,n}(I^2)$ and for any fixed *blending grid* (more precisely: (m,n)–*blending grid*)

$$G_{\mathbf{x},\mathbf{y}}^{(m,n)} := \{ (x,y) \in I^2 \ : \ \prod_{\mu=1}^{m}(x - x_\mu) \prod_{\nu=1}^{n}(y - y_\nu) = 0 \},$$

where $\mathbf{x} = (x_1,...,x_m)$, $\mathbf{y} = (y_1,...,y_n)$ with $-1 \leq x_1 < x_2 < ... < x_m \leq 1$, $-1 \leq y_1 < y_2 < ... < y_n \leq 1$, there exists a unique *blending interpolant g_f* (of Lagrange type) whose restriction to $G_{\mathbf{x},\mathbf{y}}^{(m,n)}$ satisfies

$$g_f\big|_{G_{\mathbf{x},\mathbf{y}}^{(m,n)}} = f\big|_{G_{\mathbf{x},\mathbf{y}}^{(m,n)}}.$$

An extended result holds for *blending Hermite interpolation*. Let us state this problem as follows: Given a blending grid $G_{\mathbf{x},\mathbf{y}}^{(m,n)}$ as above, we associate with every node x_μ a multiplicity i_μ, and with every y_ν a multiplicity j_ν $(1 \leq \mu \leq m, 1 \leq \nu \leq n)$ such that

$$\sum_{\mu=1}^{m} i_\mu =: M \qquad \text{and} \qquad \sum_{\nu=1}^{n} j_\nu =: N.$$

In Proposition 7 we shall show that for a given $f \in C^{M,N}(I^2)$ there exists a unique blending Hermite interpolant $g \in B^{M,N}(I^2)$ which interpolates f as follows:

$$D^{i,0}g(x_\mu,y) = D^{i,0}f(x_\mu,y) \qquad \mu = 1,...,m, \ i = 0,...,i_\mu - 1, \quad y \in [-1,1],$$

$$D^{0,j}g(x,y_\nu) = D^{0,j}f(x,y_\nu) \qquad \nu = 1,...,n, \ j = 0,...,j_\nu - 1, \quad x \in [-1,1].$$

We say that a blending interpolant g is of multiplicity 2, if $i_\mu = j_\nu = 2$ for $1 \leq \mu \leq m, 1 \leq \nu \leq n$.

In the present paper we consider *best one–sided L^1–approximation by blending functions* which is defined as follows: Let $f \in C(I^2)$ and $\Omega \subset B^{m,n}(I^2)$. A function $h_0 \in \Omega$ is defined to be a *best one–sided L^1–approximant to f from Ω* if $f - h_0 \geq 0$ and

$$\int_{I^2} (f - h_0) \;\leq\; \int_{I^2} (f - h)$$

for all $h \in \Omega$ such that $f - h \geq 0$.

In contrast to the univariate case and unlike the situation of Theorem A, we show that the best one–sided L^1–approximant with respect to the class $B^{2m,2n}(I^2)$ of blending functions is *not* a blending interpolant. This result is based on the fact that a blending function which is non–positive on some blending grid (e.g. on the set $G_{x,y}^{(m,n)}$ above) may have positive integral over I^2. Indeed, we have the following

Theorem 1. *There exists a blending function $h \in B^{2m,2n}(I^2)$ such that it is negative on the Legendre (m,n)–blending grid defined as*

$$LG_{x,y}^{(m,n)} := \{ \, (x,y) \in I^2 \;\; : \;\; P_m(x) \cdot P_n(y) = 0 \, \},$$

where P_m and P_n are the Legendre polynomials of degree m resp. n, i.e. $h|_{LG_{x,y}^{(m,n)}} < 0$, but its integral is positive:

$$\int_{I^2} h \;>\; 0.$$

By making use of this property and Armitage–Gardiner [1, Proposition 1] it follows that contrary to the univariate case for a given $f \in C^{2m,2n}(I^2)$ with $D^{2m,2n} f > 0$ the Hermite blending interpolant g of multiplicity 2 on the Legendre (m,n)–blending grid is *not* a best one–sided L^1–approximant to f with respect to $B^{2m,2n}(I^2)$. More precisely, we have

Corollary 2. *Let $f \in C^{2m,2n}(I^2)$ with $D^{2m,2n} f > 0$ on I^2. Then*

$$\inf_{h \in B^{2m,2n}(I^2), \, f - h \geq 0} \int_{I^2} (f - h) \;<\; \int_{I^2} (f - g),$$

where g is the blending Hermite interpolant of multiplicity 2 with respect to the Legendre (m,n)–blending grid.

On the other hand, if we consider a smaller class of approximating blending functions rather than $B^{2m,2n}(I^2)$, then we can characterize the best one–sided L^1–approximant by a blending interpolation condition:

Theorem 3.   *Let $f \in C^{2m,2n}(I^2)$ satisfy $D^{2m,2n}f \geq 0$ on I^2. The unique best one–sided L^1–approximant to f from*

$$B_f^{2m,2n} := \{ h \in B^{2m,2n}(I^2) \, : \, D^{2m,0}(f-h) \geq 0, \; D^{0,2n}(f-h) \geq 0, f-h \geq 0 \text{ on } I^2 \}$$

is given by the Hermite blending interpolant of multiplicity 2 based on the Legendre (m,n)–blending grid $LG_{\mathbf{x},\mathbf{y}}^{(m,n)}$.

Remark.   (i) A natural class of blending functions for one–sided approximation would be the class of all $h \in B^{2m,2n}$ that interpolate f one–sidedly with respect to some blending grid. We will show (see Section 3) that the class $B_f^{2m,2n}(I^2)$ contains all one–sided Hermite blending interpolants for f with respect to any blending grid $G_{\mathbf{x},\mathbf{y}}^{(k,\ell)}$ such that

$$\sum_{\mu=1}^{k} 2i_\mu = 2m \quad \text{and} \quad \sum_{\nu=1}^{\ell} 2j_\nu = 2n.$$

The class $B_f^{2m,2n}$ contains also non–interpolatory blending functions.

(ii) Corollary 2 and Theorem 3 show that for an $f \in C^{2m,2n}(I^2)$ satisfying $D^{2m,2n}f \geq 0$ on I^2 the best one–sided L^1–approximant to f from $B^{2m,2n}$ is *not* a blending interpolant.

If we restrict ourselves to polynomial blending functions, i.e. functions from

$$PB^{2m,2n}(I^2) := \{ h \, : \, h(x,y) = \sum_{\mu=0}^{2m} \sum_{\nu=0}^{2n} a_{\mu\nu} x^\mu y^\nu, a_{\mu\nu} \in \mathbb{R}, \; a_{2m,2n} = 0 \},$$

then a situation as in Theorem 1 cannot occur. Indeed, we have

Theorem 4.   *Let $h \in PB^{2m,2n}$ and $h|_{LG_{\mathbf{x},\mathbf{y}}^{(m,n)}} \leq 0$. Then*

$$(1) \qquad \int_{I^2} h \leq 0.$$

Equality holds in (1) *if and only if*

$$h(x, y) = P_m(x)P_n(y)\Big(P_n(y)s_{m-1}(x) + P_m(x)t_{n-1}(y) + u_{m-1,n-1}(x,y)\Big)$$

where $s_k, t_k \in \Pi_k$ *and* $u_{k,\ell} \in \Pi_{k,\ell}$, *and where* P_k *denotes the Legendre polynomial of degree* k.

Here Π_k denotes the vector space of all univariate polynomials of degree $\leq k$, and $\Pi_{k,\ell}$ is the vector space of all bivariate polynomials whose partial degrees with respect to x and y do not exceed k resp. ℓ.

Next we consider best one–sided L^1–approximation of product type functions. First we introduce some notation. For a given function u on $I = [-1, 1]$ we shall denote by $p_k^-(u; \cdot)$ the polynomial of degree k of best one–sided L^1-approximation from below to u on I. Set

$$E_k^-(u) := \int_{-1}^{1} |u(x) - p_k^-(u; x)| \, dx.$$

Similarly, $p_k^+(u; \cdot)$ and $E_k^+(u)$ are defined for approximation from above.

Theorem 5.   *Let* $u \in C^{2m}(I)$, $v \in C^{2n}(I)$ *satisfy*

$$u^{(2m)} > 0, \qquad v^{(2n)} > 0 \quad \text{on } I.$$

Then the unique best one–sided L^1-approximation from below to $u(x)v(y)$ *on* I^2 *by functions of the form*

$$u(x)f(y) + g(x)v(y) + q(x, y),$$

(where $f \in \Pi_{2n-1}$, $g \in \Pi_{2m-1}$, $q \in \Pi_{2m-1,2n-1}$) *is attained for*

$$h_0(x, y) = u(x) \cdot p_{2n-1}^-(v; y) + p_{2m-1}^-(u; x) \cdot v(y) - p_{2m-1}^-(u; x) \cdot p_{2n-1}^-(v; y).$$

The proof of Theorem 5 uses an inequality due to Brass–Schmeisser [5] and will be given in Section 5.

As a consequence of Theorem 5 we extend a result of Fromm [8] to one–sided L^1–approximation of $x^{2m}y^{2n}$ as follows:

Corollary 6. *The unique best one–sided L^1-approximant to* $x^{2m}y^{2n}$ *from the space* $PB^{2m,2n}(I^2)$ *is given by*

$$x^{2m}y^{2n} - P_m^2(x)P_n^2(y),$$

where the Legendre polynomials are normalized with leading coefficient 1.

Using Armitage–Gardiner [1, Proposition 1], Corollary 6 can be derived from Theorem 4, too.

The paper is organized as follows: In Section 2 we consider some fundamental properties of Hermite blending interpolation. Furthermore, in order to prove our results, the technique used involves blending cubature formulae of maximal degree of precision which will also be considered in Section 2.

In Section 3 we shall give the proofs of Theorems 1 and 3. Section 4 will be devoted to the proof of Theorem 4 while Theorem 5 will be proved in Section 5. Finally we will give some concluding remarks in Section 6.

Note that our space $B^{2m,2n}(I^2)$ of approximating functions is infinite–dimensional. Best one–sided L^1–approximation with respect to the infinite–dimensional space of harmonic functions was dealt with in [2]. For results and references concerning best one–sided L^1–approximation with respect to finite–dimensional spaces we refer to the book of Pinkus [13].

2   Hermite Blending Interpolation and Blending Cubature Formulae

Our proofs are based on Hermite blending interpolation and blending cubature formulae. Hence in this section we give a short overview on these concepts.

Being defined as the kernel of the differential operator $D^{m,n}$, the blending functions share several properties of the algebraic polynomials. Particularly, the basic problem of interpolation by univariate algebraic polynomials has an interesting extension to interpolation by blending functions including Hermite blending interpolation as defined in Section 1.

Proposition 7.   *For a given $f \in C^{M,N}(I^2)$ there exists a unique blending Hermite interpolant $g \in B^{M,N}(I^2)$ satisfying the interpolation conditions*

$$D^{i,0}g(x_\mu,y) \;\; = \;\; D^{i,0}f(x_\mu,y) \qquad \mu = 1,...,m, \; i = 0,...,i_\mu - 1, \quad y \in [-1,1],$$

$$D^{0,j}g(x,y_\nu) \;\; = \;\; D^{0,j}f(x,y_\nu) \qquad \nu = 1,...,n, \; j = 0,...,j_\nu - 1, \quad x \in [-1,1]$$

with respect to a given blending grid $G_{\mathbf{x},\mathbf{y}}^{(m,n)}$.

Indeed, the existence follows immediately from the representation formula

$$g(x,y) \;=\; \sum_{\mu=1}^{m}\sum_{i=0}^{i_\mu-1} D^{i,0}f(x_\mu,y)\cdot\ell^i_{\mu,M}(x) + \sum_{\nu=1}^{n}\sum_{j=0}^{j_\nu-1} D^{0,j}f(x,y_\nu)\cdot\overline{\ell}^{\,j}_{\nu,N}(y)$$

$$-\sum_{\mu=1}^{m}\sum_{i=0}^{i_\mu-1}\sum_{\nu=1}^{n}\sum_{j=0}^{j_\nu-1} D^{i,j}f(x_\mu,y_\nu)\cdot\ell^i_{\mu,M}(x)\cdot\overline{\ell}^{\,j}_{\nu,N}(y),$$

where the fundamental polynomials $\ell^i_{\mu,M}$ resp. $\overline{\ell}^{\,j}_{\nu,N}$ satisfy

$$D^{r,0}\ell^i_{\mu,M}(x_s) \;=\; \delta_{\mu s}\cdot\delta_{ir} \qquad s=1,...,m,\ r=0,...,i_s-1,$$

$$D^{0,r}\overline{\ell}^{\,j}_{\nu,N}(y_s) \;=\; \delta_{\nu s}\cdot\delta_{jr} \qquad s=1,...,n,\ r=0,...,j_s-1.$$

To prove uniqueness, consider any blending function $g \in B^{M,N}$ which solves the homogeneous Hermite interpolation problem. Then

$$D^{M,0}g(x,y) \;=\; \sum_{l=0}^{N-1} D^M b_l(x)\cdot q_l(y)\,,$$

where $D^M = \partial^M/\partial x^M$. For any fixed x_0, the right hand side is a polynomial of y which vanishes at $y_1,...,y_n$ with multiplicities $j_1,...,j_n$ which is seen from the interpolation conditions. Thus $D^{M,0}g = 0$. Hence the blending function g has the form

$$g(x,y) \;=\; \sum_{k=0}^{M-1} p_k(x)\cdot c_k(y)$$

with functions $c_k \in C^n(I)$. For fixed y_0, the polynomial $g(\cdot,y_0) \in \Pi_{M-1}$ has M zeros counting multiplicity, hence $g = 0$. $\qquad\square$

Now we consider the *remainder of the Hermite interpolation formula*. We put $(\mathbf{x},\mathbf{i}) := (x_1,...,x_m,i_1,...,i_m)$, where $(x_1,...,x_m)$ are the interpolation nodes with multiplicities $(i_1,...,i_m)$, and define

$$\omega_{\mathbf{x},\mathbf{i}}(x) \;=\; \prod_{\mu=1}^{m}(x-x_\mu)^{i_\mu}.$$

We shall write $\omega_{\mathbf{x}}$ if all multiplicities are $i_1 = ... = i_m = 1$.

In univariate Hermite interpolation we have for an M–times continuously differentiable function f the representation

$$(2) \qquad f(x) \; = \; H^M(f) \; + \; \omega_{\mathbf{x},\mathbf{i}}(x) \int_{-1}^{1} B_{M-1}((\mathbf{x},\mathbf{i}), x; \xi) \cdot f^{(M)}(\xi) \, d\xi,$$

where $H^M(f)$ denotes the Hermite interpolant corresponding to $(\mathbf{x},\mathbf{i})$ and B_{M-1} is the normalized B–spline (with $\int_I B_{M-1}((\mathbf{x},\mathbf{i}), x; \xi) \, d\xi = 1/M!$) of degree $M-1$ with knots $(\mathbf{x}, x) = (x_1, x_2, ..., x_m, x)$ and multiplicities $(i_1, i_2, ..., i_m, 1)$. For details about B–splines, we refer to the book by Bojanov–Hakopian–Sahakian [3].

Applying (2) first to $f \in C^{(M,N)}(I^2)$ with respect to x and then to $f^{(M,0)}$ with respect to y, we get a representation of f in terms of the Hermite blending interpolant

$$f(x,y) \; = \; H_x^M(f) \; + \; H_y^N(f) \; - \; H_x^M H_y^N(f) \; + \; R_{M,N}^H(f; \, x, y).$$

Here and later the indices x and y mean that the univariate operator is applied with respect to the variable x resp. y.

The remainder is given by

$$R_{M,N}^H(f; \, x, y)$$

$$= \omega_{\mathbf{x},\mathbf{i}}(x) \cdot \omega_{\mathbf{y},\mathbf{j}}(y) \int_{-1}^{1} \int_{-1}^{1} B_{M-1}((\mathbf{x},\mathbf{i}), x; \xi) \cdot B_{N-1}((\mathbf{y},\mathbf{j}), y; \eta) f^{(M,N)}(\xi, \eta) d\xi d\eta.$$

By the mean value theorem we can rewrite the remainder in the form

$$R_{M,N}^H(f; \, x, y) \; = \; \frac{f^{(M,N)}(u_x, v_y)}{M! \, N!} \; \omega_{\mathbf{x},\mathbf{i}}(x) \cdot \omega_{\mathbf{y},\mathbf{j}}(y).$$

This remainder representation can be also obtained by the method used in Haußmann–Zeller [9, p. 548].

Now we consider *cubature formulae* which are exact for classes of blending functions. Given two quadratures Q_x and Q_y of the form

$$\int_a^b f(t) \, dt \; = \; Q(f) \; + \; R(f) \; := \; \sum_{\mu=1}^{m} a_\mu L_\mu(f) \; + \; \int_a^b M(t) f^{(r)}(t) \, dt,$$

where the L_μ denote linear functionals (point evaluation, derivatives) and where M is the corresponding Peano kernel, we construct a cubature formula by

$$\int_a^b \int_c^d f(x,y)\,dxdy \;=\; \int_c^d Q_x(f)\,dy + \int_a^b Q_y(f) - \sum_{\mu=1}^m \sum_{\nu=1}^n a_\mu^{(x)} \cdot a_\nu^{(y)} \cdot L_\mu^{(x)} L_\nu^{(y)}(f)$$

$$+ \int_a^b \int_c^d M_x(t) M_y(u) \cdot f^{(r,s)}(t,u)\,dtdu.$$

For easy reference, we shall call any cubature of this type a *blending cubature formula*. As a first example we mention the compound rectangular blending cubature formula

$$\int_{I^2} f \;=\; \frac{2}{m} \int_{G_{\mathbf{x,y}}^{(m,n)}} f \;-\; \frac{4}{m^2} \sum_{\mu=0}^{m-1} \sum_{\nu=0}^{m-1} f\left(x_\mu, y_\nu\right) \;+\; \frac{D^{2,2} f(\xi,\eta)}{9m^4}\,,$$

where $G_{\mathbf{x,y}}^{(m,m)}$ is the (m,m)–blending grid based on the node

$$x_\mu = y_\mu = -1 + \frac{2\mu+1}{m},\ \mu = 0,...,m-1.$$

Another example is the trapezoidal blending cubature formula

$$\int_{I^2} f \;=\; \frac{2}{m} \int_{G_{\mathbf{x,y}}^{(m+1,n+1)}} f \;-\; \frac{1}{m} \int_{\partial I^2} f \;-\; \frac{4}{m^2} \sum_{\mu=1}^{m-1} \sum_{\nu=1}^{m-1} f(x_\mu, y_\nu)$$

$$- \frac{2}{m^2} \sum_{\mu=1}^{m-1} \left(f(-1,y_\mu) + f(1,y_\mu) + f(x_\mu,-1) + f(x_\mu,1) \right)$$

$$- \frac{1}{m^2} \left(f(-1,-1) + f(-1,1) + f(1,-1) + f(1,1) \right) + \frac{4}{9m^4} D^{2,2} f(\xi,\eta),$$

where $G_{\mathbf{x,y}}^{(m+1,m+1)}$ is the $(m+1,m+1)$–blending grid based on the points

$$x_\mu \;=\; y_\mu \;=\; -1 + \frac{2\mu}{m} \qquad\qquad (\mu = 0,1,...,m).$$

For our purpose, we turn now to *Gaussian blending cubature*. First we define the blending degree of precision of such a formula. We say that a cubature formula Q has a *blending degree of precision* $\mathrm{BDP}(Q) = (m,n)$ if Q integrates all blending functions from $B^{m,n}(I^2)$ to the exact value, but not from $B^{m_1,n_1}(I^2)$ with $m_1 > m$ or $n_1 > n$.

Given any (m,n)–blending grid $G_{\mathbf{x},\mathbf{y}}^{(m,n)}$ based on $\mathbf{x} : x_1 < x_2 < ... < x_m$ and $\mathbf{y} : y_1 < y_2 < ... < y_n$ we can construct a cubature formula of $\mathrm{BDP}(Q) = (m,n)$; indeed, take the interpolating blending cubature.

If $\mathbf{x}$ and $\mathbf{y}$ are taken as the zeros of the *Legendre polynomials* P_m and P_n of degree m resp. n, then we achieve a cubature formula with $\mathrm{BDP} = (2m, 2n)$.

This cubature formula will be referred to as *Gaussian blending cubature*. Its existence can be seen by integrating the Hermite blending interpolant (with multiplicities 2) on the Legendre (m,n)–blending grid. Hence Gaussian blending cubature can be expressed as follows:

$$\int_{I^2} f \; = \; \sum_{\mu=1}^{m} A_\mu^m \int_I f(x_\mu, y)\, dy \; + \; \sum_{\nu=1}^{n} A_\nu^n \int_I f(x, y_\nu)\, dx$$

$$- \; \sum_{\mu=1}^{m} \sum_{\nu=1}^{n} A_\mu^m A_\nu^n \cdot f(x_\mu, y_\nu) \; + \; R_{m,n}(f),$$

where A_μ^m are the coefficients of the corresponding univariate Gaussian quadrature. Note that $A_\mu^m > 0$ for $\mu = 1, ..., m$.

The degree $(2m, 2n)$ is a maximal BDP in the sense that for any (m,n)–blending grid $G_{\mathbf{x},\mathbf{y}}^{(m,n)}$ there exists a blending function of degree $(2m + 1, j)$ for every $j \geq 0$ for which no blending cubature based on $G_{\mathbf{x},\mathbf{y}}^{(m,n)}$ is exact. To see this, take (for $j = 0$)

$$f(x,y) \; = \; \psi(y) \cdot \prod_{\mu=1}^{m} (x - x_\mu)^2,$$

where $\psi(y_\nu) = 0$, $\nu = 1, ..., n$, and $\psi(y) > 0$ on $I \setminus \{y_1, ..., y_n\}$.

The Legendre (m,n)–blending grid (based on the zeros of the corresponding Legendre polynomials) is unique with the property that the corresponding blending cubature has $(2m, 2n)$ as BDP. Indeed, if a cubature formula has $\mathrm{BDP} = (2m, 2n)$, it is exact for

$$f_i(x,y) \; = \; \psi(y) \cdot x^i \prod_{\mu=1}^{m} (x - x_\mu), \qquad\qquad 0 \leq i \leq m - 1,$$

ψ as above. This determines the nodes $x_1, ..., x_m$ uniquely as the zeros of the Legendre polynomial of degree m. From this it follows that the Gaussian blending cubature is the unique (m, n)–blending cubature with maximal BDP $= (2m, 2n)$.

The remainder $R_{m,n}(f)$ of the Gaussian blending cubature can be expressed in different forms, e.g.

$$R_{m,n}(f) \; = \; \frac{D^{2m,2n} f(\xi, \eta)}{(2m)! \, (2n)!} \int_{I^2} \omega_{\mathbf{x}}^2(x) \cdot \omega_{\mathbf{y}}^2(y) \, dx dy.$$

Note that the Gaussian blending cubature can be obtained also by integrating the *Lagrange blending interpolant* with respect to the Legendre (m, n)–blending grid.

3   Proofs of Theorems 1 and 3

3.1. *Proof of Theorem 1.*

For any fixed (x_{μ_0}, y_{ν_0}) in $LG_{\mathbf{x,y}}^{(m,n)}$ and sufficiently small $\varepsilon > 0$, $\delta > 0$ we consider the auxiliary function

$$H_{\varepsilon, \delta} \quad := \quad H_\varepsilon - \delta,$$

where $H_\varepsilon(x, y)$ is given by

$$\begin{cases} -e^2 \cdot \exp\left(-\dfrac{\varepsilon^2}{\varepsilon^2 - (x - x_{\mu_0})^2}\right) \cdot \exp\left(-\dfrac{\varepsilon^2}{\varepsilon^2 - (y - y_{\nu_0})^2}\right) & \begin{aligned} &|x - x_{\mu_0}| < \varepsilon, \\ &|y - y_{\nu_0}| < \varepsilon, \end{aligned} \\[2em] 0 & \text{elsewhere on } I^2, \end{cases}$$

with ε such that $H_\varepsilon(x_\mu, y_\nu) = 0$ for $(\mu, \nu) \neq (\mu_0, \nu_0)$. Then $H_{\varepsilon, \delta} \in C^\infty(I^2)$ and

$$D^{1,0} H_{\varepsilon, \delta}(x_\mu, y) = D^{0,1} H_{\varepsilon, \delta}(x, y_\nu) = 0 \quad \text{for } 1 \leq \mu \leq m, \; 1 \leq \nu \leq n.$$

Hence the Hermite blending interpolant uses only function values and has the form

$$h_{\varepsilon, \delta}(x, y) = \sum_{\mu=1}^{m} H_{\varepsilon, \delta}(x_\mu, y) \, \ell_{\mu, 2m}^0(x) + \sum_{\nu=1}^{n} H_{\varepsilon, \delta}(x, y_\nu) \, \overline{\ell}_{\nu, 2n}^0(y)$$

$$- \sum_{\mu=1}^{m} \sum_{\nu=1}^{n} H_{\varepsilon, \delta}(x_\mu, y_\nu) \, \ell_{\mu, 2m}^0(x) \, \overline{\ell}_{\nu, 2n}^0(y).$$

By integration on I^2 we get

$$\int_{I^2} h_{\varepsilon,\delta} \;=\; \sum_{\mu=1}^{m} A_{\mu}^{m} \int_{-1}^{1} H_{\varepsilon,\delta}\left(x_{\mu},y\right) dy \;+\; \sum_{\nu=1}^{n} A_{\nu}^{n} \int_{-1}^{1} H_{\varepsilon,\delta}\left(x,y_{\nu}\right) dx$$

$$-\sum_{\mu=1}^{m}\sum_{\nu=1}^{n} A_{\mu}^{m} A_{\nu}^{n} H_{\varepsilon,\delta}\left(x_{\mu},y_{\nu}\right)$$

$$\geq (1+\delta) A_{\mu_0}^{m} A_{\nu_0}^{n} \;+\; \delta \sum_{\substack{\mu=1 \\ (\mu,\nu)\neq(\mu_0,\nu_0)}}^{m}\sum_{\nu=1}^{n} A_{\mu}^{m} A_{\nu}^{n}$$

$$-2\delta \sum_{\substack{\mu=1 \\ \mu\neq\mu_0}}^{m} A_{\mu}^{m} \;-\; 2\delta \sum_{\substack{\nu=1 \\ \nu\neq\nu_0}}^{n} A_{\nu}^{n} \;-\; \left(A_{\mu_0}^{m} + A_{\nu_0}^{n}\right)\left(\,2\varepsilon(1+\delta)+(2-2\varepsilon)\delta\,\right)$$

$$\geq A_{\mu_0}^{m} A_{\nu_0}^{n} + 4\delta - 8\delta - 8\varepsilon \;=\; A_{\mu_0}^{m} A_{\nu_0}^{n} - 4\delta - 8\varepsilon > 0$$

for ε and δ sufficiently small. Hence it is obvious that $h := h_{\varepsilon,\delta}$ satisfies the properties stated in Theorem 1. $\qquad\square$

3.2. *Proof of Corollary 2.*

Since $D^{2m,2n} f > 0$, the zero set $Z(f-g)$ is exactly the Legendre (m,n)–blending grid $LG_{\mathbf{x,y}}^{(m,n)}$. Thus Corollary 2 is an immediate consequence of Theorem 1 and Armitage–Gardiner [1, Proposition 1], since property (b) of that proposition is violated in our situation. $\qquad\square$

3.3. *Proof of Theorem 3.*

Let $h \in B_{f}^{2m,2n}$, and let g be the Hermite blending interpolant with multiplicities 2 on the Legendre (m,n)–blending grid. The best one–sided approximation property will follow from

$$(3) \qquad\qquad \int_{I^2}(g-h) \;\geq\; 0.$$

Indeed, for any $h \in B_{f}^{2m,2n}$ we have

$$\int_{I^2}(f-h) \;=\; \int_{I^2}(f-g) \;+\; \int_{I^2}(g-h) \;\geq\; \int_{I^2}(f-g).$$

Hence we have to prove (3). We compute the integral in (3) by integration of the Hermite blending interpolant representation for $g - h$ and get

$$\int_{I^2}(g - h) = \int_{I^2}\left(H_x(g - h) + H_y(g - h) - 2H_xH_y(g - h) \right) + \int_{I^2} H_xH_y(g - h)$$

$$= \sum_{\nu=1}^{n} A_\nu^n \int_{-1}^{1}\int_{-1}^{1} B_{2m-1}(\mathbf{x}, x; t)\,(g - h)^{(2m,0)}(t, y_\nu)\,dt \cdot \omega_{\mathbf{x}}^2(x)\,dx$$

$$+ \sum_{\mu=1}^{m} A_\mu^m \int_{-1}^{1}\int_{-1}^{1} B_{2n-1}(\mathbf{y}, y; u)\,(g - h)^{(0,2n)}(x_\mu, u)\,du \cdot \omega_{\mathbf{y}}^2(y)\,dy$$

$$+ \sum_{\mu=1}^{m}\sum_{\nu=1}^{n} A_\mu^m A_\nu^n (g - h)\,(x_\mu, y_\nu)$$

where the B–splines involved have knots $\mathbf{x}$ and $\mathbf{y}$ with multiplicities 2. Since g interpolates f on $LG_{\mathbf{x},\mathbf{y}}^{(m,n)}$, it follows that $g^{(2m,0)}(\cdot, y_\nu) \geq h^{(2m,0)}(\cdot, y_\nu)$, $g^{(0,2n)}(x_\mu, \cdot) \geq h^{(0,2n)}(x_\mu, \cdot)$ and $g(x_\mu, y_\nu) \geq h(x_\mu, y_\nu)$, hence (3) holds, i.e. g is a best one–sided L^1-approximant to f.

Equality in (3) occurs only if

(i) $h(x_\mu, y_\nu) = g(x_\mu, y_\nu)$ $\qquad\qquad\qquad \mu = 1, ..., m,\ \nu = 1, ..., n,$

(ii) $h^{(2m,0)}(x, y_\nu) = g^{(2m,0)}(x, y_\nu)$ $\qquad\qquad \nu = 1, ..., n,$

(iii) $h^{(0,2n)}(x_\mu, y) = g^{(0,2n)}(x_\mu, y)$ $\qquad\qquad \mu = 1, ..., m.$

We shall show that (i)–(iii) imply $h = g$ and thus uniqueness. In order to prove this, we shall use a simple lemma whose proof follows immediately from Taylor's formula.

Lemma 8. *Assume that $v, w \in C^1[a - \varepsilon, a + \varepsilon]$ for some $\varepsilon > 0$. Let*

$$v(a) = w(a), \text{ and } v(x) \geq w(x) \text{ on } [a - \varepsilon, a + \varepsilon].$$

Then

$$v'(a) = w'(a).$$

Now we turn to the proof of the uniqueness in Theorem 3. Assume that $h \in B_f^{2m,2n}$ satisfies (i)–(iii). Since $g|_{LG_{\mathbf{x,y}}^{(m,n)}} = f|_{LG_{\mathbf{x,y}}^{(m,n)}}$, we can assume equivalently that (i)–(iii) hold with g replaced by f. Without loss of generality, we can assume that $g = 0$.

The function h is represented as

$$h(x,y) \;=\; \sum_{k=0}^{2m-1} \varphi_k(y) \cdot x^k \;+\; \sum_{\ell=0}^{2n-1} \psi_\ell(x) \cdot y^\ell,$$

hence

$$(4) \qquad\qquad h^{(2m,0)}(x,y) \;=\; \sum_{\ell=0}^{2n-1} \psi_\ell^{(2m)}(x) \cdot y^\ell.$$

Now fix $x_0 \in I$. Since $h \in B_f^{2m,2n}$, we have $D^{2m,0}(f - h) \geq 0$ and since $D^{2m,0}(f - h)(x_0, y_\nu) = 0$ (by (ii)), Lemma 8 implies

$$D^{2m,1} h(x_0, y_\nu) \;=\; D^{2m,1} f(x_0, y_\nu) \;=\; 0$$

for $\nu = 1, ..., n$. From (4) it is seen that $h^{2m,0}(x_0, \cdot)$ is a polynomial of degree $\leq 2n - 1$; it vanishes at the points $y_1, ..., y_n$ with multiplicities ≥ 2, hence $h^{2m,0}(x_0, \cdot) = 0$. This is true for any $x_0 \in I$, thus $h^{(2m,0)}$ vanishes identically on I^2. This implies that $\psi_\ell^{(2m)} = 0$ on I, i.e. ψ_ℓ is a polynomial of degree $2m - 1$. Analogously, the functions φ_k are polynomials of degree $2n - 1$, hence h has the form

$$h(x,y) \;=\; \sum_{k=0}^{2m-1} \sum_{\ell=0}^{2n-1} d_{k\ell} x^k y^\ell \,.$$

According to (i), h vanishes on $\{x_1, ..., x_m\} \times \{y_1, ..., y_n\}$. Now by Lemma 8, it follows that

$$D^{1,0} h(x_\mu, y_\nu) \;=\; D^{0,1} h(x_\mu, y_\nu) \;=\; 0,$$

hence $h = 0$ on $G_{\mathbf{x,y}}^{(m,n)}$. For fixed x_0, the polynomial $h(x_0, \cdot)$ vanishes at the points $y_1, ..., y_n$ with multiplicity ≥ 2 by Lemma 8. Hence $h = 0$ on I^2. $\qquad\square$

Remark. (i) In order to see that the one–sided Hermite blending interpolants are included in $B_f^{2m,2n}$, we note that if $h \in B^{2m,2n}(I^2)$ interpolates f on some blending grid $G_{\mathbf{x,y}}^{(k,\ell)}$ with multiplicities $2i_\mu$, $\mu = 1, ..., k$ resp. $2j_\nu$, $\nu = 1, ..., \ell$, then

$$D^{2m,0} h(x,y) \;=\; \sum_{\nu=1}^{k} \sum_{j=0}^{2j_\nu - 1} D^{2m,j} f(x, y_\nu) \, \overline{\ell}_{\nu,2n}^{\,j}(y),$$

i.e. for any fixed $x_0 \in I$, the derivative $D^{2m,0}h$ interpolates $D^{2m,0}f$ with respect to y. Since $D^{2m,2n}f \geq 0$, the univariate even order Hermite interpolation yields $D^{2m,0}(f-h) \geq 0$. In a similar way, one concludes $D^{0,2n}(f-h) \geq 0$. The even order interpolation yields $f - h \geq 0$.

(ii) Our proof of Theorem 3 uses only the assumption that $D^{2m,0}(f-h) \geq 0$ and $D^{0,2n}(f-h) \geq 0$ holds on some neighbourhood of the grid $G_{\mathbf{x,y}}^{(m,n)}$.

4 $\quad$ Proof of Theorem 4

Each $h \in PB^{2m,2n}(I^2)$ can be represented in the form

$$(5) \qquad h(x,y) = P_n^2(y)s_{2m-1}(x) + P_m^2(x)t_{2n-1}(y) + q_{2m-1,2n-1}(x,y)$$

where $s_{2m-1} \in \Pi_{2m-1}, t_{2n-1} \in \Pi_{2n-1}, q \in \Pi_{2m-1,2n-1}$.

By assumption, we have

$$h(x_\mu, y_\nu) \;=\; q_{2m-1,2n-1}(x_\mu, y_\nu) \;\leq\; 0 \quad \mu = 1,\ldots,m, \; \nu = 1,\ldots,n,$$

and from here using Gaussian blending cubature we get

$$(6) \qquad \int_{I^2} q_{2m-1,2n-1}(x,y)\,dx dy \;\leq\; 0.$$

From

$$h(x_\mu, y) \;\leq\; 0 \qquad \mu = 1,\ldots,m,$$

by using univariate Gaussian quadrature we conclude that

$$0 \geq \sum_{\mu=1}^{m} A_\mu^m h(x_\mu, y)$$

$$= P_n^2(y) \int_{-1}^{1} s_{2m-1}(x)\,dx + \int_{-1}^{1} q_{2m-1,2n-1}(x,y)\,dx$$

for each $y \in I$. Similarly

$$0 \;\geq\; P_m^2(x) \int_{-1}^{1} t_{2n-1}(y)\,dy + \int_{-1}^{1} q_{2m-1,2n-1}(x,y)\,dy.$$

Let L_m be the Lobatto quadrature formula with $m-1$ interior nodes. Note that L_m has positive coefficients and algebraic degree of precision $2m-1$ (see Brass

[4], Davis–Rabinowitz [6]). From the above inequalities, since $L_m(P_m^2) > 0$, we have

$$(7) \qquad \int_{-1}^{1} t_{2n-1}(y)\, dy + \frac{1}{L_m(P_m^2)} \cdot \int_{I^2} q_{2m-1,2n-1}(x,y)\, dx\, dy \;\leq\; 0$$

resp.

$$(8) \qquad \int_{-1}^{1} s_{2m-1}(x)\, dx + \frac{1}{L_n(P_n^2)} \int_{I^2} q_{2m-1,2n-1}(x,y)\, dx\, dy \;\leq\; 0.$$

Now we express $\int_{I^2} h$ in the following form:

$$\int_{I^2} h = \int_{-1}^{1} P_m^2(x)\, dx \left[\int_{-1}^{1} t_{2n-1}(y)\, dy + \frac{1}{L_m(P_m^2)} \int_{I^2} q_{2m-1,2n-1} \right]$$

$$+ \int_{-1}^{1} P_n^2(y)\, dy \left[\int_{-1}^{1} s_{2m-1}(x)\, dx + \frac{1}{L_n(P_n^2)} \int_{I^2} q_{2m-1,2n-1} \right]$$

$$+ \left[1 - \frac{\int_{-1}^{1} P_m^2(x)\, dx}{L_m(P_m^2)} - \frac{\int_{-1}^{1} P_n^2(y)\, dy}{L_n(P_n^2)} \right] \cdot \int_{I^2} q_{2m-1,2n-1}.$$

According to (7), (8) and (6) it suffices to show that

$$\Delta \;:=\; 1 - \frac{\int_{-1}^{1} P_m^2(x)\, dx}{L_m(P_m^2)} - \frac{\int_{-1}^{1} P_n^2(y)\, dy}{L_n(P_n^2)} \;\geq\; 0.$$

Note that P_m satisfies the differential equation (Natanson [12])

$$(1 - x^2) P_m''(x) - 2x P_m'(x) + m(m+1) P_m(x) = 0$$

and that the interior nodes of L_m are the zeros of P_m'. Hence (note that the leading coefficient of P_m is 1)

$$\int_{-1}^{1} P_m^2(x)\, dx = L_m(P_m^2) + \frac{(P_m^2)^{(2m)}}{(2m)!} \cdot \frac{1}{m^2} \int_{-1}^{1} (x^2 - 1)(P_m'(x))^2\, dx$$

$$= L_m(P_m^2) + \frac{1}{m^2} \int_{-1}^{1} (x^2 - 1)(P_m'(x))^2\, dx.$$

Using the differential equation for P_m, the integral on the right hand side is

$$\int_{-1}^{1} (x^2 - 1)(P_m'(x))^2\, dx = \int_{-1}^{1} (x^2 - 1) P_m'(x)\, dP_m(x)$$

$$= -\int_{-1}^{1} P_m(x)\Big[(x^2 - 1)P_m''(x) + 2x\,P_m'(x)\Big]\,dx$$

$$= -m(m+1)\int_{-1}^{1} P_m^2(x)\,dx,$$

thus

$$L_m(P_m^2) = \Big[1 + \frac{m(m+1)}{m^2}\Big]\int_{-1}^{1} P_m^2(x)\,dx.$$

In the same way we get

$$L_n(P_n^2) = \Big[1 + \frac{n(n+1)}{n^2}\Big]\int_{-1}^{1} P_n^2(x)\,dx\Big].$$

So $\Delta = 1 - \frac{1}{2+\frac{1}{m}} - \frac{1}{2+\frac{1}{n}} > 0$. In order to get $\int_{I^2} h = 0$, we need to have (i)

$h(x_\mu, y_\nu) = 0$ $\quad \mu = 1,\ldots,m,\ \nu = 1,\ldots,n,$which is $q_{2m-1,2n-1}(x_\mu, y_\nu) = 0$, for $\mu = 1,\ldots,m,\ \nu = 1,\ldots,n$, further (ii) that $L_n(h(x_\mu,\cdot)) = 0$, i.e. $h(x_\mu, -1) = h(x_\mu,1) = h(x_\mu,\tau_\nu) = 0$ for $1 \le \nu \le n-1$, where the τ_ν are the zeros of P_n' For a fixed x_μ, $h(x_\mu,y)$ is a polynomial of degree $2n$ having $2n+1$ zeros, hence $h(x_\mu,\cdot) = 0$, thus $s_{2m-1}(x_\mu) = 0$ for $1 \le \mu \le m$. From (5) it follows that

$$q_{2m-1,2n-1}(x_\mu,\cdot) = 0 \qquad \text{for} \quad 1 \le \mu \le m.$$

Similarly

$$q_{2m-1,2n-1}(\cdot,y_\nu) = 0 \qquad \text{for} \quad 1 \le \nu \le n.$$

In summary we have

$$\left\{ \begin{array}{rcl} s_{2m-1}(x_\mu) & = & 0 \\ t_{2n-1}(y_\nu) & = & 0 \\ q_{2m-1,2n-1}(x, y_\nu) & = & q_{2m-1,2n-1}(x_\mu, y) = 0 \end{array} \right.$$

for $1 \le \mu \le m,\ 1 \le \nu \le n$. Thus h must be of the desired form. Conversely, if h is of the form

$$h(x,y) = P_m(x)P_n(y)\Big(P_m(x)t_{n-1}(y) + P_n(y)s_{m-1}(x) + u_{m-1,n-1}(x,y)\Big),$$

then $\int_{I^2} h = 0$. $\hfill \square$

5   One–Sided L^1–Approximation of Products

Consider the Chebyshev polynomials U_m of second kind. Fromm [8] showed that the best L^1-approximation to $x^m y^n$ with respect to $PB^{m,n}(I^2)$ is given by $q(x,y) = x^m y^n - 2^{-m-n} \cdot U_m(x)U_n(y)$. This means that the approximation error can be written as a product, namely

$$x^m \cdot y^n - q(x,y) = 2^{-m-n} \cdot U_m(x) \cdot U_n(y).$$

Similar results hold for other norms, see e.g. Ehlich–Zeller [7]. In addition, corresponding results hold when $x^m y^n$ is replaced by a function $u(x)v(y)$, see Shapiro [14] for the uniform norm and Haußmann–Zeller [10] for more general norms. Now we turn to the proof of Theorem 5.

Proof of Theorem 5. Put

$$U(x) := u(x) - p^-_{2m-1}(u;x), \quad V(y) := v(y) - p^-_{2n-1}(v;y).$$

Note that U and V vanish only on the zeros of the Legendre polynomials of degree m resp. n with mulitplicities 2.

Clearly, the problem in the theorem is equivalent to the problem of best approximation of $U(x)V(y)$ by functions of the form

$$\beta(x,y) := U(x) \cdot f(y) + g(x) \cdot V(y) + q(x,y)$$

such that

(9) $$U(x)V(y) \geq \beta(x,y) \text{ on } I.$$

We claim that the unique best one–sided approximant is given by $\beta = 0$. To this end we need

Proposition 9.   *If we have*

$$\beta(x,y) := U(x) \cdot f(y) + g(x) \cdot V(y) + q(x,y) \leq 0$$

on the Legendre blending grid $LG^{(m,n)}_{\mathbf{x},\mathbf{y}}$, *then* $\int_{I^2} \beta \leq 0$.

Proof. We make use of the Gauss quadrature formula and represent the integral of β in the following form:

$$\int_{I^2} \beta = \int_{-1}^{1} U(x)\,dx \cdot \sum_{j=1}^{n} A_j^n f(y_j) + \sum_{i=1}^{m} A_i^m g(x_i) \cdot \int_{-1}^{1} V(y)\,dy + \int_{I^2} q$$

$$= \sum_{j=1}^{n} A_j^n \cdot \left[f(y_j) \cdot \int_{-1}^{1} U(x)\,dx + \frac{1}{2} \int_{-1}^{1} q(x, y_j)\,dx \right]$$

$$+ \sum_{i=1}^{m} A_i^m \cdot \left[g(x_i) \cdot \int_{-1}^{1} V(y)\,dy + \frac{1}{2} \int_{-1}^{1} q(x_i, y)\,dy \right].$$

Since the Gaussian coefficients are positive we need to show that every expression in the square brackets is non–positive. In order to do this, note that (9) yields

$$(10)\qquad\qquad f(y_j) \cdot U(x) + q(x, y_j) \leq 0 \quad \text{for each } x \in I.$$

In particular, taking $x = x_i$ we get $q(x_i, y_j) \leq 0$ and then, by the Gauss quadrature formula, we conclude that

$$\int_{-1}^{1} q(x, y_j)\,dx \leq 0.$$

Therefore, if $f(y_j) \leq 0$, then clearly

$$(11)\qquad\qquad f(y_j) \int_{-1}^{1} U(x)\,dx + \frac{1}{2} \int_{-1}^{1} q(x, y_j)\,dx \leq 0.$$

If, on the other hand, $f(y_j) > 0$, then, in view of (10)

$$-q(x, y_j) \geq f(y_j) \cdot U(x) \text{ for all } x \in I.$$

Therefore

$$-\int_{-1}^{1} q(x, y_j)\,dx \geq f(y_j) \int_{-1}^{1} p_{2m-1}^{+}(U; x)\,dx$$

$$= f(y_j)\left\{ \int_{-1}^{1} [p_{2m-1}^{+}(U; x) - U(x)]\,dx + \int_{-1}^{1} U(x)\,dx \right\}$$

$$= f(y_j) \left\{ E_{2m-1}^{+}(U) + E_{2m-1}^{-}(U) \right\}.$$

According to Brass–Schmeisser [5, Theorem 6] it follows from the assumption $u^{(2m)} > 0$ that

$$E_{2m-1}^{+}(U) > E_{2m-1}^{-}(U) = \int_{-1}^{1} U(x)\,dx.$$

Hence

$$-\int_{-1}^{1} q(x, y_j)\,dx > 2f(y_j) \int_{-1}^{1} U(x)\,dx.$$

The latter implies (11) with strict inequality. Thus we showed that $\int_{I^2} \beta \leq 0$, provided that β satisfies (9). Thus Proposition 9 is settled, and hence 0 is a best approximant. $\qquad\square$

Next we show the uniqueness assertion in Theorem 5. Assume $\int_{I^2} \beta = 0$. Then $f(y_j) \leq 0$, since otherwise the integral $\int_{I^2} \beta$ would be strictly negative. This, together with (11) and the observation $\int_{-1}^{1} q(x, y_j)dx \leq 0$, implies

$$q(x_i, y_j) = 0 \quad \text{for } i = 1,\ldots,m, \quad \text{and} \quad f(y_j) = 0.$$

Then in view of (10), $q(x, y_j) \leq 0$ on I. By Lemma 8 all $x_i's$ are double zeros and consequently $q(\cdot, y_j) = 0$ for each fixed $y_j, j = 1,\ldots,n$. Hence q is identically zero, and therefore

$$\beta(x, y) \quad = \quad U(x)f(y) + g(x)V(y).$$

Since $\beta(x, y_j) \leq 0$, we apply Lemma 8 to deduce that

$$f'(y_j) \quad = \quad 0 \quad \text{for } j = 1,\ldots,n$$

and from this $f = 0$. The same reasoning yields $g = 0$, thus $\beta = 0$. $\qquad\square$

6   Concluding Remarks

We have presented our results in the case of approximation from below. Following the same scheme, we can get corresponding results for approximation from above which involve Lobatto type quadratures rather than Gaussian ones. We also can get results putting restrictions on odd order derivatives in which Radau quadratures will play a role.

Acknowledgement. This work was supported by Volkswagen Foundation under grant I/70523. We acknowledge the information given to us by K. Jetter [11] concerning a previous version of Theorem 3.

References

[1] Armitage, D. H., Gardiner, S. J.: *Best one–sided L^1–approximation by harmonic and subharmonic functions.* In: *Advances in Multivariate Approximation*, W. Haußmann et al. (eds.), Wiley–VCH, Berlin 1999, pp. 43–56.

[2] Armitage, D. H., Gardiner, S. J., Haußmann W., Rogge, L.: *Best one–sided L^1–approximation by harmonic functions.* manuscripta math. **96**, 181–194 (1998).

[3] Bojanov, B. D., Hakopian, H. A., Sahakian, A. A.: *Spline Functions and Multivariate Interpolations.* Dordrecht–Boston–London: Kluwer, 1993.

[4] Brass, H.: *Quadraturverfahren.* Göttingen–Zürich: Vandenhoeck & Ruprecht, 1977.

[5] Brass, H., Schmeisser, G.: *Error estimates for interpolatory quadrature formulae.* Numer. Math. **37**, 371–386 (1981).

[6] Davis, P. J., Rabinowitz, P.: *Methods of Numerical Integration.* New York–London: Academic Press, 1975.

[7] Ehlich, H., Zeller, K.: *Čebyšev Polynome in mehreren Veränderlichen.* Math. Z. **93**, 142–143 (1966).

[8] Fromm, J.: *L_1–approximation to zero.* Math. Z. **151**, 31–33 (1976).

[9] Haußmann, W., Zeller, K.: *Blending interpolation and best L^1–approximation.* Arch. Math. (Basel) **40**, 545–552 (1983).

[10] Haußmann, W., Zeller, K.: *Mixed norm multivariate approximation with blending functions.* In: Constructive Theory of Functions, Bl. Sendov et. al. (eds.), Publ. House Bulg. Acad. Sci., Sofia 1984, pp. 403–408.

[11] Jetter, K.: Private communication.

[12] Natanson, I. P.: *Constructive Function Theory, Vol. II.* New York: Ungar, 1965.

[13]   Pinkus, A.: *On L^1-approximation.* Cambridge University Press, 1989.

[14]   Shapiro, H. S.: *Topics in Approximation Theory.* Berlin–Heidelberg–New York: Springer, 1971.

Addresses:

BORISLAV D. BOJANOV, DIMITER P. DRYANOV,
GENO P. NIKOLOV
Department of Mathematics
University of Sofia
James Boucher Blvd. No. 5
BG–1164 Sofia
Bulgaria

WERNER HAUSSMANN
Department of Mathematics
Gerhard–Mercator–University
Lotharstrasse 65
D–47048 Duisburg
Germany

On the Asymptotics of Points which Maximize Determinants of the Form $det\,[g(|x_i - x_j|)]$

Len Bos and Ulrike Maier

Dedicated to Prof. Dr. M. Reimer on the occasion of his 65th birthday

Abstract

The question of how to choose good point systems for interpolation in general is a difficult task. For functions of the form $f(x) = g(|x|)$ as occur in the theory of radial basis functions we inspect which point systems maximize the determinants $\det(g(|x_i - x_j|)_{1 \le i,j \le n}$ for several choices of the functions g.

We give several examples where the optimal points are precisely equally spaced. Numerical experiments indicate that this is asymptotically true for a wide class of functions and we begin here a program to show that this is indeed the case.

1 Introduction

The general problem in curve and surface fitting is, given data $(x_i, y_i) \in \mathrm{I\!R}^d \times \mathrm{I\!R}$, $1 \le i \le n$, find a "simple" function $p(x)$ which interpolates the data, i.e. $p(x_i) = y_i$, $1 \le i \le n$, in some consistent manner. One of the most successful methods proposed for the solution of such problems is the so-called method of radial basis functions, [3, 6, 7]. Here we fix a basis function $f : \mathrm{I\!R}^d \to \mathrm{I\!R}$, chosen for simplicity to be radial, i.e. $f(x) = g(|x|)$ for some function $g : \mathrm{I\!R}^+ \to \mathrm{I\!R}$, and seek an interpolant (essentially) of the form $p(x) = \sum_{i=1}^n a_i f(x - x_i)$. The interpolation conditions, $p(x_i) = y_i$, result in the linear system

$$[f(x_i - x_j)]_{1 \le i,j \le n}\, \mathbf{a} = \mathbf{y},$$

$$\text{where} \quad \mathbf{a} = \begin{pmatrix} a_1 \\ a_2 \\ \vdots \\ a_n \end{pmatrix} \quad \text{and} \quad \mathbf{y} = \begin{pmatrix} y_1 \\ y_2 \\ \vdots \\ y_n \end{pmatrix},$$

Advances in Multivariate Approximation; W. Haußmann, K. Jetter and M. Reimer (eds.)
Mathematical Research, Vol. 107, pp. 107–128, ISBN 3-527-40236-5
© WILEY-VCH, Berlin 1999

or, in other words,

$$[g(|x_i - x_j|)]_{1 \le i,j \le n}\, \mathbf{a} = \mathbf{y}.$$

For practical purposes it is therefore of some interest to understand for which
points such matrices are best conditioned.

Now, finding points which minimize the condition number of $[g(|x_i - x_j|)]_{1 \le i,j \le n}$
is likely a forbiddingly difficult problem, and thus we were led to consider the
slightly easier, but yet relevant, problem of maximizing $\det[g(|x_i - x_j|)]_{1 \le i,j \le n}$
where the interpolation nodes x_i are allowed to range over a fixed compact set
$K \subset \mathbb{R}^d$. Moreover, it is perhaps encouraging to note that there is a beautiful,
analogous theory of such optimal points in the theory of polynomial interpolation. The points which maximize the corresponding determinant, specifically
the Vandermonde determinant, are known as the Fekete points. They have
been much studied, especially as they provide a link between polynomial interpolation and logarithmic potential theory in $\mathbb{C}$. Specifically, the asymptotic
distribution of the Fekete points for a compact $K \subset \mathbb{C}$ is the same as the
so-called equilibrium measure of K (see e. g. [10, pp. 141 ff]).
This then is the most general form of the problem we wish to consider: for
$K \subset \mathbb{R}^d$ and $g : \mathbb{R}^d \to \mathbb{C}$ find the asymptotic distribution as $n \to \infty$ of the
points $x_1, \ldots, x_n \in K$ for which $|\det[g(|x_i - x_j|)]_{1 \le i,j \le n}|$ is a maximum. We
will make this more precise in the sequel.

This also appears to be a lengthly and perhaps difficult program and hence, in
this paper, we begin with the univariate case, $d = 1$, $K = [-1, 1]$.

2   Some Simple Examples

The simplest interesting case is when $g(x) = x$, i.e. we maximize $\det[|x_i - x_j|]$.
If we order the points $-1 \le x_1 < x_2 < \ldots < x_n \le 1$ and let $h_i := x_{i+1} - x_i$,
$1 \le i \le n - 1$, we have the determinant

$$D_n := \begin{vmatrix} 0 & h_1 & h_1 + h_2 & \cdots & \cdots & h_1 + h_2 + \ldots + h_{n-1} \\ h_1 & 0 & h_2 & \cdots & \cdots & h_2 + \ldots + h_{n-1} \\ h_1 + h_2 & h_2 & 0 & h_3 & \cdots & \vdots \\ \vdots & & \ddots & \ddots & & h_{n-1} \\ h_1 + h_2 + \ldots + h_{n-1} & \cdots & & h_{n-1} & & 0 \end{vmatrix}$$

which may be evaluated by elementary means to be

$$D_n = (-1)^{n-1} 2^{n-2} \left(\prod_{i=1}^{n-1} h_i\right) \left(\sum_{i=1}^{n-1} h_i\right),$$

so that

$$|D_n| = 2^{n-2} \left(\prod_{i=1}^{n-1} h_i\right) \left(\sum_{i=1}^{n-1} h_i\right).$$

Clearly, this latter value is monotonically increasing in each h_i and hence, in particular h_1 and h_{n-1} must be as large as possible in order to maximize $|D_n|$. It follows that $x_1 = -1$ and $x_n = +1$, and consequently $\sum_{i=1}^{n-1} h_i = 1 - (-1) = 2$, and we are left with the problem of maximizing $\prod_{i=1}^{n-1} h_i$ subject to the constraint that $\sum_{i=1}^{n-1} h_i = 2$. But, as is well-known, this maximum is uniquely attained for

$$h_i = \frac{2}{n-1}, \quad 1 \le i \le n-1,$$

i.e. precisely for the equally spaced points

$$x_i = -1 + (i-1)\frac{2}{n-1}, \quad 1 \le i \le n,$$

which are clearly uniformly distributed on $[-1, 1]$.

(In contrast, the Fekete points for $[-1, 1]$ have nearly the same distribution as the (extended) Chebyshev points, and have asymptotically the so-called arcsin distribution.)

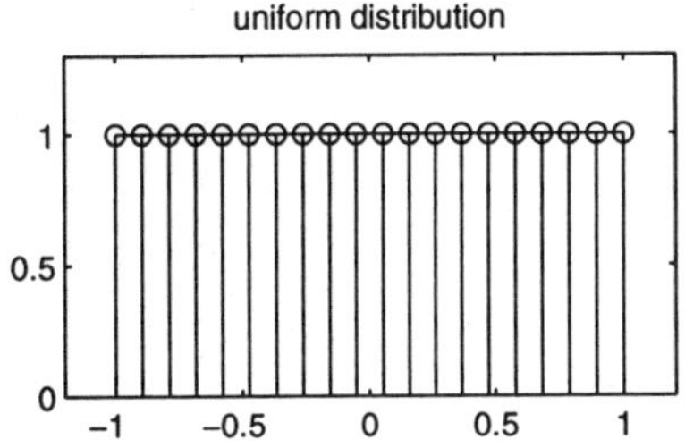
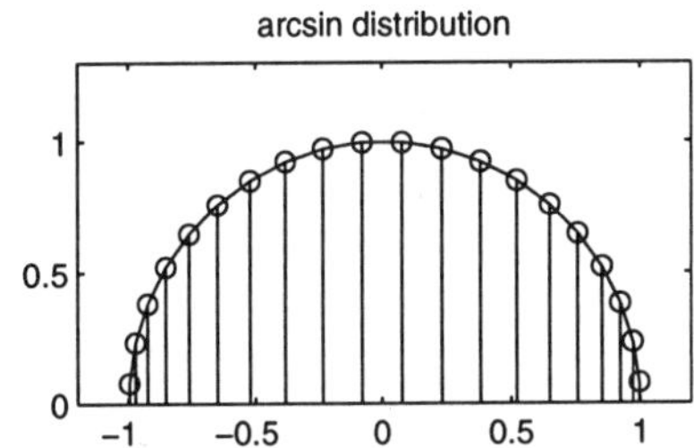

Figure 1: point distributions on $[-1,1]$

Consider now a second, seemingly very different, example, $g(x) = e^x$. With the same notation as above, we now have the $n \times n$–determinant $(n \geq 2)$

$$D_n := \begin{vmatrix} 1 & e^{h_1} & e^{h_1+h_2} & \cdots & e^{h_1+h_2+\ldots+h_{n-1}} \\ e^{h_1} & 1 & e^{h_2} & \cdots & e^{h_2+\ldots+h_{n-1}} \\ e^{h_1+h_2} & e^{h_2} & 1 & \cdots & \\ \vdots & \ddots & \ddots & \ddots & \vdots \\ e^{h_1+h_2+\ldots+h_{n-1}} & e^{h_2+\ldots+h_{n-1}} & \cdots & e^{h_{n-1}} & 1 \end{vmatrix}.$$

This determinant may also easily be evaluated:

Subtracting $e^{h_{n-1}} \times$ (column $n-1$) from column n yields

$$\begin{matrix} 0 \\ 0 \\ \vdots \\ 1 - e^{2h_{n-1}} \end{matrix}$$

for the last column. It follows by induction that

$$D_n = \prod_{i=1}^{n-1} (1 - e^{2h_i}) \quad \text{and} \quad |D_n| = \prod_{i=1}^{n-1} (e^{2h_i} - 1).$$

Again, $|D_n|$ is monotone in each h_i separately so that for a maximal $|D_n|$ we have $x_1 = -1$, $x_n = +1$ and $\sum_{i=1}^{n-1} h_i = 2$. Thus we may write

$$|D_n| = \prod_{i=1}^{n-1} e^{h_i} 2^{n-1} \prod_{i=1}^{n-1} \left(\frac{e^{h_i} - e^{-h_i}}{2} \right)$$

$$= e^2 \, 2^{n-1} \prod_{i=1}^{n-1} \sinh(h_i)$$

and we wish to maximize $\prod_{i=1}^{n-1} \sinh(h_i)$ subject to the constraint that $\sum_{i=1}^{n-1} h_i = 2$. An easy application of Lagrange multipliers shows that the unique maximum is attained again for $h_i = \frac{2}{n-1}$, so that the optimal points for $g(x) = e^x$ are also precisely uniformly distributed.

More generally, by the same calculation, we may show that for $g(x) = a^x$, $a \in \mathbb{C}$, $a \neq 0$,

$$D_n = (-1)^{n-1} \prod_{i=1}^{n-1} \left(a^{2h_i} - 1 \right)$$

which again is maximized in absolute value for $h_1 = h_2 = \ldots = h_{n-1} = \dfrac{2}{n-1}$.

Another class of examples is given by

$$g(x) = k\left(e^{ax} - e^{-ax}\right) \quad (\text{or} \quad k\sinh(ax)) \tag{2.1}$$

where $k, a \in \mathbb{C}$ are arbitrary (a particular case would then be $g(x) = \sin(bx)$).

Proposition 2.1. *For $g(x)$ given by (2.1),*

$$\det[g(|x_i - x_j|)] = (-1)^{n+1}2^{n-2}k^n g\left(\sum_{i=1}^{n-1} h_i\right) \prod_{i=1}^{n-1} g(h_i).$$

Proof : Since k is simple a factor of each row in the determinant, we may suppose that $k = 1$. As usual, we set $h_j := x_{j+1} - x_j$, so that the determinant has the form

$$\begin{vmatrix}
0 & g(h_1) & g(h_1 + h_2) & \cdots & g(h_1 + \ldots + h_{n-1}) \\
g(h_1) & 0 & g(h_2) & \cdots & g(h_2 + \ldots h_{n-1}) \\
g(h_1 + h_2) & g(h_2) & 0 & & \vdots \\
\vdots & \ddots & \ddots & \ddots & \\
g(h_2 + \ldots h_{n-1}) & \cdots & g(h_{n-2}) & 0 & g(h_{n-1}) \\
g(h_1 + \ldots + h_{n-1}) & \cdots & & g(h_{n-1}) & 0
\end{vmatrix}.$$

Now let $z_j := e^{ah_j}$, $j = 1, \ldots, n-1$, so that

$$g(h_1) = z_1 - \frac{1}{z_1}$$

$$g(h_1 + h_2) = e^{a(h_1+h_2)} - e^{-a(h_1+h_2)} = z_1 z_2 - \frac{1}{z_1 z_2}$$

$$\vdots$$

$$g(h_1 + \ldots + h_{n-1}) = z_1 \ldots z_{n-1} - \frac{1}{z_1 \ldots z_{n-1}}.$$

Then the determinant may be expressed as

$$\begin{vmatrix}
0 & z_1 - \frac{1}{z_1} & z_1 z_2 - \frac{1}{z_1 z_2} & \cdots & z_1 \ldots z_{n-1} - \frac{1}{z_1 \ldots z_{n-1}} \\
z_1 - \frac{1}{z_1} & 0 & z_2 - \frac{1}{z_2} & \cdots & z_2 \ldots z_{n-1} - \frac{1}{z_2 \ldots z_{n-1}} \\
z_1 z_2 - \frac{1}{z_1 z_2} & & & & \\
\vdots & \ddots & \ddots & \ddots & \vdots \\
& & & & z_{n-1} - \frac{1}{z_{n-1}} \\
z_1 \ldots z_{n-1} - \frac{1}{z_1 \ldots z_{n-1}} & & \cdots & z_{n-1} - \frac{1}{z_{n-1}} & 0
\end{vmatrix}.$$

Now subtract z_j times the $(j+1)$st column from the jth column, $1 \le j \le n-1$, to obtain

$$
\begin{vmatrix}
-(z_1^2-1) & -z_1(z_2^2-1) & \cdots & -z_1\ldots z_{n-2}(z_{n-1}^2-1) & z_1\ldots z_{n-1} - \frac{1}{z_1\ldots z_{n-1}} \\
\frac{(z_1^2-1)}{z_1} & -(z_2^2-1) & \cdots & -z_2\ldots z_{n-2}(z_{n-1}^2-1) & z_2\ldots z_{n-1} - \frac{1}{z_2\ldots z_{n-1}} \\
\frac{(z_1^2-1)}{z_1 z_2} & \frac{(z_2^2-1)}{z_2} & & & \\
\vdots & \frac{(z_2^2-1)}{z_2 z_3} & \ddots & \vdots & \vdots \\
& \vdots & & -(z_{n-1}^2-1) & z_{n-1} - \frac{1}{z_{n-1}} \\
\frac{(z_1^2-1)}{z_1\ldots z_{n-1}} & \frac{(z_2^2-1)}{z_2\ldots z_{n-1}} & \cdots & \frac{(z_{n-1}^2-1)}{z_{n-1}} & 0
\end{vmatrix}
$$

$$
= \prod_{i=1}^{n-1}(z_i^2-1)
\begin{vmatrix}
-1 & -z_1 & -z_1 z_2 & \cdots & -z_1\ldots z_{n-2} & z_1\ldots z_{n-1} - \frac{1}{z_1\ldots z_{n-1}} \\
\frac{1}{z_1} & -1 & -z_2 & \cdots & -z_2\ldots z_{n-2} & z_2\ldots z_{n-1} - \frac{1}{z_2\ldots z_{n-1}} \\
\frac{1}{z_1 z_2} & \frac{1}{z_2} & -1 & \cdots & & \\
\vdots & \vdots & \frac{1}{z_3} & \ddots & & \vdots \\
& & & \ddots & -1 & z_{n-1} - \frac{1}{z_{n-1}} \\
\frac{1}{z_1\ldots z_{n-1}} & \frac{1}{z_2\ldots z_{n-1}} & \frac{1}{z_3\ldots z_{n-1}} & \cdots & \frac{1}{z_{n-1}} & 0
\end{vmatrix} .
$$

Now expand down the last column. Notice that the first $n-1$ elements of the first row are just $-z_1\ldots z_{n-1}$ times those of the last row. Hence the minors for the second through the second last entries of the nth column are zero (as first and last rows of the minors are multiples of each other). The nth entry in the last column is 0 and so our determinant simplifies to

$$
\prod_{i=1}^{n-1}(z_i^2-1)(-1)^{n+1}\left(z_1\ldots z_{n-1} - \frac{1}{z_1\ldots z_{n-1}}\right)
$$

$$
\times
\begin{vmatrix}
\frac{1}{z_1} & -1 & -z_2 & \cdots & -z_2\ldots z_{n-1} \\
\frac{1}{z_1 z_2} & \frac{1}{z_2} & -1 & & -z_3\ldots z_{n-1} \\
& & \frac{1}{z_3} & \ddots & \vdots \\
\vdots & & & \ddots & -1 \\
\frac{1}{z_1\ldots z_{n-1}} & \frac{1}{z_2\ldots z_{n-1}} & \frac{1}{z_3\ldots z_{n-1}} & \cdots & \frac{1}{z_{n-1}}
\end{vmatrix} .
$$

To evaluate this last determinant, subtract $\frac{1}{z_j}$ times the $(j+1)$st column from the jth column, $1 \le j \le n-1$, to obtain

$$(-1)^{n+1} \prod_{i=1}^{n-1} (z_i^2 - 1) \left(z_1 \ldots z_{n-1} - \frac{1}{z_1 \ldots z_{n-1}} \right)$$

$$\times \begin{vmatrix} \frac{2}{z_1} & 0 & 0 & \ldots & 0 & -z_2 \ldots z_{n-1} \\ 0 & \frac{2}{z_2} & 0 & \ldots & & -z_3 \ldots z_{n-1} \\ \vdots & 0 & \frac{2}{z_3} & & & \vdots \\ \vdots & & \ddots & \ddots & \vdots & \\ & & & \ddots & \frac{2}{z_{n-2}} & -1 \\ 0 & 0 & 0 & \ldots & 0 & \frac{1}{z_{n-1}} \end{vmatrix}$$

$$= (-1)^{n+1} \prod_{i=1}^{n-1} (z_i^2 - 1) \left(z_1 \ldots z_{n-1} - \frac{1}{z_1 \ldots z_{n-1}} \right) \cdot \frac{2^{n-2}}{z_1 \ldots z_{n-1}}$$

$$= (-1)^{n+1} 2^{n-2} \prod_{i=1}^{n-1} (z_i - \frac{1}{z_i}) \left(z_1 \ldots z_{n-1} - \frac{1}{z_1 \ldots z_{n-1}} \right)$$

$$= (-1)^{n+1} 2^{n-2} \left(\prod_{i=1}^{n-1} (e^{ah_i} - e^{-ah_i}) \right) \left(\prod_{i=1}^{n-1} e^{ah_i} - \prod_{i=1}^{n-1} e^{-ah_i} \right)$$

$$= (-1)^{n+1} 2^{n-2} \left(\prod_{i=1}^{n-1} g(h_i) \right) g\left(\sum_{i=1}^{n-1} h_i \right). \qquad \square$$

For our particular problem, we wish to maximize the absolute value of this expression subject to the constraint that $\sum_{i=1}^{n-1} h_i = 2$, which provided $g\left(\sum_{i=1}^{n-1} h_i\right) \ne 0$, is equivalent to maximizing the absolute value of $\prod_{i=1}^{n-1} g(h_i)$ subject to the same constraint. Lagrange multipliers inform us that the critical points are given by

$$\mathrm{Re}\frac{g'(h_1)}{g(h_1)} = \mathrm{Re}\frac{g'(h_2)}{g(h_2)} = \ldots = \mathrm{Re}\frac{g'(h_{n-1})}{g(h_{n-1})}$$

which <u>includes</u> the equally spaced points, but there <u>may be</u> others.

A closer examination of the particular case $g(x) = \sin(bx)$ is instructive. First note that, if $b = \frac{m\pi}{2}$ for some integer m, then $\sin(b\sum_{i=1}^{n-1} h_i) = \sin(\frac{m\pi}{2} \cdot 2) = 0$ so that our determinant is zero for <u>any</u> choice of h_i.

If $b = 3$, then the determinant for three equally spaced points (with $h_1 = h_2$

$= 1$) is

$$2\sin(3)\sin(3)\sin(6) = 2\sin^2(3)\sin(6) = -0.01112903421,$$

but for $h_1 = \frac{(3-\frac{\pi}{2})}{3}$, $h_2 = \frac{(3+\frac{\pi}{2})}{3}$, the determinant is

$$2\sin(3 - \frac{\pi}{2})\sin(3 + \frac{\pi}{2})\sin(6) = 0.5477019572,$$

so that the equally spaced points do <u>not</u> maximize the (absolute value of the) determinant.

If, however, $0 < b < \frac{\pi}{4}$, then $0 < bh_i < \frac{\pi}{2}$ and the solution of $\sin(bh_1) = \sin(bh_2) = \ldots = \sin(bh_{n-1})$ is <u>uniquely</u> $h_1 = h_2 = \ldots = h_{n-1} = \frac{2}{n-1}$.

It is interesting to note, that by the same means as for Proposition 2.1, we may prove

Proposition 2.2. *If $g(x) = k\cosh(ax)$, where $k, a \in \mathbb{C}$ are arbitrary parameters, then*

$$\det[g(|x_i - x_j|)] \equiv 0. \qquad \square$$

3 $\quad g(x) = x^{2k+1}$

The examples of Section 2 lead one naturally to ask if it should always be the case that there exist optimal points that are uniformly distributed. Let us therefore consider $g(x) = x^{2k+1}$. The translates $g(|x_i - x_j|) = |x_i - x_j|^{2k+1}$ are but certain splines of degree $2k + 1$ (with single knot at x_i) and thus the (univariate) radial basis interpolant is just a spline interpolant. But, with the assistance of a computer algebra system, one easily finds that for $k = 1$, the four optimal points for $n = 4$, are $-1, -1 + h, 1 - h, 1$ where h is the real root of

$$p(x) = 15x^5 - 96x^4 + 332x^3 - 536x^2 + 384x - 96.$$

The value of h is certainly not $2/3$ since $p(2/3) = 256/81 \neq 0$ (numerically $h = 0.5432617730$), hence these points are <u>not</u> equally spaced. However, numerical experiments indicate that the optimal points are asymptotically uniformly distributed. Specifically, let $x_1 < \ldots < x_n$ be these optimal points and $t_j = -1 + \frac{2(j-1)}{n-1}$, $1 \leq j \leq n$, be the corresponding equally spaced points.

Further, let $L_{n,k}(x)$ be the piecewise affine linear function with knots at the t_j such that $L_{n,k}(t_j) = x_j, 1 \leq j \leq n$.

The figures 2–5 show the graphs of $L_{n,k}$ for various values of n (and k).

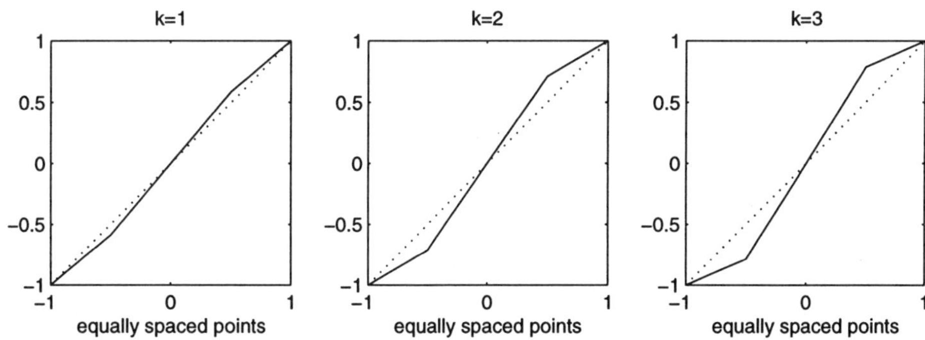

Figure 2: 5 optimal points for $g(x) = x^{2k+1}$, $k = 1, 2, 3$, vs. equally spaced points on $[-1,1]$

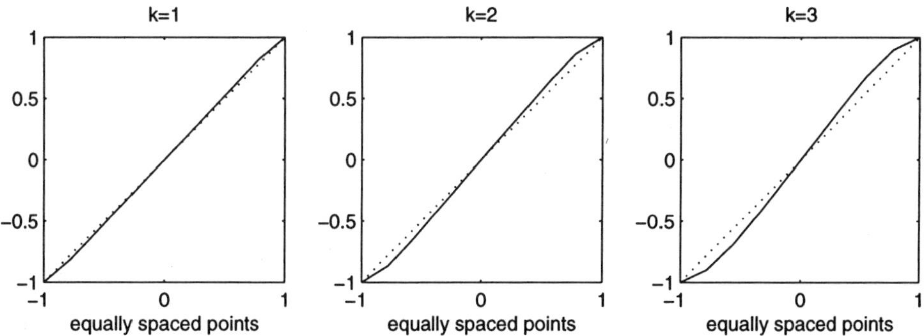

Figure 3: 10 optimal points for $g(x) = x^{2k+1}$, $k = 1, 2, 3$, vs. equally spaced points on $[-1,1]$

Conjecture 3.1. *For each $k = 1, 2, 3, \ldots$ the points which maximize the determinants* $\det \left[|x_i - x_j|^{2k+1} \right]_{1 \leq i,j \leq n}$ *are asymptotically uniformly distributed, in the sense that* $\lim\limits_{n \to \infty} L_{n,k}(x) = x.$

We also experimented with $g(x) = e^{-x^2}$, $g(x) = \sqrt{x + c}$ as well as with $g(x) = x^2 \log(x + c)$, where c is a constant. Let $L_n(x)$ be the piecewise affine linear function with knots at the equally spaced points t_j such that $L_n(t_j) = x_j$, $j = 1, \ldots, n$, for the optimal points x_j of g. On this basis we make the further

Conjecture 3.2.

The points which maximize the determinants $\det [g(|x_i - x_j|)]_{1 \leq i,j \leq n}$ *are asymptotically uniformly distributed, in the sense that* $\lim\limits_{n \to \infty} L_n(x) = x$, *for a "broad" class of functions $g(x)$.*

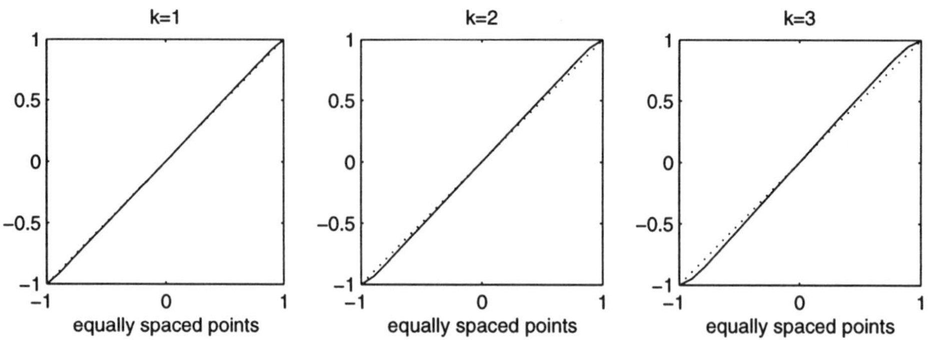

Figure 4: 20 optimal points for $g(x) = x^{2k+1}$, $k = 1, 2, 3$, vs. equally spaced points on $[-1,1]$

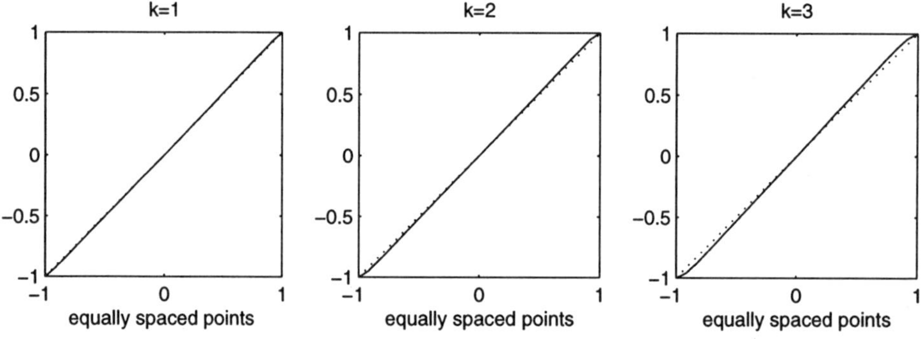

Figure 5: 25 optimal points for $g(x) = x^{2k+1}$, $k = 1, 2, 3$, vs. equally spaced points on $[-1,1]$

We are not able at this point to prove these conjectures. Hence, we consider the special case of $g(x) = x^{2k+1}$ for $k = 1$, i.e. $g(x) = x^3$ in more detail. To begin, we note that there is a close connection between the matrix $[|x_i - x_j|^{2k+1}]_{1 \le i,j \le n}$, $k = 1, 2, ...$, and certain associated Gram matrices.

Proposition 3.3. *Let* $s_i(x) = |x - x_i|^{2k+1}$, $x_1 < x_2 < ... < x_n$. *Then*

$$2(-1)^{k+1}(2k+1)! \left[|x_i - x_j|^{2k+1}\right]_{1 \le i,j \le n}$$

$$= G\left(s_1^{(k+1)}, s_2^{(k+1)}, \ldots, s_n^{(k+1)}\right) - \left(\frac{(2k+1)!}{k!}\right)^2 G\left((x - x_1)^k, \ldots, (x - x_n)^k\right).$$

Here we use the notation

$$G(f_1, \ldots, f_n) = \left[\int_{x_1}^{x_n} f_i(x)f_j(x)dx \right]_{1 \leq i,j \leq n} .$$

Proof: Since each of the matrices is symmetric we need only show that the corresponding elements agree for $j \geq i$. Thus, let $j \geq i$ and consider

$$\int_{x_1}^{x_n} s_i^{(k+1)}(x)s_j^{(k+1)}(x)dx.$$

But

$$s_i^{(k+1)}(x) = \frac{d^{k+1}}{dx^{k+1}}|x - x_i|^{2k+1} = \frac{(2k+1)!}{k!}\text{sign}(x - x_i)(x - x_i)^k$$

so that

$$\int_{x_1}^{x_n} s_i^{(k+1)}(x)s_j^{(k+1)}(x)dx$$

$$= \left(\frac{(2k+1)!}{k!}\right)^2 \left\{\int_{x_1}^{x_i} - \int_{x_i}^{x_j} + \int_{x_j}^{x_n} (x - x_i)^k(x - x_j)^k dx\right\}$$

$$= \left(\frac{(2k+1)!}{k!}\right)^2 \left\{\int_{x_1}^{x_n}(x - x_i)^k(x - x_j)^k dx - 2\int_{x_i}^{x_j}(x - x_i)^k(x - x_j)^k dx\right\}.$$

Now, letting $u = (x - x_i)/(x_j - x_i)$,

$$\int_{x_i}^{x_j}(x - x_i)^k(x - x_j)^k dx = (x_j - x_i)^{2k+1}\int_0^1 u^k(u - 1)^k du$$

$$= (-1)^k(x_j - x_i)^{2k+1}\int_0^1 u^k(1 - u)^k du$$

$$= (-1)^k\frac{k!\,k!}{(2k+1)!}(x_j - x_i)^{2k+1}.$$

Consequently,

$$G\left(s_1^{(k+1)}, \ldots, s_n^{(k+1)}\right)_{ij} = \left(\frac{(2k+1)!}{k!}\right)^2 \left\{G\left((x - x_1)^k, \ldots, (x - x_n)^k\right)_{ij}\right.$$

$$\left. +2(-1)^{k+1}\frac{(k!)^2}{(2k+1)!}|x_i - x_j|^{2k+1}\right\}$$

$$= \left(\frac{(2k+1)!}{k!}\right)^2 G\left((x - x_1)^k, \ldots, (x - x_n)^k\right)_{ij}$$

$$+2(-1)^{k+1}(2k+1)!|x_j - x_i|^{2k+1}$$

and the result follows. $\square$

But now note that since $(x - x_1)^k, \ldots, (x - x_n)^k$ are all polynomials of fixed degree k, the matrix $G\left((x - x_1)^k, \ldots, (x - x_n)^k\right)$ is of low rank, $k + 1$, compared to the size of the matrix, n, which we let tend to infinity. It thus seems natural to ignore this term and consider first the problem of maximizing the Gram determinant

$$g\left(s_1^{(k+1)}, \ldots, s_n^{(k+1)}\right) := \det G\left(s_1^{(k+1)}, \ldots, s_n^{(k+1)}\right).$$

But for $k = 1$ (the case we wish to consider),

$$s_i^{(k+1)}(x) = s_i''(x) = 6|x - x_i|.$$

A further simplification results by writing

$$|x - x_i| = \sum_{j=1}^{n} |x_i - x_j| H_j(x),$$

where $H_j(x)$ is the so-called hat function (or linear B-spline) defined by

$$H_j(x) = \begin{cases} 0 & \text{if} \quad x < x_{j-1} \\ \dfrac{x - x_{j-1}}{x_j - x_{j-1}} & \text{if} \quad x_{j-1} \leq x \leq x_j \\ \dfrac{x_{j+1} - x}{x_{j+1} - x_j} & \text{if} \quad x_j \leq x \leq x_{j+1} \\ 0 & \text{if} \quad x_{j+1} < x \end{cases} \tag{3.1}$$

with $x_0 := x_1$, $x_{n+1} := x_n$. For then it follows (see e.g. Davis [4, p.177] and [1]) that

$$G(|x-x_1|, |x-x_2|, \ldots, |x-x_n|) = [|x_i-x_j|]_{1 \leq i,j \leq n} G(H_1, \ldots, H_n)[|x_i-x_j|]_{1 \leq i,j \leq n}$$

and

$$g(|x - x_1|, |x - x_2|, \ldots, |x - x_n|) = \det[|x_i - x_j|]^2 g(H_1, \ldots, H_n).$$

Our first example showed that the points that maximize $\det[|x_i - x_j|]$ are precisely the equally spaced ones and thus we are finally led to the Gram determinant

$$\det\left[\int_{-1}^{1} H_i(x) H_j(x) dx\right]$$

where we have scaled the points so that $x_1 = -1$, and $x_n = +1$.

Now we may easily compute this to be explicitly the tridiagonal matrix

$$
T_n \; := \; \left[\int_{-1}^{1} H_i(x)H_j(x)dx\right]
$$

$$
= \; \frac{1}{3}
\begin{bmatrix}
h_1 & h_1/2 & 0 & \cdots & & & 0 \\
h_1/2 & h_1 + h_2 & h_2/2 & 0 & & \cdots & 0 \\
0 & h_2/2 & h_2 + h_3 & h_3/2 & & \cdots & 0 \\
\vdots & & \ddots & \ddots & & \ddots & \vdots \\
\vdots & & & & h_{n-2}/2 & h_{n-2}+h_{n-1} & h_{n-1}/2 \\
0 & \cdots & & & 0 & h_{n-1}/2 & h_{n-1}
\end{bmatrix}
$$

where, as usual, $h_i := x_{i+1} - x_i$.

Theorem 3.4. *The points which maximize* $\det(T_n)$ *are asymptotically equally spaced in the sense that* $\exists\, c, c' > 0$ *s.t.* $c'/n \le h_j + h_{j-1} \le c/n$.

The proof of the Theorem depends on a sequence of (sometimes technical) propositions. Note that maximizing $\det(T_n)$ over $-1 = x_1 \le x_2 \le \ldots \le x_n = 1$ is equivalent to maximizing $\det(T_n)$ as a function of $h_1, \ldots, h_{n-1}$ over the simplex $h_1 \ge 0, \ldots, h_{n-1} \ge 0, \sum_{i=1}^{n-1} h_i = 2$. We would like to emphasize also that the proof is complete except for a demonstration of the following conjecture which we believe to be true, in particular on the basis of extensive numerical testing.

Technical Conjecture 3.5. *The* h_i, $i = 1, \ldots, n-1$, *which maximize* $\det(T_n)$ *are all strictly positive.*

Then, assuming this Conjecture, we may characterize the optimal h_i's as follows.

Theorem 3.6. *Let* $\mathbf{h} = (h_1, \ldots, h_{n-1})^\top \in \mathbb{R}^{n-1}$ *and suppose that* $\mathbf{h}^{opt}$ *is the vector which maximizes* $\det(T_n)$. *Then for any other competing choice of* $\mathbf{h}$ *with* $h_i \ge 0$, $i = 1, \ldots n-1$, $\sum_{i=1}^{n-1} h_i = 2$,

$$
\operatorname{trace}\left(T_n(\mathbf{h}^{opt})^{-1} T_n(\mathbf{h})\right) = n.
$$

Moreover, $\mathbf{h}^{opt}$ *is unique. (Here, as usual,* trace *denotes the sum of the diagonal entries of a matrix.)*

Proof : (adapted from [2, Theorem 2]).
Suppose that $\mathbf{h}^{(1)}$ and $\mathbf{h}^{(2)}$ are two vectors of allowable $\mathbf{h}$'s. Then since the

entries $T_n(\mathbf{h})$ are linear in $\mathbf{h}$, we have that for $t \in [0,1]$

$$T_n\left((1-t)\mathbf{h}^{(1)} + t\mathbf{h}^{(2)}\right) = (1-t)T_n\left(\mathbf{h}^{(1)}\right) + tT_n\left(\mathbf{h}^{(2)}\right).$$

But $T_n\left(\mathbf{h}^{(1)}\right)$ and $T_n\left(\mathbf{h}^{(2)}\right)$ are symmetric and positive definite and so (see e.g. [8, p. 469]) there exists a non–singular matrix A such that $A^\top T_n\left(\mathbf{h}^{(1)}\right) A = \mathrm{diag}(a_1,\ldots,a_n)$ and $A^\top T_n\left(\mathbf{h}^{(2)}\right) A = \mathrm{diag}(b_1,\ldots,b_n)$ are both diagonal, with $a_j \geq 0$, $b_j \geq 0$, $1 \leq j \leq n$.

Hence,

$$
\begin{aligned}
\det A^\top T_n&\left((1-t)\mathbf{h}^{(1)} + t\mathbf{h}^{(2)}\right) A \\
&= \det\left[(1-t)\,\mathrm{diag}(a_1,\ldots,a_n) + t\,\mathrm{diag}(b_1,\ldots,b_n)\right]
\end{aligned}
$$

so that

$$
\begin{aligned}
\det T_n\left((1-t)\mathbf{h}^{(1)} + t\mathbf{h}^{(2)}\right) &= (\det A)^{-2}\det[\,\mathrm{diag}((1-t)a_i + tb_i)] \\
&= (\det A)^{-2}\prod_{i=1}^{n}\left((1-t)a_i + tb_i\right).
\end{aligned}
$$

But by the arithmetic-geometric-mean inequality,

$$(1-t)a_i + tb_i \geq a_i^{1-t}b_i^{t}$$

so that

$$
\begin{aligned}
\det T_n&\left((1-t)\mathbf{h}^{(1)} + t\mathbf{h}^{(2)}\right) \\
&\geq (\det A)^{-2}\left(\prod_{i=1}^{n}a_i\right)^{1-t}\left(\prod_{i=1}^{n}b_i\right)^{t} \\
&= \left(\det\left(A^{-\top}\,\mathrm{diag}(a_1,\ldots,a_n)A^{-1}\right)\right)^{1-t} \cdot \left(\det\left(A^{-\top}\,\mathrm{diag}(b_1,\ldots,b_n)A^{-1}\right)\right)^{t} \\
&= \left(\det T_n\left(\mathbf{h}^{(1)}\right)\right)^{1-t}\left(\det T_n\left(\mathbf{h}^{(2)}\right)\right)^{t} \quad\quad\quad (3.2)
\end{aligned}
$$

with equality iff $a_i = b_i$, $1 \leq i \leq n$, i.e., $T_n\left(\mathbf{h}^{(1)}\right) = T_n\left(\mathbf{h}^{(2)}\right)$, i.e. iff $\mathbf{h}^{(1)} = \mathbf{h}^{(2)}$.

It follows that if $\mathbf{h}^{(1)}$ and $\mathbf{h}^{(2)}$ both optimize $\det T_n(\mathbf{h})$, then, on the one hand

$$\det T_n\left(\frac{\mathbf{h}^{(1)} + \mathbf{h}^{(2)}}{2}\right) \leq \det T_n(\mathbf{h}^{(1)}) = \det T_n(\mathbf{h}^{(2)}),$$

but on the other hand by (3.2) (with $t = 1/2$)

$$\det T_n\left(\frac{\mathbf{h}^{(1)} + \mathbf{h}^{(2)}}{2}\right) \geq \left(\det T_n(\mathbf{h}^{(1)})\right)^{1/2} \times \left(\det T_n(\mathbf{h}^{(2)})\right)^{1/2}$$

$$= \det T_n(\mathbf{h}^{(1)})$$

and hence $\mathbf{h}^{(1)} = \mathbf{h}^{(2)}$. In other words, $\mathbf{h}^{opt}$ is $\underline{\text{unique}}$.

Now if $\mathbf{h}^{opt}$ optimizes $\det T_n(\mathbf{h}^{opt})$ it also optimizes $\det T_n((1-t)\mathbf{h}^{opt} + t\mathbf{h}^{(2)})$ for any given $\mathbf{h}^{(2)}$, as a function of $t \in [0,1]$. Consequently, it also optimizes $\log \det T_n\left((1-t)\mathbf{h}^{opt} + t\mathbf{h}^{(2)}\right)$ as a function of $t \in [0,1]$. From the assumption that $h_i^{opt} > 0$, $1 \leq i \leq n-1$, we have then that

$$\frac{d}{dt} \log \det T_n\left((1-t)\mathbf{h}^{opt} + t\mathbf{h}^{(2)}\right)\Big|_{t=0} = 0.$$

But we may calculate that

$$\frac{d}{dt}\log \det T_n\left((1-t)\mathbf{h}^{opt} + t\mathbf{h}^{(2)}\right)$$

$$= \frac{d}{dt}\operatorname{trace}\left(\log T_n\left((1-t)\mathbf{h}^{opt} + t\mathbf{h}^{(2)}\right)\right)$$

$$= \operatorname{trace}\left(T_n^{-1}\left((1-t)\mathbf{h}^{opt} + t\mathbf{h}^{(2)}\right) \cdot \frac{d}{dt}T_n\left((1-t)\mathbf{h}^{opt} + t\mathbf{h}^{(2)}\right)\right).$$

Therefore, we have

$$\begin{aligned}
0 &= \operatorname{trace}\left(T_n^{-1}\left(\mathbf{h}^{opt}\right)\left(T_n\left(-\mathbf{h}^{opt} + \mathbf{h}^{(2)}\right)\right)\right) \\
&= \operatorname{trace}\left(T_n^{-1}\left(\mathbf{h}^{opt}\right)\left(T_n\left(\mathbf{h}^{(2)}\right) - T_n\left(\mathbf{h}^{opt}\right)\right)\right) \\
&= \operatorname{trace}\left(T_n^{-1}\left(\mathbf{h}^{opt}\right)T_n\left(\mathbf{h}^{(2)}\right)\right) - \operatorname{trace}\left(T_n^{-1}\left(\mathbf{h}^{opt}\right)T_n\left(\mathbf{h}^{opt}\right)\right) \\
&= \operatorname{trace}\left(T_n^{-1}\left(\mathbf{h}^{opt}\right)T_n\left(\mathbf{h}^{(2)}\right)\right) - \operatorname{trace}(I_n)
\end{aligned}$$

and consequently,

$$\operatorname{trace}\left(T_n^{-1}\left(\mathbf{h}^{opt}\right)T_n\left(\mathbf{h}^{(2)}\right)\right) = n. \qquad \square$$

Corollary 3.7. *If $T_n^{-1}(\mathbf{h}^{opt}) = [a_{ij}]_{1 \leq i,j \leq n}$ then*

$$a_{i,i} + a_{i,i+1} + a_{i+1,i+1} = \frac{3n}{2}, \qquad 1 \leq i \leq n-1.$$

Proof : Take $\mathbf{h} = (h_1, \ldots, h_{n-1})^\mathsf{T}$ with $h_1 = h_2 = \ldots = h_{i-1} = 0$, $h_i = 2$, $h_{i+1} = \ldots = h_{n-1} = 0$. Then

$$T_n(\mathbf{h}) = \frac{1}{3}
\begin{bmatrix}
0 & 0 & & & \cdots & 0 \\
0 & \ddots & & & & \\
\vdots & & 2 & 1 & & \vdots \\
\vdots & & 1 & 2 & & \vdots \\
& & & & \ddots & \\
0 & \cdots & & & \cdots & 0
\end{bmatrix}.$$

Hence, for this particular $\mathbf{h}$, we have

$$
\begin{aligned}
n &= \operatorname{trace}\left(T_n^{-1}(\mathbf{h}^{opt})T_n(\mathbf{h})\right) \\
&= \frac{1}{3}\operatorname{trace}[a_{ij}]
\begin{bmatrix}
0 & 0 & & & \cdots & 0 \\
0 & \ddots & & & & \\
\vdots & & 2 & 1 & & \vdots \\
\vdots & & 1 & 2 & & \vdots \\
& & & & \ddots & \\
0 & \cdots & & & \cdots & 0
\end{bmatrix} \\
&= \frac{1}{3}(2a_{ii} + 2a_{i,i+1} + 2a_{i+1,i+1})
\end{aligned}
$$

and so, indeed,

$$a_{i,i} + a_{i,i+1} + a_{i+1,i+1} = \frac{3n}{2}. \qquad \square$$

There is also a useful equivalent characterization of $\mathbf{h}^{opt}$ in terms of the (reciprocal of the) Christoffel function for the space of piecewise linear functions with breaks at the x_i determined by $\mathbf{h}^{opt}$, or, in other words, the space of the associated hat functions H_i.

In fact, consider the alternate basis $\{p_1(x), \ldots, p_n(x)\}$ given by

$$\begin{pmatrix} p_1(x) \\ \vdots \\ p_n(x) \end{pmatrix} = \left(T_n(\mathbf{h}^{opt})\right)^{-1/2} \begin{pmatrix} H_1(x) \\ \vdots \\ H_n(x) \end{pmatrix}.$$

Then

$$\begin{aligned} G(p_1, \ldots, p_n) &= \int_{\mathbb{R}} \begin{pmatrix} p_1(x) \\ \vdots \\ p_n(x) \end{pmatrix} (p_1(x), \ldots, p_n(x)) dx \\ &= \int_{\mathbb{R}} T_n(\mathbf{h}^{opt})^{-1/2} \begin{pmatrix} H_1(x) \\ \vdots \\ H_n(x) \end{pmatrix} (H_1(x), \ldots, H_n(x)) \left(T_n(\mathbf{h}^{opt})\right)^{-1/2} dx \\ &= \left(T_n(\mathbf{h}^{opt})\right)^{-1/2} T_n(\mathbf{h}^{opt}) \left(T_n(\mathbf{h}^{opt})\right)^{-1/2} \\ &= I. \end{aligned}$$

Hence, $\{p_1(x), \ldots, p_n(x)\}$ is in fact an orthonormal basis.
Now, let

$$K_n(x) := \sum_{k=1}^{n} p_k^2(x)$$

be the diagonal of the reproducing kernel

$$K_n(x, y) := \sum_{k=1}^{n} p_k(x) p_k(y).$$

The reciprocal of $K_n(x)$ is known as the Christoffel function and plays an important role in the theory of orthogonal polynomials (see e.g. Nevai [9]).
Then

$$\begin{aligned} K_n(x) &= \sum_{k=1}^{n} p_k^2(x) = (p_1(x), \ldots, p_n(x)) \begin{pmatrix} p_1(x) \\ \vdots \\ p_n(x) \end{pmatrix} \\ &= (H_1(x), \ldots, H_n(x)) \left(T_n(\mathbf{h}^{opt})\right)^{-1} \begin{pmatrix} H_1(x) \\ \vdots \\ H_n(x) \end{pmatrix}. \end{aligned}$$

We may calculate then,

$$\int_{x_k}^{x_{k+1}} K_n(x)dx = \sum_{i,j=1}^{n} a_{ij} \int_{x_k}^{x_{k+1}} H_i(x)H_j(x)dx$$

where we have again the notation $T_n^{-1}(\mathbf{h}^{opt}) = [a_{ij}]_{1\leq i,j\leq n}$.

But $H_i \not\equiv 0$ on $[x_k, x_{k+1}]$ only for $i = k$ and $i = k + 1$. Thus

$$\begin{aligned}
\int_{x_k}^{x_{k+1}} K_n(x)dx &= a_{k,k} \int_{x_k}^{x_{k+1}} H_k^2(x)dx + 2a_{k,k+1} \int_{x_k}^{x_{k+1}} H_k(x)H_{k+1}(x)dx \\
&\quad + a_{k+1,k+1} \int_{x_k}^{x_{k+1}} H_{k+1}^2(x)dx \\
&= \frac{h_k}{3}(a_{k,k} + a_{k,k+1} + a_{k+1,k+1})
\end{aligned}$$

by an easy evaluation of the integrals.

Hence we have shown that

Proposition 3.8. *For the optimal points*

$$\int_{x_k}^{x_{k+1}} K_n(x)dx = \frac{h_k}{3}\frac{3n}{2} = \frac{nh_k}{2}, \quad 1 \leq k \leq n-1. \qquad \square$$

This latter may be used to give bounds on nh_k, $1 \leq k \leq n - 1$.

Proposition 3.9. *For $k = 1,\ldots,n - 1$ we have $nh_k \leq 8$.*

Proof : Nikolskii's inequality (see e.g. [5, p. 102, Theorem 2.6]) informs us that for a polynomial $p(x)$

$$\max_{a\leq x\leq b} |p(x)| \leq \left(2(2 + 1)(\deg p)^2\right)^{1/2} \left(\frac{1}{b - a} \int_a^b p^2(x)dx\right)^{1/2}$$

or, for $x \in [a, b]$,

$$|p(x)|^2 \leq 6(\deg p)^2\frac{1}{b - a} \int_a^b p^2(x)dx.$$

In case $p(x)$ is affine linear, we have

$$|p(x)|^2 \leq \frac{6}{b - a} \int_a^b p^2(x)dx,$$

but the constant 6 may be improved to 4, as is shown in Lemma 3.11 below. Consequently, if $L(x)$ is piecewise affine linear with breaks at $x_1 < x_2 < \ldots < x_n$, then for $x \in [x_k, x_{k+1}]$,

$$L^2(x) \leq \frac{4}{h_k} \int_{x_k}^{x_{k+1}} L^2(y)dy \leq \frac{4}{h_k} \int_{-1}^{1} L^2(y)dy.$$

In other words, for $x \in [x_k, x_{k+1}]$,

$$\frac{L^2(x)}{\int_{-1}^{1} L^2(y)dy} \leq \frac{4}{h_k}.$$

But, as is well known,

$$K_n(x) = \sup_L \frac{L^2(x)}{\int_{-1}^{1} L^2(y)dy}$$

so that we have, in fact,

$$K_n(x) \leq \frac{4}{h_k} \quad \text{for} \quad x \in [x_k, x_{k+1}].$$

From Proposition 3.8 we have

$$\frac{nh_k}{2} = \int_{x_k}^{x_{k+1}} K_n(x)dx \leq \int_{x_k}^{x_{k+1}} \frac{4}{h_k}dx = 4$$

so that, indeed, $nh_k \leq 8$. $\qquad\square$

Proposition 3.10. *For $k = 2, 3, \ldots, n - 1$, $nh_k + nh_{k-1} \geq 2$, $nh_1 \geq 2$ and $nh_{n-1} \geq 2$.*

Proof: Again using the fact that

$$K_n(x) = \sup_L \frac{L^2(x)}{\int_{-1}^{1} L^2(y)dy},$$

we have, for $L(x) = H_k(x)$ (H_k being the hat function defined in (3.1)),

$$H_k^2(x) \leq K_n(x) \int_{-1}^{1} H_k^2(y)dy.$$

Hence,

$$\int_{x_k}^{x_{k+1}} H_k^2(x)dx \leq \int_{x_k}^{x_{k+1}} K_n(x)dx \int_{-1}^{1} H_k^2(y)dy$$

or

$$\frac{h_k}{3} \leq \frac{nh_k}{2} \frac{(h_k + h_{k-1})}{3}.$$

Since $h_k \neq 0$ it follows that $n(h_k + h_{k-1}) \geq 2$. When $k = 1$, this specializes to $nh_1 \geq 2$ and when $k = n$ to $nh_{n-1} \geq 2$. $\qquad\qquad\square$

Lemma 3.11. *(Improved Version of Nikolskii's inequality for linear polynomials) If* $p(x) = Ax + B$ *then for* $x \in [a,b]$

$$p^2(x) \leq \frac{4}{b-a} \int_a^b p^2(y)dy.$$

Proof : By a linear change of variables, we may assume that $[a,b] = [0,1]$. Hence we wish to show that for $x \in [0,1]$,

$$p^2(x) \leq 4 \int_0^1 p^2(y)dy.$$

We may further simplify our task by noting that if $A = 0$, there is nothing to do and otherwise dividing by A^2 allows us to assume that $A = 1$.
Now

$$\max_{x\in[0,1]}(x + B)^2 = \max\{B^2, (B+1)^2\}$$

and

$$4 \int_0^1 (y + B)^2 dy = \frac{4}{3}\{3B^2 + 3B + 1\}.$$

Hence we must show that

$$3\max\{B^2, (B+1)^2\} \leq 4\{3B^2 + 3B + 1\} \quad \forall B \in \mathbb{R}.$$

But

$$4\{3B^2 + 3B + 1\} - 3B^2 = 9B^2 + 12B + 4 = 9(B + \frac{2}{3})^2 \geq 0$$

and

$$4\{3B^2 + 3B + 1\} - 3(B+1)^2 = 9B^2 + 6B + 1 = 9(B + \frac{1}{3})^2 \geq 0$$

and we are done. $\qquad\qquad\square$

We remark that the constant 4 is best possible since equality is obtained for $p(x) = x - \frac{1}{3}$.

Proof of Theorem 3.4.

Supposing that the Technical Conjecture 3.5 holds, Theorem 3.4 would then follow from Propositions 3.9 and 3.10. $\qquad\square$

We now also give the (weak) asymptotics of the Christoffel function $K_n(x)$.

Proposition 3.12. *For the points which maximize* $\det(T_n)$, *the weak*–limit of* $\frac{1}{n}K_n(x)$ *on* $[-1,1]$ *as* $n \to \infty$ *is* $\lim\limits_{n\to\infty} \frac{1}{n}K_n(x) = \frac{1}{2}$.

Proof : For $f \in C[-1,1]$, let

$$
\begin{aligned}
\mu_n(f) \;&:=\; \int_{-1}^{1} f(x)\frac{K_n(x)}{n}dx \\
&=\; \sum_{i=1}^{n-1} \int_{x_k}^{x_{k+1}} f(x)\frac{K_n(x)}{n}dx.
\end{aligned}
$$

Then if $m_k := \min\limits_{x\in[x_k,x_{k+1}]} f(x)$ and $M_k := \max\limits_{x\in[x_k,x_{k+1}]} f(x)$, with the aid of Proposition 3.8

$$
\begin{aligned}
\frac{m_k h_k}{2} \;&=\; m_k \int_{x_k}^{x_{k+1}} \frac{K_n(x)}{n}dx \leq \int_{x_k}^{x_{k+1}} f(x)\frac{K_n(x)}{n}dx \\
&\leq\; M_k \int_{x_k}^{x_{k+1}} \frac{K_n(x)}{n}dx = \frac{M_k h_k}{2}.
\end{aligned}
$$

Hence,

$$
\frac{1}{2}\sum_{k=1}^{n-1} h_k m_k \leq \mu_n(f) \leq \frac{1}{2}\sum_{k=1}^{n-1} h_k M_k.
$$

But as $f(x)$ is continuous on $[-1,1]$, both $\frac{1}{2}\sum\limits_{k=1}^{n-1} h_k m_k$ and $\frac{1}{2}\sum\limits_{k=1}^{n-1} h_k M_k$ are certain Riemann sums for $\frac{1}{2}\int_{-1}^{1} f(x)dx$. Therefore,

$$
\lim_{n\to\infty} \mu_n(f) = \frac{1}{2}\int_{-1}^{1} f(x)dx,
$$

or, in other words, the weak*–limit of $\frac{1}{n}K_n(x)$ is $\frac{1}{2}$. $\qquad\square$

References

[1] L. P. Bos, K. Salkauskas: *On the matrix $[|x_i - x_j|^3]$ and the cubic spline continuity equations*, J. Approx. Theory **51** (1987), 81–88.

[2] L. Bos: *Some Remarks on the Féjer problem for Lagrange interpolation in several variables*, J. Approx. Theory **60** (1990), 133–140.

[3] M. D. Buhmann: *Multivariate interpolation with radial basis functions*, Constr. Approx. **6** (1990), 103–130.

[4] Ph. J. Davis: *Interpolation and Approximation*, Dover Publications, New York 1975.

[5] R. deVore, G. G. Lorentz: *Constructive Approximation*, Grundlehren der math. Wissenschaften **303**, Springer, Berlin–Heidelberg 1993.

[6] N. Dyn, W. A. Light, E. W. Cheney: *Interpolation by piecewise-linear radial basis functions I*, J. Approx. Theory **59** (1989), 202–223.

[7] N. Dyn: *Interpolation and Approximation by radial and related functions*, in: *Approximation Theory VI*, C. K. Chui et al. (eds.), 211–234, Academic Press, New York 1989.

[8] G. H. Golub, Ch. F. VanLoan: *Matrix Computations*, 2nd edition, John Hopkins Univ. Press, Baltimore 1989.

[9] P. G. Nevai: *Orthogonal Polynomials*, Memoirs Amer. Math. Soc. Vol. **18**, Nr. 1, Providence, R. I., 1979.

[10] E. B. Saff, V. Totik: *Logarithmic Potentials with External Fields*, Grundlehren der math. Wissenschaften **316**, Springer, Berlin–Heidelberg 1997.

Addresses:

LEN BOS
Department of Mathematics and Statistics
University of Calgary
Calgary, Alberta T2N 1N4
Canada

ULRIKE MAIER
Fachbereich Mathematik
Universität Dortmund
D–44221 Dortmund
Germany

Besov Regularity for the Stokes Problem

Stephan Dahlke

Dedicated to Prof. M. Reimer on the occasion of his 65th birthday

Abstract

This paper is concerned with regularity estimates for the solutions to the Stokes problem in polygonal domains in $\mathrm{I\!R}^2$. Especially, we derive regularity results in specific scales of Besov spaces which determine the efficiency of adaptive numerical schemes. The proofs of the main results are based on representations of the solution spaces which were given by Osborn [20] and on characterizations of Besov spaces by wavelet expansions.

1   Introduction

In recent years, much effort has been spent to design and to analyze adaptive schemes for the numerical treatment of elliptic boundary value problems of the form

$$
\begin{aligned}
Lu &= f &&\text{on}\quad \Omega \subset \mathrm{I\!R}^d, \\
u &= 0 &&\text{on}\quad \partial\Omega,
\end{aligned}
\tag{1.1}
$$

where L denotes a second order elliptic differential operator and Ω is a Lipschitz domain. Especially, the investigation of adaptive algorithms based on *wavelet expansions* has become a field of increasing importance [1,7]. The general idea of adaptive schemes is to improve the performance of the numerical algorithm by using nonuniform grid or space refinements, respectively, i.e., the underlying approximation space is refined only in regions where the current approximation is still 'far away' from the exact solution u to (1.1). Although this strategy seems to be very plausible at first glance, the principal question arises if an adaptive scheme indeed provides some gain of efficiency when compared with uniform (nonadaptive) methods. It turns out that the answer to this question is related with the regularity properties of the solution u in (1.1) as we shall now explain very briefly. In general, an adaptive scheme can be interpreted as some kind of *nonlinear approximation*. It can be shown that for a function F in $L_2(\Omega)$ the order of approximation that can be achieved by a nonlinear wavelet method

Advances in Multivariate Approximation; W. Haußmann, K. Jetter and M. Reimer (eds.)
Mathematical Research, Vol. 107, pp. 129–138, ISBN 3–527–40236–5
© WILEY–VCH, Berlin 1999

is determined by its regularity in the specifc scale $B^s_\tau(L_\tau(\Omega))$, $1/\tau = s/d + 1/2$, of *Besov* spaces (see, e.g., [12,22] for the definition and the main properties of Besov spaces). For a detailed description of these fundamental relationships and of its consequences for numerical schemes, the reader is referred, e.g., to [6,10,11]. In contrast to this, the efficiency of *uniform* methods is determined by the regularity of F in the usual Sobolev scale $H^s(\Omega)$. Therefore the following question arises: Does the solution u to (1.1) have a higher regularity in the scale $B^s_\tau(L_\tau(\Omega))$, $1/\tau = s/d + 1/2$, of Besov spaces compared to the corresponding Sobolev scale? For then, adaptive algorithms can indeed perform better than uniform schemes, at least in principle.

Quite recently, several results in this direction have been shown [2,3,4,5,8,16]. It has turned out that for many important cases the Besov regularity of the solution u is high enough to justify the use of adaptive schemes. The deepest results were obtained for problems on general Lipschitz domains where the Sobolev regularity decreases significantly due to singularities near the boundary [8]. This note can be interpreted as a continuation of the above studies. We shall be concerned with an important special case, i.e., with the *2D–Stokes problem*. Let Ω be a bounded, simply connected, polygonal domain in $\mathrm{I\!R}^2$. Then, given a vector field $f \in H^{-1}(\Omega)^2$ and a function $g \in L_{2,0}(\Omega) := \{q \in L_2(\Omega) \; : \; \int_\Omega q(x)dx = 0\}$, one has to determine the velocity $u \in H^1_0(\Omega)^2$ and the pressure $p \in L_{2,0}(\Omega)$ such that

$$\begin{aligned} -\Delta u + \nabla p &= f \quad \text{in } \Omega, \\ -\nabla \cdot u &= g \quad \text{in } \Omega. \end{aligned} \tag{1.2}$$

In the *mixed formulation*, the problem reads as follows: find a pair

$$(u,p) \in H^1_0(\Omega)^2 \times L_{2,0}(\Omega)$$

such that

$$\begin{aligned} a(u,v) \;+\; b(v,p) &= \langle f,v \rangle \quad && \text{for all } v \in H^1_0(\Omega)^2, \\ b(u,q) \quad\quad &= \langle g,q \rangle \quad && \text{for all } q \in L_{2,0}(\Omega), \end{aligned} \tag{1.3}$$

where

$$a(u,v) \;:=\; (\nabla u, \nabla v) = \sum_{i,j=1}^{2} \int_\Omega \frac{\partial u_i}{\partial x_j}(x)\frac{\partial v_i}{\partial x_j}(x)dx,$$

$$b(v,q) \;:=\; -(\nabla \cdot v, q) = -\sum_{i=1}^{2} \int_\Omega q(x)\frac{\partial}{\partial x_i}v_i(x)dx.$$

For further information concerning the theory and the numerical treatment of the Stokes equations, the reader is referred, e.g., to Girault and Raviart [14] and to Teman [21].

The main result of this paper shows that the Besov regularity of u and p, respectively, is again much higher than the Sobolev regularity, so that the use of adaptive schemes is completely justified. More precisely, it turns out that under some further technical conditions u and p have the optimal regularity in the interesting Besov scale, i.e., for $f \in H^m(\Omega)^2$, $g \in H^{m+1}(\Omega)$, one has $u \in B_\tau^s(L_\tau(\Omega))^2$, $s < m + 2$, $p \in B_\tau^s(L_\tau(\Omega))$, $s < m + 1$, $1/\tau = s/2 + 1/2$.

2   A New Regularity Theorem

Our aim is to investigate the dependence of the regularity of the pair (u, p) in the scale $B_\tau^s(L_\tau(\Omega))$, $1/\tau = s/2 + 1/2$, of Besov spaces on the smoothness of f and g and on the shape of the domain Ω. Before we can state our main result, some preparations are necessary. Let the segments of $\partial\Omega$ be denoted by $\overline{\Gamma}_l$, Γ_l open, $l = 1, \ldots, N$, numbered in positive orientation. Furthermore, let S_l denote the endpoint of Γ_l. Let us now suppose that $f \in H^m(\Omega)^2$ and $g \in H^{m+1}(\Omega)$ for some $m \in \mathrm{I\!N}$. By using the regularity theory for smooth domains, see, e.g., [18] for details, we first observe that $u \in H^{m+2}(\tilde{\Omega})^2$, $p \in H^{m+1}(\tilde{\Omega})$ for any subdomain $\tilde{\Omega}$ of Ω with smooth boundary not containing a vertex of Ω. Then the well–known embeddings of Besov spaces $H^\alpha(\Omega) = B_2^\alpha(L_2(\Omega)) \hookrightarrow B_\tau^s(L_\tau(\Omega))$, $s < \alpha$, $\tau < 2$, give the estimates

$$u \in B_\tau^s(L_\tau(\tilde{\Omega}))^2, \qquad 1/\tau = s/2 + 1/2, \; s < m + 2, \qquad (2.1)$$
$$p \in B_\tau^s(L_\tau(\tilde{\Omega})), \qquad 1/\tau = s/2 + 1/2, \; s < m + 1.$$

Therefore it remains to study the regularity of u and p near the vertices. By the usual decomposition technique using suitable C^∞ truncation functions, it turns out that u and p can be written as

$$u \;=\; u_I + u_B, \qquad u_B = \sum_{l=1}^{N} u_l, \qquad (2.2)$$

$$p \;=\; p_I + p_B, \qquad p_B = \sum_{l=1}^{N} p_l, \qquad (2.3)$$

where the functions u_l and p_l are supported in the neighbourhood of the vertex S_l and are solutions to a modified Stokes problem, see Osborn [20] for details. Since (u,p) equals (u_l,p_l) in the vicinity of S_l, we see that the study of p and u near the vertex S_l is reduced to the study of the Stokes problem in a sector. Therefore the remaining results in this paper will all be stated for the Stokes equation in a sector.

We need some further notations. By a change of coordinates, we may assume that the vertex S_l is placed at 0 and that one of the sides of the corresponding sector V lies on the positive x_1–axis. Let ω denote the measure of the interior angle of V. Furthermore, let λ_j denote one of the roots of the transcendental equation

$$v(z) := \sinh^2(z^2\omega) - z^2 \sin^2(\omega) = 0, \qquad (2.4)$$

which lie in the upper half plane. Moreover, m_j is defined to be the order of λ_j as a zero of $v(z)$. Finally, we define the weighted Sobolev space $W^{m,\alpha}(V)$ to be the set of all functions for which the following norm is finite:

$$\|w\|_{W^{m,\alpha}(V)} := \sum_{\nu=0}^{m} \int_V r^{\alpha-2(m-\nu)} \Big(\sum_{|\mu|=\nu} |D^\mu w|^2 \Big) dx, \qquad r := (x_1^2 + x_2^2)^{1/2}. \quad (2.5)$$

Then the main result reads as follows.

Theorem 2.1. *Suppose that $f \in W_0^{m,0}(V)^2$, $g \in W_0^{m+1,0}(V)$ and that no λ_j lies on the line $\Im z = m+1$ in the complex plane. Let (u,p) denote the solution to the Stokes problem*

$$\begin{aligned} -\Delta u + \nabla p &= f \quad in\ V, \\ -\nabla \cdot u &= g \quad in\ V. \end{aligned} \qquad (2.6)$$

Then

$$\begin{aligned} u &\in B_\tau^s(L_\tau(V))^2, \quad for\ all\ \ s < m+2,\ 1/\tau = s/2 + 1/2, \\ p &\in B_\tau^s(L_\tau(V)), \quad for\ all\ \ s < m+1,\ 1/\tau = s/2 + 1/2. \end{aligned}$$

The proof is based on the following characterization of the solution space to (2.6) which was derived by Osborn [20]. Similar results have also been obtained by Grisvard [15] and Kondrat'ev [17].

Theorem 2.2. *Suppose that the conditions of Theorem 2.1 are satisfied. Let (r, θ) denote polar coordinates in V. Then u and p have expansions $u = u_R + u_S$, $p = p_R + p_S$, where $u_R \in W_0^{m+2,0}(V)^2$, $p_R \in W^{m+1,0}(V)$ and*

$$u_S \;=\; \sum_{0<\Im\lambda_j<m+1}\;\sum_{l=0}^{m_j-1} C_{j,l}^u(\theta) r^{-i\lambda_j}\log^l(r), \qquad (2.7)$$

$$p_S \;=\; \sum_{0<\Im\lambda_j<m+1}\;\sum_{l=0}^{m_j-1} C_{j,l}^p(\theta) r^{-i\lambda_j-1}\log^l(r), \qquad (2.8)$$

where $C_{j,l}^u(\theta)$ and $C_{j,l}^p(\theta)$ are C^∞–functions of θ.

We have to establish Besov regularity for u_R, u_S, p_R and p_S. The functions u_R and p_R can be treated as above by using suitable embeddings. It remains to study the singular parts u_S and p_S. It turns out that these parts, although not very smooth in the usual Sobolev scale, have arbitrary high regularity in the specific scale of Besov spaces we are interested in.

Theorem 2.3. *Suppose that the conditions of Theorem 2.1 are satisfied. Then for the functions u_S and p_S according to (2.7) and (2.8), respectively, the following holds:*

$$u_S \in B_\tau^s(L_\tau(V))^2, \qquad 1/\tau = s/2 + 1/2, \text{ for all } s > 0,$$
$$p_S \in B_\tau^s(L_\tau(V)), \qquad 1/\tau = s/2 + 1/2, \text{ for all } s > 0.$$

By employing Theorem 2.3 which will be proved in Section 3, the result follows.

Remark 2.4. *The reader should observe that, in contrary to the usual Sobolev regularity, the Besov regularity of u and p is independent of the shape of the domain and depends only on the smoothness of the functions f and g.*

3 Proof of Theorem 2.3

The proof can be performed by employing the ideas developed in [4]. We shall briefly discuss the most important steps. We only present the arguments for the function p_S according to (2.8), the function u_S can be treated analogously. It is sufficient to establish Besov regularity for a function $h(r, \theta)$ of the form

$$h(r, \theta) = C(\theta) r^{-i\gamma-1}\log^l(r), \qquad (3.1)$$

where $C(\theta)$ is a C^∞-function and $\Im(\gamma) > 0$. We want to use the fact that function spaces such as Besov spaces can be characterized by wavelet expansions. Let Ψ be the set of $2^d - 1$ functions built in the usual way by tensor products from the univariate, compactly supported, orthonormal Daubechies wavelets, see [9], [19]. Then the functions

$$\eta_I := \eta_{j,k} := 2^{jd/2}\eta(2^j \cdot -k), \quad I = 2^{-j}k + 2^{-j}[0,1]^d, \quad k \in \mathbb{Z}^d, j \in \mathbb{Z}, \eta \in \Psi, \tag{3.2}$$

form an orthonormal basis for $L_2(\mathbb{R}^d)$. If the functions $\eta \in \Psi$ are sufficiently smooth (which can always be achieved, see [9] for details), then a function F is in the Besov space $B_\tau^\alpha(L_\tau(\mathbb{R}^d)), 1/\tau = \alpha/d + 1/2$, if and only if

$$\|P_0(F)\|_{L_\tau(\mathbb{R}^d)} + \left(\sum_{\eta \in \Psi} \sum_{I \in \mathcal{D}^+} |\langle F, \eta_I \rangle|^\tau \right)^{1/\tau} < \infty, \tag{3.3}$$

where $\mathcal{D}^+$ denotes the set of all dyadic cubes of measure < 1 and P_0 is a projector onto a suitable subspace of $L_2(\mathbb{R}^d)$, see, e.g., [19] for the case $\tau > 1$ and [13] for the general case. According to (3.3), we have to estimate the wavelet coefficients of a function h of the form (3.1). By employing a suitable extension technique, we may view $h(\theta, r)$ as a function on all of $\mathbb{R}^2$, see [4] for details. It can be shown that for this extended function the term $\|P_0(h)\|_{L_\tau(\mathbb{R}^d)}$ is always finite, see [8]. Therefore it remains to estimate the second term in (3.3), i.e., we have to show that

$$\sum_{(I,\eta) \in \Lambda} |\langle h, \eta_I \rangle|^\tau < \infty, \tag{3.4}$$

where Λ denotes the set of all pairs (I, η), $I \in \mathcal{D}^+$, $\eta \in \Psi$ for which $Q(I) \cap V \neq \emptyset$. Here $Q(I)$ denotes a suitable cube which contains the support of η_I. Let us start by estimating one wavelet coefficient. By using the vanishing moment property of wavelets, see again [9] for details, and employing a classical Whitney–type estimate for the error of approximation by polynomials on cubes, it turns out that there exists a polynomial P_I of total degree $< n$ such that

$$\begin{aligned}
|\langle h, \eta_I \rangle| &\leq \|h - P_I\|_{L_2(Q(I))} \|\eta_I\|_{L_2(Q(I))} \tag{3.5} \\
&\lesssim |Q(I)|^{(n+1)/2} |h|_{W^n(L_\infty(Q(I)))} \\
&\lesssim 2^{-j(n+1)} |h|_{W^n(L_\infty(Q(I)))}.
\end{aligned}$$

(By ' $\lesssim$ ' we clearly indicate inequality up to constants). Now we have to sum these expressions. First, we fix a refinement level j by considering the set

$$\Lambda_j := \{(I,\eta) \in \Lambda \ : \ |I| = 2^{-2j}\}. \tag{3.6}$$

For each level, we cover V by layers, i.e., we define

$$\Lambda_{j,k} \ := \ \{(I,\eta) \in \Lambda_j \ : \ k2^{-j} \le \delta_I < (k+1)2^{-j}\}, \tag{3.7}$$

where δ_I denotes the distance of the cube $Q(I)$ to zero,

$$\delta_I := \inf_{x \in Q(I)} r(x).$$

We first consider the sets

$$\Lambda_j^o := \Lambda_j \backslash \Lambda_{j,c}, \ \Lambda_{j,c} := \{(I,\eta) \in \Lambda_j \ : \ \delta_I < c2^{-j}\} \tag{3.8}$$

for some suitable constant c and estimate $|h|_{W^n(L_\infty(Q(I)))}$ for a typical cube $Q(I)$, $(I,\eta) \in \Lambda_j^o$. By using polar coordinates and Leibniz' rule, we obtain for $|\beta| = n$ on $Q(I)$

$$
\begin{aligned}
|D^\beta h| &\lesssim \sum_{\nu=0}^{n} r^{-(n-\nu)} \left| \left(\frac{d}{dr}\right)^\nu (r^{-i\gamma-1}\log^l(r)) \right| \\
&\lesssim \sum_{\nu=0}^{n} r^{-(n-\nu)} \sum_{\mu=0}^{\nu} r^{\Im\gamma-1-(\nu-\mu)} \left| \left(\frac{d}{dr}\right)^\mu \log^l(r) \right| \\
&\lesssim r^{-n+\Im\gamma-1} \sum_{\nu=0}^{\min(n,l-1)} |\log^{l-\nu}(r)| \\
&\lesssim r^{-n+\Im\gamma-1-\varepsilon}
\end{aligned}
$$

for some suitable small $\varepsilon > 0$. Hence

$$|h|_{W^n(L_\infty(Q(I)))} \lesssim \delta_I^{\Im\gamma-n-1-\varepsilon} \qquad \text{for } (I,\eta) \in \Lambda_j^o. \tag{3.9}$$

Furthermore, one has

$$|\Lambda_{j,k}| \lesssim k \tag{3.10}$$

so that, by combining (3.5), (3.9) and (3.10), we obtain

$$
\begin{aligned}
\sum_{(I,\eta)\in\Lambda_j^o} |\langle h,\eta_I\rangle|^\tau &\lesssim \sum_{k=k_1}^{\infty} \sum_{(I,\eta)\in\Lambda_{j,k}} 2^{-j(n+1)\tau} \delta_I^{(\Im\gamma-n-1-\varepsilon)\tau} \\
&\lesssim \sum_{k=k_1}^{\infty} k \cdot 2^{-j(n+1)\tau} (k \cdot 2^{-j})^{(\Im\gamma-n-1-\varepsilon)\tau} \\
&\lesssim 2^{-j(-\varepsilon+\Im\gamma)\tau} \sum_{k=k_1}^{\infty} k^{1+(\Im\gamma-n-1-\varepsilon)\tau}, \tag{3.11}
\end{aligned}
$$

where k_1 depends on the constant c in (3.8). If we choose n large enough, the sum involving k is clearly finite. Summing over all refinement levels, we are left with a geometric series which is convergent if we choose $\varepsilon < \Im\gamma$ which is clearly possible.

It remains to study the sets $\Lambda_{j,c}$. It follows from (2.8) that $h \in H^s(V)$, for some sufficiently small s. Using this fact and following the lines of the proof of Theorem 3.2 in [8], we obtain the condition

$$-2s\tau/(2 - \tau) < 0$$

which is clearly satisfied. The theorem is proved. $\qquad\square$

Acknowledgment. The author feels grateful to the Deutsche Forschungsgemeinschaft which has supported this work (DA 117/13–1). He also wants to thank K. Urban for several helpful remarks.

References

[1]   A. Cohen, W. Dahmen, R. DeVore: *Adaptive wavelet methods for elliptic operator equations - convergence rates*, preprint, 1998.

[2]   S. Dahlke: *Wavelets: Construction Principles and Applications to the Numerical Treatment of Operator Equations*, Habilitation thesis, Shaker Verlag, Aachen 1997.

[3]   S. Dahlke: *Besov regularity for elliptic boundary value problems with variable coefficients*, Manuscripta Math. **95** (1998), 59–77.

[4]   S. Dahlke: *Besov regularity for elliptic boundary value problems on polygonal domains*, Appl. Math. Lett., in press (1999).

[5]   S. Dahlke: *Besov regularity for interface problems*, Z. Angew. Math. Mech., in press (1999).

[6]   S. Dahlke, W. Dahmen, R. DeVore: *Nonlinear approximation and adaptive techniques for solving elliptic equations*, in: *Multiscale Wavelet Methods for Partial Differential Equations*, W. Dahmen et al. (eds.), 237–283, Academic Press, San Diego 1997.

[7] S. Dahlke, W. Dahmen, R. Hochmuth, R. Schneider: *Stable multiscale bases and local error estimation for elliptic problems*, Appl. Numer. Math. **23** (1997), 21–47.

[8] S. Dahlke, R. DeVore: *Besov regularity for elliptic boundary value problems*, Comm. Partial Differential Equations **22**(1&2) (1997), 1–16.

[9] I. Daubechies: *Ten Lectures on Wavelets*, CBMS–NSF Regional Conference Series in Applied Math. **61**, SIAM, Philadelphia 1992.

[10] R. DeVore: *Nonlinear approximation*, Acta Numerica **7**, 51–150, Cambridge University Press, Cambridge 1998.

[11] R. DeVore, B. Jawerth, V. Popov: *Compression of wavelet decompositions*, Amer. J. Math. **114** (1992), 737–785.

[12] R. DeVore, V. Popov: *Interpolation of Besov spaces*, Trans. Amer. Math. Soc. **305** (1988), 397–414.

[13] M. Frazier, B. Jawerth: *Decomposition of Besov spaces*, Indiana Univ. Math. J. **34** (1985), 777–799.

[14] V. Girault, P.–A. Raviart: *Finite Element Methods for Navier–Stokes–Equations*, Springer, Berlin 1986.

[15] P. Grisvard: *Singularities in Boundary Value Problems*, Research Notes in Applied Mathematics **22**, Springer, Berlin 1992.

[16] R. Hochmuth: *Restricted nonlinear approximation and singular solutions of boundary integral equations*, preprint, 1998.

[17] V. A. Kondrat'ev: *Asymptotic of a solution of the Navier–Stokes equation near the angular part of the boundary*, Prikl. Mat. Meh. **31** (1967), 119–123 (Russian); J. Appl. Math. Mech. **31** (1967), 125–129.

[18] O. A. Ladyzhenskaya: *The Mathematical Theory of Viscous Impressible Flow*, Gordan and Breach, New York 1962.

[19] Y. Meyer: *Wavelets and Operators*, Cambridge Studies in Advanced Mathematics **37**, Cambridge 1992.

[20]   J. Osborn: *Regularity of solutions of the Stokes problem in a poly-gonal domain*, in *Symposium on Numerical Solutions of Partial Differential Equations III*, B. Hubbart (ed.), 393–411, Academic Press, New York 1975.

[21]   R. Teman: *Navier–Stokes Equations. Theory and Numerical Analysis*, North Holland, Amsterdam 1977.

[22]   H. Triebel: *Interpolation Theory, Function Spaces, Differential Operators*, North–Holland, Amsterdam 1978.

Address:

STEPHAN DAHLKE
Institut für Geometrie und
Praktische Mathematik
RWTH Aachen
Templergraben 55
D–52056 Aachen
Germany

Optimal Periodic Interpolation in Multivariate Periodic Hilbert Spaces

Franz–Jürgen Delvos

Dedicated to Prof. Dr. Manfred Reimer on the occasion of his 65th birthday

Abstract

Minimum norm interpolation in multivariate periodic Hilbert spaces are considered and error bounds in appropriate subspaces of the multivariate Wiener algebra are derived.

Introduction

The concept of one–dimensional periodic Hilbert spaces (Babuška [1, 2]) was introduced to investigate universally optimal linear functionals in these spaces. The relation to optimal periodic interpolation was established by Prager [9], see also Locher [8].

The approximation properties of optimal periodic interpolation both in the uniform norm as well as in the the mean square norm were investigated in the papers [3, 4, 5] using the ideas of Golomb [7] for periodic splines.

In the paper [6] we generalized the concept of periodic Hilbert space to the multivariate setting and investigated approximation properties of multivariate trigonometric polynomials in these spaces.

In this paper we will study optimal periodic interpolation in multivariate periodic Hilbert spaces and derive error bounds in appropriate subspaces of the multivariate Wiener algebra.

1 Periodic Hilbert Spaces

A multivariate periodic Hilbert space is constructed with the aid of a real nonnegative defining sequence

$$D \in l_1(\mathbf{Z}^n) \ .$$

Advances in Multivariate Approximation; W. Haußmann, K. Jetter and M. Reimer (eds.)
Mathematical Research, Vol. 107, pp. 139–150, ISBN 3–527–40236–5
© WILEY–VCH, Berlin 1999

Its Fourier series

$$d(x) = \sum_{k \in \mathbf{Z}^n} D_k e_k(x) \tag{1.1}$$

with $e_k(x) = e^{ik \cdot x}$ defines the generating function which is an element of the Wiener algebra

$$A(\mathbf{T}^n) = \left\{ f \in L_2(\mathbf{T}^n) : \sum_{k \in Z^n} |F_k| < \infty \right\} \quad (\mathbf{T} = [0, 2\pi[\ ).$$

The *finite Fourier transform* of $f \in L_2(\mathbf{T}^n)$ is given by

$$F_k = \frac{1}{(2\pi)^n} \int_{\mathbf{T}^n} f(x) e_k(-x) \, dx. \tag{1.2}$$

It is used to define the periodic Hilbert space as the linear space of functions

$$H_D(\mathbf{T}^n) = \left\{ f : \sum_{k \in \mathbf{Z}^n} |F_k|^2 / D_k < \infty \right\}.$$

Its inner product is given by

$$(f, g)_D = \sum_{k \in \mathbf{Z}^n} F_k \overline{G_k} / D_k. \tag{1.3}$$

For $x \in \mathbf{R}^n$ the translate of a function $h \in L_2(\mathbf{T}^n)$ is defined by

$$h_x(y) = h(y - x).$$

Any $f, g \in H_D(\mathbf{T}^n)$ satisfy

$$(f_x, g)_D = (f, g_x)_D = (f, g)_D. \tag{1.4}$$

$H_D(\mathbf{T}^n)$ is a *reproducing Hilbert space* by

$$(f, d_x)_D = f(x). \tag{1.5}$$

Recall that the Wiener algebra $A(\mathbf{T}^n)$ is a Banach algebra with respect to the norm

$$\|f\|_a = \sum_{k \in \mathbf{Z}^n} |F_k|.$$

The Wiener algebra $A(\mathbf{T}^n)$ is also a subalgebra of the algebra of continuous periodic functions $C(\mathbf{T}^n)$ which norm is given by

$$\|f\|_\infty = \sup\left\{|f(x)| : x \in \mathbf{T}^n\right\}.$$

Note that

$$\|f\|_\infty \leq \|f\|_a \tag{1.6}$$

with equality if the finite Fourier transform of f is nonnegative.
The topological inclusions

$$H_D(\mathbf{T}^n) \subset A(\mathbf{T}^n) \subset C(\mathbf{T}^n)$$

are true in view of the inequalities

$$\|f\|_\infty \leq \|f\|_a \leq \|f\|_D \sqrt{d(0)}.$$

It follows from the reproducing kernel properties that the translates of the generating function form a dense subset of the periodic Hilbert space

$$\mathrm{clin}\left\{d(\cdot - a) : a \in \mathbf{T}^n\right\} = H_D(\mathbf{T}^n).$$

Here "clin" means closed linear span.

2 Multivariate Defining Sequences

We discuss two methods of construction of multivariate defining sequences from univariate ones. For this purpose we use the geometric and harmonic mean. The geometric mean of a vector $a = (a_1, ..., a_n)$ of positive numbers is defined by

$$G(a_1, ..., a_n) = [a_1...a_n]^{1/n} \tag{2.1}$$

while the harmonic mean of a vector $a = (a_1, ..., a_n)$ of positive numbers is defined by

$$H(a_1, ..., a_n) = \frac{n}{\frac{1}{a_1} + ...\frac{1}{a_n}}. \tag{2.2}$$

The *anisotropic method* is based on (simple forming) tensor products of univariate defining sequences:

$$(D_P)_{k_1,...,k_n} = D_{k_1}...D_{k_n}. \tag{2.3}$$

This method is related to the geometric mean by

$$(D_P)_{k_1,\dots,k_n} = G(D_{k_1},\dots,D_{k_n})^n. \tag{2.4}$$

The *isotropic method* is based on the harmonic mean via

$$(D_S)_{k_1,\dots,k_n} = \left(\frac{n}{\frac{1}{D_{k_1}} + \dots + \frac{1}{D_{k_n}}} \right)^n. \tag{2.5}$$

A typical example of a univariate defining sequence is given by

$$D_k = 1/(1 + k^2)\ .$$

In this case,

$$H_{D_P}(\mathbf{T}^n) = W^{(1,\dots,1)}(\mathbf{T}^n)$$

is a tensor product Sobolev space of periodic functions while

$$H_{D_S}(\mathbf{T}^n) = W^n(\mathbf{T}^n)$$

is a multivariate Sobolev space of periodic functions.
For the univariate defining sequence

$$D_k = \exp(-\,|k|),$$

we get similar constructions for multivariate periodic analytic functions

$$H_{D_P}(\mathbf{T}^n) = H^{(1,\dots,1)}(\mathbf{T}^n),\ \ H_{D_S}(\mathbf{T}^n) = H^n(\mathbf{T}^n).$$

We conclude with recalling the fundamental inequality between the geometric and harmonic mean

$$H(a_1,\dots,a_n) \le G(a_1,\dots,a_n) \tag{2.6}$$

which implies

$$(D_S)_{k_1,\dots,k_n} \le (D_P)_{k_1,\dots,k_n} \tag{2.7}$$

and

$$H_{D_S}(\mathbf{T}^n) \subset H_{D_P}(\mathbf{T}^n). \tag{2.8}$$

3 Optimal Periodic Interpolation

Optimal periodic interpolation is defined as minimum norm interpolation in a periodic Hilbert space. We consider first a vector of positive integers

$$b = (b_1, ..., b_n), \quad b_j \in \mathbb{N} \tag{3.1}$$

and the associated *translation vector*

$$M = (M_1, ..., M_n), \quad M_j = 2b_j + 1. \tag{3.2}$$

The *interpolation points* are defined by

$$t_{j,M} = (2\pi j_1/M_1, ..., 2\pi j_n/M_n) = 2\pi j/M, \ j \in \mathbf{Z}^n. \tag{3.3}$$

The *associated closed subspace* of the periodic Hilbert space is defined by

$$N_b = \{f \in H_D(\mathbf{T}^n) : f(2\pi j/M) = 0, \ j \in [-b, b] \cap \mathbf{Z}^n\}. \tag{3.4}$$

The projection theorem in Hilbert spaces yields the existence of a unique *associated orthogonal projector* satisfying

$$\ker(Q_b) = N_b . \tag{3.5}$$

Proposition 1 . *Let* $f + N_b$ *be linear a manifold associated with a function* $f \in H_D(\mathbf{T}^n)$. *The* minimum norm interpolant $Q_b(f)$ *of* f *is defined as the unique solution of the variational problem*

$$\|Q_b(f)\|_D = \min\{\|g\|_D : g \in f + N_b\}. \tag{3.6}$$

The *characterizing equations* of the minimum norm interpolant are given by

$$g = Q_b(f) \Leftrightarrow (f - g, d(\cdot - t_{j,M}))_D = 0, \ j \in [-b, b] \cap \mathbf{Z}^n . \tag{3.7}$$

Thus the linear space of minimum norm interpolants can be described as follows:

$$\mathrm{rng}(Q_b) = \ker(Q_b)^\perp = \lin\{d(\cdot - t_{j,M}) : j \in [-b, b] \cap \mathbf{Z}^n\}.$$

The abbreviation "rng" means range while "lin" means linear span.

We obtain the representation of the minimum norm interpolant in the following form

$$Q_b(f)(x) = \sum_{j \in [-b,b] \cap \mathbf{Z}^n} c_j d(x - t_{j,M}) \ . \tag{3.8}$$

Using the Fourier series of the generating function, we obtain the Fourier series characterization of minimum norm interpolant

$$Q_b(f)(x) = \sum_{r \in \mathbf{Z}^n} a_r D_r e_r(x), \quad a_{r+kM} = a_r, \quad k \in \mathbf{Z}^n \ . \tag{3.9}$$

We use periodization of the finite Fourier transform to obtain Fourier series representation of the interpolants. Let

$$f(x) = \sum_{r \in \mathbf{Z}^n} F_r e_r(x).$$

Then we can conclude

$$f(t_{j,M}) = \sum_{r \in \mathbf{Z}^n} F_r e_r(t_{j,M})$$

$$= \sum_{k \in [-b,b] \cap \mathbf{Z}^n} \sum_{s \in \mathbf{Z}^n} F_{k+sM} e_{k+sM}(t_{j,M}) = \sum_{k \in [-b,b] \cap \mathbf{Z}^n} F_k^M e_k(t_{j,M}) \ .$$

Here

$$F_k^M = \sum_{s \in \mathbf{Z}^n} F_{k+sM} \tag{3.10}$$

denotes the *periodization* of the finite Fourier transform. This yields the following representation of the *trigonometric interpolation projector*

$$T_b(f)(x) = \sum_{k \in [-b,b] \cap \mathbf{Z}^n} F_k^M e_k(x) \ . \tag{3.11}$$

It follows from (3.9) that the Fourier series of the optimal periodic interpolant $Q_b(f)$ appears as an attenuated form of the trigonometric interpolant $T_b(f)$:

$$Q_b(f)(x) = \sum_{r \in \mathbf{Z}^n} \frac{D_r}{D_r^M} F_r^M e_r(x) \ . \tag{3.12}$$

Proposition 2. *The projector Q_b of optimal periodic interpolation extends to the Wiener algebra $A(\mathbf{T}^n)$ satisfying*

$$\|Q_b(f)\|_a = \|T_b(f)\|_a \leq \|f\|_a \ . \tag{3.13}$$

Proof.   We have

$$\|Q_b(f)\|_a = \sum_{r\in\mathbf{Z}^n}(D_r/D_r^M)|F_r^M| = \sum_{k\in[-b,b]\cap\mathbf{Z}^n}|F_k^M|/D_k^M\sum_{s\in\mathbf{Z}^n}D_{k+sM}$$

$$= \sum_{k\in[-b,b]\cap\mathbf{Z}^n}|F_k^M| = \|T_b(f)\|_a \le \sum_{k\in[-b,b]\cap\mathbf{Z}^n}\sum_{s\in\mathbf{Z}^n}|F_{k+sM}| = \|f\|_a\,.$$

$\square$

4   Convergence in the Wiener Algebra

The exponentials e_k, $k\in\mathbf{Z}^n$, generate a dense subspace of the Wiener algebra $A(\mathbf{T}^n)$. We will apply the Banach–Steinhaus principle to the family of uniformly bounded projectors Q_b (by Proposition 2) by establishing convergence for the exponentials.

It follows from (3.12) that

$$Q_b(e_s)(x) = \sum_{r\in\mathbf{Z}^n}(D_{s+rM}/D_s^M)\,e_{s+rM}(x)\,. \tag{4.1}$$

This implies

$$\|e_s - Q_b(e_s)\|_a = (1-(D_s/D_s^M)) + \sum_{r\in\mathbf{Z}^n,r\neq 0}(D_{s+rM}/D_s^M)\,. \tag{4.2}$$

We introduce the quantities

$$\tilde{D}_s^M := D_s^M - D_s = \sum_{r\in\mathbf{Z}^n,r\neq 0}D_{s+rM} \tag{4.3}$$

which describe the quality of approximation by optimal periodic interpolation. Clearly,

$$\tilde{D}_s^M \to 0 \quad \text{for} \quad \min\{b_j : 1\le j\le n\} \to \infty\,.$$

With respect to the norm of the Wiener algebra, we obtain the estimate

$$\|e_s - Q_b(e_s)\|_a \le 2\tilde{D}_s^M/D_s \to 0 \quad \text{for} \quad \min\{b_j : 1\le j\le n\} \to \infty\,. \tag{4.4}$$

Thus the Banach–Steinhaus principle is applicable and we obtain

Proposition 3. *Given* $f\in A(\mathbf{T}^n)$, *we have*

$$\|f - Q_b(f)\|_a \to 0 \quad \text{for} \quad \min\{b_j : 1\le j\le n\} \to \infty. \tag{4.5}$$

5   Convergence in Smooth Subspaces of the Wiener Algebra

Smooth subspaces of the Wiener algebra are related to the defining sequence D by

$$A_D(\mathbf{T}^n) = \left\{ f : \sum_{k \in \mathbf{Z}^n} |F_k| / D_k < \infty \right\} .$$

The associated linear operator is defined by

$$T_D(f)(x) = \sum_{k \in \mathbf{Z}^n} (F_k / D_k) e_k(x) .$$

We use the approach of Golomb to establish quantitative error bounds for optimal periodic interpolation in terms of the *error measure*

$$\tilde{D}^M := \sup_{k \in [-b,b]} \tilde{D}_k^M . \tag{5.1}$$

The Golomb approach is based on three steps:

- approximation by the Fourier partial sum,

- approximation of the Fourier partial sum by its optimal periodic interpolant,

- boundedness of the optimal periodic interpolation projector.

The *Fourier partial sum projector S_b* is given by

$$S_b(f)(x) = \sum_{k \in \mathbf{Z}^n} \chi_{[-b,b]}(k) F_k e_k(x) . \tag{5.2}$$

It is easily seen that

$$\|f - S_b(f)\|_a \leq \sup_{k \in \mathbf{Z}^n \setminus [-b,b]} D_k \, \|T_D(f) - S_b(T_D(f))\|_a , \; f \in A_D(\mathbf{T}^n) . \tag{5.3}$$

Since

$$\tilde{D}^M \geq \sup_{k \in \mathbf{Z}^n \setminus [-b,b]} D_k , \tag{5.4}$$

we obtain

$$\|f - S_b(f)\|_a \leq \tilde{D}^M \, \|T_D(f) - S_b(T_D(f))\|_a , \qquad f \in A_D(\mathbf{T}^n) . \tag{5.5}$$

Next we investigate the approximation of the Fourier partial sum by its optimal periodic interpolation. We have in view of (4.4)

$$\|S_b(f) - Q_b(S_b(f))\|_a = \left\|\sum_{k\in\mathbf{Z}^n}\chi_{[-b,b]}(k)F_k e_k - Q_b\Big(\sum_{k\in\mathbf{Z}^n}\chi_{[-b,b]}(k)F_k e_k\Big)\right\|_a$$

$$\leq \sum_{k\in[-b,b]}|F_k|\,\|e_k - Q_b(e_k)\|_a$$

$$\leq \sum_{k\in[-b,b]}|F_k|\,2\tilde{D}_k^M/D_k \ \leq 2\sup_{s\in[-b,b]}\tilde{D}_s^M\sum_{k\in[-b,b]}|F_k|/D_k \ \ .$$

This shows

$$\|S_b(f) - Q_b(S_b(f))\|_a \leq 2\cdot\tilde{D}^M\,\|T_D(S_b(f))\|_a\,, \quad f\in A_D(\mathbf{T}^n). \tag{5.6}$$

Taking Proposition 2 into account, we can conclude

$$\|f - Q_b(f)\|_a \leq \|f - S_b(f)\|_a + \|S_b(f) - Q_b(S_b(f))\|_a + \|Q_b(f) - Q_b(S_b(f))\|_a$$

$$\leq \tilde{D}^M\,\|T_D(f) - S_b(T_D(f))\|_a + \|S_b(f) - Q_b(S_b(f))\|_a + \|f - S_b(f)\|_a$$

$$\leq 2\tilde{D}^M\,\|T_D(f) - S_b(T_D(f))\|_a + \|S_b(f) - Q_b(S_b(f))\|_a$$

$$\leq 2\tilde{D}^M\,\|T_D(f) - S_b(T_D(f))\|_a + 2\tilde{D}^M\,\|T_D(S_b(f))\|_a \,,$$

i.e., we have shown (see also (1.6))

Proposition 4. *For any $f\in A_D(\mathbf{T}^n)$ we have*

$$\|f - Q_b(f)\|_\infty \leq \|f - Q_b(f)\|_a \leq 4\tilde{D}^M\,\|T_D(f)\|_a\,, \qquad f\in A_D(\mathbf{T}^n). \tag{5.7}$$

6 Multivariate Error Measures

For notational simplicity we consider mainly the bivariate case. Recall first that
$$M_1 = 2b_1 + 1,\ M_2 = 2b_2 + 1.$$

Then we have for $s_1\in\{-b_1,...,b_1\}, s_2\in\{-b_2,...,b_2\}$:
$$D_{s_1,s_2}^{M_1,M_2} = \sum_{r_1}\sum_{r_2}D_{s_1+r_1M_1,\,s_2+r_2M_2} \ \ .$$

Since

$$\tilde{D}^M_{s_1,s_2} = D^{M_1,M_2}_{s_1,s_2} - D_{s_1,s_2}, \tag{6.1}$$

we obtain

$$\tilde{D}^M_{s_1,s_2} = \sum_{r_1\neq 0} D_{s_1+r_1M_1,s_2} + \sum_{r_2\neq 0} D_{s_1,s_2+r_2M_2} + \sum_{r_1\neq 0}\sum_{r_2\neq 0} D_{s_1+r_1M_1,s_2+r_2M_2}. \tag{6.2}$$

To simplify this expression, we assume the following properties for the defining sequence:

- $1 \geq D_{k_1,k_2} = D_{k_2,k_1} > 0,$

- $D_{-k_1,k_2} = D_{k_1,k_2},\ D_{k_1,-k_2} = D_{k_1,k_2},$

- $D_{k_1+1,k_2} \leq D_{k_1,k_2},\quad k_1 \geq 0.$

This implies

$$D^{M_1,M_2}_{s_1,s_2} = D^{M_1,M_2}_{-s_1,s_2} = D^{M_1,M_2}_{s_1,-s_2},\quad \tilde{D}^{M_1,M_2}_{s_1,s_2} = \tilde{D}^{M_1,M_2}_{-s_1,s_2} = \tilde{D}^{M_1,M_2}_{s_1,-s_2}. \tag{6.3}$$

Thus we assume for the sequel

$$s_1 \in \{0,...,b_1\},\, s_2 \in \{0,...,b_2\}.$$

Using monotonicity, we get

$$\sum_{r_1\neq 0} D_{s_1+r_1M_1,s_2} \leq \sum_{r_1>0} D_{r_1b_1,0}, \tag{6.4}$$

$$\sum_{r_2\neq 0} D_{s_1,s_2+r_2M_2} \leq \sum_{r_2>0} D_{0,r_2b_2}, \tag{6.5}$$

$$\sum_{r_1\neq 0}\sum_{r_2\neq 0} D_{s_1+r_1M_1,s_2+r_2M_2} \leq \sum_{r_1>0}\sum_{r_2>0} D_{r_1b_1,r_2b_2}. \tag{6.6}$$

Since

$$\tilde{D}^M = \sup_{s\in[-b,b]} \tilde{D}^M_s = \sup_{s\in[0,b]} \tilde{D}^M_s,$$

we obtain the estimation of the error measure

$$\tilde{D}^M \leq \sum_{r_1>0} D_{r_1b_1,0} + \sum_{r_2>0} D_{0,r_2b_2} + \sum_{r_1>0}\sum_{r_2>0} D_{r_1b_1,r_2b_2} =: D^*_b. \tag{6.7}$$

A first application is given for *anisotropic periodic Hilbert spaces*. In this case we have

$$(D_P)_{k_1,k_2} = D_{k_1} D_{k_2}$$

and we can conclude

$$(D_P)_b^* = \sum_{r_1>0} D_{r_1 b_1} + \sum_{r_2>0} D_{r_2 b_2} + \sum_{r_1>0} D_{r_1 b_1} \sum_{r_2>0} D_{r_2 b_2}. \tag{6.8}$$

Thus the anisotropic error measure is determined by the univariate error measure. As a special case we get for $D_k = 1/(1+k^2)$

Proposition 5. *For any $f \in A_{D_P}(\mathbf{T}^2)$ we have*

$$\|f - Q_b(f)\|_\infty \leq \|f - Q_b(f)\|_a = O(b_1^{-2}) \cdot b_1 = b_2 . \tag{6.9}$$

Next we consider *isotropic periodic Hilbert spaces* . In this case we have

$$(D_S)_{k_1,k_2} = \left(\frac{2}{D_{k_1}^{-1} + D_{k_2}^{-1}}\right)^2 = \left(\frac{2 D_{k_1} D_{k_2}}{D_{k_1} + D_{k_2}}\right)^2 \leq D_{k_1} D_{k_2}$$

$$(D_S)_{k_1,0} \leq 4 D_{k_1}^2, \ (D_S)_{0,k_2} \leq 4 D_{k_2}^2,$$

and we can conclude

$$(D_S)_b^* = 4 \sum_{r_1>0} D_{r_1 b_1}^2 + 4 \sum_{r_2>0} D_{r_2 b_2}^2 + \sum_{r_1>0} D_{r_1 b_1} \sum_{r_2>0} D_{r_2 b_2}.$$

Thus the isotropic error measure is bounded by the "squared" univariate error measure. As a special case we get for $D_k = 1/(1+k^2)$

Proposition 6. *For any $f \in A_{D_S}(\mathbf{T}^2)$ we have*

$$\|f - Q_b(f)\|_\infty \leq \|f - Q_b(f)\|_a = O(b_1^{-4}) \cdot b_1 = b_2 . \tag{6.10}$$

References

[1]   I. Babuška: *Über universal optimale Quadraturformeln I*, Apl. Math. **13** (1968), 304–308.

[2]   I. Babuška: *Über universal optimale Quadraturformeln II*, Apl. Math. **13** (1968), 388–404.

[3]   F.-J. Delvos: *Approximation by optimal periodic interpolation,* Apl. Math. **35** (1990), 451–457.

[4]   F.-J. Delvos: *Approximation properties of periodic interpolation by translates of one function,* Modélisation Mathematique et Analyse Numérique **28** (1994), 177–188.

[5]   F.-J. Delvos: *Mean square approximation by optimal periodic interpolation,* Apl. Math. **40** (1995), 451–457.

[6]   F.-J. Delvos: *Trigonometric approximation in multivariate periodic Hilbert spaces,* in: *Multivariate Approximation: Recent Trends and Results,* W. Haußmann, K. Jetter, M. Reimer (eds.), Mathematical Research, Vol. **101**, pp. 35–44, Akademie Verlag, Berlin 1997.

[7]   M. Golomb: *Approximation by periodic spline interpolation on uniform meshes,* J. Approx. Theory **1** (1968), 26–65.

[8]   F. Locher: *Interpolation on uniform meshes by translates of one function and related attenuation factors,* Math. Comp. **37** (1981), 403–416.

[9]   M. Prager: *Universally optimal approximation of functionals,* Apl. Math. **24** (1979), 406–420.

Address:

Franz–Jürgen Delvos
University of Siegen
FB Mathematik I
Walter–Flex–Strasse 3
D–57068 Siegen
Germany

Multiscale Modelling
from Geopotential Data

Willi Freeden, Oliver Glockner and Rolf Litzenberger

Dedicated to Prof. Dr. M. Reimer on the occasion of his 65th birthday

Abstract

An approach to harmonic wavelets is presented within a framework of a Sobolev-like Hilbert space $\mathcal{H}$. Basic tool is the construction of $\mathcal{H}$-product kernels in terms of an (outer harmonics) orthonormal basis in $\mathcal{H}$. Scaling function and wavelet are introduced by means of $\mathcal{H}$-product kernels. Harmonic wavelets are shown to be "building blocks" that decorrelate geopotential data. A pyramid scheme provides fast computations. Finally, multiscale modelling of the earth's gravitational potential from NASA EGM96–model data is illustrated by use of harmonic wavelets.

Introduction

The earth is an almost perfect sphere. Deviations from its spherical shape are less than 0.4% of its radius (6371 km) and arise from its rotation. All equipotential surfaces are nearly spherical too. Thus spherical tools play an important role in the mathematical treatment of earth's gravitational potential determination. At present time the use of spherical harmonics is a well established technique in all geosciences for the purpose of representing geopotentials. Reference models for the earth's gravitational or magnetic potential are widely given by tables of coefficients for the spherical harmonic expansion. However, the fundamental difficulty of spherical harmonic approximation of a geopotential is that one has to deal in one way or other with functions not showing any phenomenon of space localization. This is the reason why any local change of a geopotential under consideration affects the whole table of potential coefficients. As a matter of fact, local modelling cannot be performed adequately within the framework of spherical harmonics. For future purposes a new component of approximation has to come into play, viz. harmonic wavelets. The power of wavelets is based on a multiresolution analysis which enables both (angular) momentum and space localization. Wavelets may be used as mathematical means for breaking up the complicated structure of a geopotential into

Advances in Multivariate Approximation; W. Haußmann, K. Jetter and M. Reimer (eds.)
Mathematical Research, Vol. 107, pp. 151–174, ISBN 3-527-40236-5
© WILEY-VCH, Berlin 1999

many simple pieces of different scales and positions. The representation of data in terms of wavelets is somehow "more compact" than the original representation. Basically this is achieved by forming convolutions of the geopotential against "dilated" and "shifted" versions of one fixed harmonic function with vanishing zeroth moment, i.e. the mother wavelet. A geopotential is represented by a two-parameter family reflecting the different space localizations and different levels of resolution. The definition of wavelets has to be given in close connection with a so-called scaling function approaching in the limit case the Dirac functional. Wavelet and scaling function are related via their symbols by scaling equations which are difference equations in the case of scale discrete wavelet theory as presented here. The scale parameter serves as a measure for decreasing momentum localization or equivalently as a measure for increasing space localization. The multiresolution analysis provides the approximation scheme. Another advantage in comparison to the outer harmonic approach is that, in terms of filtering, the sequence of convolutions of scaling functions and wavelets may be interpreted as low-pass filter and band-pass filter, respectively. The main importance of wavelet multiresolution, however, lies in the fact that only a small number of wavelet (potential) coefficients is required to deal with large data sets. In fact, the fast decorrelation power of wavelets is the key to applications such as data compression, fast data transmission, noise cancellation, etc.

The paper represents a generalization of spherical results in [4, 5, 6, 7, 8].

1   Basic Settings

Let Ω_R be the sphere in the three-dimensional Euclidean space $\mathbb{R}^3$ around the origin with radius R, Ω_R^{ext} denotes the outer space of Ω_R in $\mathbb{R}^3$ and $\overline{\Omega_R^{ext}}$ is the topological closure of Ω_R^{ext}, i.e. $\overline{\Omega_R^{ext}} = \Omega_R \cup \Omega_R^{ext}$. $Pot^{(\infty)}(\overline{\Omega_R^{ext}})$ denotes the space consisting of all functions $F \in C^{(\infty)}(\overline{\Omega_R^{ext}})$ satisfying $\Delta F = 0$ in $\Omega_R{}^{ext}$ and $F(x) = O(1/|x|)$ as $|x| \to \infty$. On $Pot^{(\infty)}(\overline{\Omega_R^{ext}})$ we are able to impose an inner product $(\cdot, \cdot)_{\mathcal{H}}$ (cf. [3])) as follows:

$$(F, G)_{\mathcal{H}} = \int_{\Omega_R} F(x)G(x)\, d\omega_R(x), \qquad (d\omega_R : \text{surface element}).$$

The Sobolev space $\mathcal{H}$ is the completion of $Pot^{(\infty)}(\overline{\Omega_R^{ext}})$ under the norm induced by $(\cdot,\cdot)_{\mathcal{H}}$:

$$\mathcal{H} = \overline{Pot^{(\infty)}(\overline{\Omega_R^{ext}})}^{\|\mathcal{H}\|}.$$

If $\{Y_{n,k}\}$, $n = 0,1,\ldots$, $k = 1,\ldots,2n+1$ is an $\mathcal{L}^2$-orthonormal system of (surface) spherical harmonics in the sense that $\int_{|x|=1} Y_{n,j}(x)Y_{k,l}(x)\,d\omega_1(x) = \delta_{nk}\delta_{jl}$, then the system $\{H_{n,k}\}$, $n = 0,1,\ldots$, $k = 1,\ldots,2n+1$ of outer harmonics defined by

$$H_{n,k}(x) = \frac{1}{R}\left(\frac{R}{|x|}\right)^{n+1} Y_{n,k}\left(\frac{x}{|x|}\right), \qquad x \in \overline{\Omega_R^{ext}}$$

satisfies the following conditions:

(i) $H_{n,k}$ is of class $Pot^{(\infty)}(\overline{\Omega_R^{ext}})$,

(ii) $H_{n,k}\,|_{\Omega_R} = (1/R)Y_{n,k}$,

(iii) $\int_{\Omega_R} H_{n,j}(x)H_{k,l}(x)\,d\omega_R(x) = \delta_{nk}\delta_{jl}$.

It is known (cf. [2, 3]) that

$$\mathcal{H} = \overline{span\,_{\substack{n=0,1,\ldots \\ k=1,\ldots,2n+1}} (H_{n,k})}^{\|\cdot\|_{\mathcal{H}}}.$$

Any potential $F \in \mathcal{H}$ can be represented as orthogonal Fourier series in terms of outer harmonics (in $\mathcal{H}$-topology) as follows:

$$\lim_{N \to \infty} \left\| F - \sum_{n=0}^{N} \sum_{k=1}^{2n+1} F^{\wedge}(n,k)H_{n,k} \right\|_{\mathcal{H}} = 0,$$

where the Fourier coefficients $F^{\wedge}(n,k)$ are given by

$$F^{\wedge}(n,k) = (F, H_{n,k})_{\mathcal{H}} = \int_{\Omega_R} F(x)H_{n,k}(x)\,d\omega_R(x).$$

The idea we follow in this paper is to present (in the $\mathcal{H}$-framework) a J-level approximation of a function (signal) $F \in \mathcal{H}$ by means of wavelet analysis. Essential tools are the concept of $\mathcal{H}$-product kernels and $\mathcal{H}$-convolutions which should be discussed now.

1.1 Product Kernels

Any function $\Gamma : \overline{\Omega_R^{ext}} \times \overline{\Omega_R^{ext}} \to \mathbb{R}$ of the form

$$\Gamma(x,y) = \sum_{n=0}^{\infty} \sum_{k=1}^{2n+1} \Gamma^{\wedge}(n) H_{n,k}(x) H_{n,k}(y), \tag{1}$$

$x, y \in \overline{\Omega_R^{ext}}$ with $\Gamma^{\wedge}(n) \in \mathbb{R}$, $n \in \mathbb{N}_0$, is called an $\mathcal{H}$-product kernel (briefly called $\mathcal{H}$-kernel). The sequence $\{\Gamma^{\wedge}(n)\}_{n=0,1,\ldots}$ is said to be the *symbol* of the $\mathcal{H}$-kernel (1). By virtue of the well-known addition theorem of outer harmonics (cf., for example, [4]), the $\mathcal{H}$-kernel (1) can be rewritten as

$$\Gamma(x,y) = \sum_{n=0}^{\infty} \Gamma^{\wedge}(n) \frac{2n+1}{4\pi R^2} \left(\frac{R^2}{|x||y|} \right)^{n+1} P_n \left(\frac{x}{|x|} \cdot \frac{y}{|y|} \right), \qquad x, y \in \overline{\Omega_R^{ext}},$$

where P_n denotes the Legendre polynomial of degree n.

Definition 1.1. A symbol $\{\Gamma^{\wedge}(n)\}_{n=0,1,\ldots}$ of an $\mathcal{H}$-product kernel (1) is called $\mathcal{H}$-admissible if it satisfies the following condition:

$$\sum_{n=0}^{\infty} \frac{2n+1}{4\pi R^2} \left(\Gamma^{\wedge}(n) \right)^2 < \infty.$$

$\mathcal{H}$-convolutions will be introduced in the following way.

Definition 1.2. Let F be of class $\mathcal{H}$. Suppose that Γ is an $\mathcal{H}$-kernel of the form (1) with $\mathcal{H}$-admissible symbol $\{\Gamma^{\wedge}(n)\}_{n=0,1,\ldots}$, then the convolution $\Gamma *_{\mathcal{H}} F$ of Γ against F is defined by

$$(\Gamma *_{\mathcal{H}} F)(x) = (\Gamma(x,\cdot), F)_{\mathcal{H}} = \sum_{n=0}^{\infty} \sum_{k=1}^{2n+1} \Gamma^{\wedge}(n) F^{\wedge}(n,k) H_{n,k}(x), \tag{2}$$

$x \in \overline{\Omega_R^{ext}}$.

If Γ is an $\mathcal{H}$-kernel with $\mathcal{H}$-admissible symbol $\{\Gamma^{\wedge}(n)\}_{n=0,1,\ldots}$ and F is of class $\mathcal{H}$, then $\Gamma *_{\mathcal{H}} F$ is a member of class $\mathcal{H}$, and we have

$$(\Gamma *_{\mathcal{H}} F)^{\wedge}(n,k) = \Gamma^{\wedge}(n) F^{\wedge}(n,k), \tag{3}$$

$n = 0,1,\ldots$, $k = 1,\ldots,2n+1$. The convolution of two $\mathcal{H}$-product kernels with $\mathcal{H}$-admissible symbols leads to the following statement.

Theorem 1.3. *Let Γ_1 and Γ_2 be $\mathcal{H}$-kernels with $\mathcal{H}$-admissible symbols $\{\Gamma_1^{\wedge}(n)\}_{n=0,1,\ldots}$ and $\{\Gamma_2^{\wedge}(n)\}_{n=0,1,\ldots}$, respectively. Then*

$$
\begin{aligned}
(\Gamma_1 *_{\mathcal{H}} \Gamma_2)(x,y) &= (\Gamma_1 *_{\mathcal{H}} \Gamma_2(\cdot,y))(x) \\
&= (\Gamma_1(x,\cdot), \Gamma_2(\cdot,y))_{\mathcal{H}} \\
&= \sum_{n=0}^{\infty} \sum_{k=1}^{2n+1} \Gamma_1^{\wedge}(n)\Gamma_2^{\wedge}(n)H_{n,k}(x)H_{n,k}(y)
\end{aligned}
\tag{4}
$$

holds for all $x,y \in \overline{\Omega_R^{ext}}$, and $\{(\Gamma_1 *_{\mathcal{H}} \Gamma_2)^{\wedge}(n)\}_{n=0,1,\ldots}$ given by

$$
(\Gamma_1 *_{\mathcal{H}} \Gamma_2)^{\wedge}(n) = \Gamma_1^{\wedge}(n)\Gamma_2^{\wedge}(n)
\tag{5}
$$

forms the $\mathcal{H}$-admissible symbol of the $\mathcal{H}$-kernel $\Gamma_1 *_{\mathcal{H}} \Gamma_2$.

1.2 $\mathcal{H}$-scaling Functions

After having explained the convolution between two $\mathcal{H}$-kernels with $\mathcal{H}$-admissible symbols we are now interested in developing countable families $\{\Gamma_J\}$, $J \in \mathbb{Z}$, of $\mathcal{H}$-product kernels Γ_J which may be understood as scaling functions in our $\mathcal{H}$-wavelet concept.

As preparation we introduce a dilation operator acting on these families in the following way: Let Γ_J be a member of a family of admissible $\mathcal{H}$-kernels. Then the *dilation operator* D_k, $k \in \mathbb{Z}$, is defined by $D_k\Gamma_J = \Gamma_{J+k}$. In particular, we have $\Gamma_J = D_J\Gamma_0$. Thus we refer to Γ_0 as the "mother kernel". Moreover, we define a *shifting operator* S_x, $x \in \overline{\Omega_R^{ext}}$, by $S_x\Gamma_J = \Gamma_J(x,\cdot)$, $J \in \mathbb{Z}$. In doing so we consequently get by composition of the operators D_J and S_x the kernel $\Gamma_J(x,\cdot) = S_x D_J\Gamma_0$ for all $x \in \overline{\Omega_R^{ext}}$ and all $J \in \mathbb{Z}$. Note that all kernels Γ_J are symmetric, so that $\Gamma_J(x,y) = \Gamma_J(y,x)$, $x,y \in \overline{\Omega_R^{ext}}$ for all $J \in \mathbb{Z}$.

We are now able to introduce scaling functions.

Definition 1.4. Let $\{(\Phi_J)^\wedge(n)\}_{n=0,1,\dots}$, $J \in \mathbb{Z}$, define an $\mathcal{H}$-admissible symbol of a family $\{\Phi_J\}$ of $\mathcal{H}$-kernels satisfying additionally the following properties:

$$(i) \quad \lim_{J\to\infty} \left((\Phi_J)^\wedge(n)\right)^2 = 1, \qquad n \in \mathbb{N},$$

$$(ii) \quad \left((\Phi_J)^\wedge(n)\right)^2 \geq \left((\Phi_{J-1})^\wedge(n)\right)^2, \qquad J \in \mathbb{Z}, n \in \mathbb{N},$$

$$(iii) \quad \lim_{J\to-\infty} \left((\Phi_J)^\wedge(n)\right)^2 = 0, \qquad n \in \mathbb{N},$$

$$(iv) \quad \left((\Phi_J)^\wedge(0)\right)^2 = 1, \qquad J \in \mathbb{Z}.$$

Then $\left\{(\Phi_J)^\wedge(n)\right\}_{n=0,1,\dots}$ is called the generating symbol of an $\mathcal{H}$-scaling function. The family of $\mathcal{H}$-kernels $\{\Phi_J\}$, $J \in \mathbb{Z}$, given by

$$\Phi_J(x,y) = \sum_{n=0}^{\infty} \sum_{k=1}^{2n+1} (\Phi_J)^\wedge(n) H_{n,k}(x) H_{n,k}(y), \qquad x,y \in \overline{\Omega_R^{ext}}, \tag{6}$$

is called *$\mathcal{H}$-scaling function.*

From the results of the previous section it follows immediately that $\Phi_J(x,\cdot) \in \mathcal{H}$ for all $x \in \overline{\Omega_R^{ext}}$ and $J \in \mathbb{Z}$. Furthermore it is easily seen that $\Phi_J *_{\mathcal{H}} \Phi_J$ is an $\mathcal{H}$-kernel with $\mathcal{H}$-admissible symbol $\{((\Phi_J)^\wedge(n))^2\}$, $n = 0, 1, \dots$.

This leads us to the following result:

Theorem 1.5. *Let $\{(\Phi_J)^\wedge(n)\}_{n=0,1,\dots}$, $J \in \mathbb{Z}$, be the generating symbol of a scaling function $\{\Phi_J\}$. Then*

$$\lim_{J\to\infty} \|F_J - F\|_{\mathcal{H}} = 0, \tag{7}$$

holds for all $F \in \mathcal{H}$, where F_J given by

$$F_J = (\Phi_J *_{\mathcal{H}} \Phi_J) *_{\mathcal{H}} F, \qquad F \in \mathcal{H} \tag{8}$$

is said to be the J-level approximation of F.

Proof. We introduce the operator $T_J : \mathcal{H} \to \mathcal{H}$, $J \in \mathbb{Z}$, by

$$F_J = T_J F = (\Phi_J *_{\mathcal{H}} \Phi_J) *_{\mathcal{H}} F. \tag{9}$$

From the definition of the convolution and the fact that $\Phi_J *_{\mathcal{H}} \Phi_J$ is an $\mathcal{H}$-kernel with $\mathcal{H}$-admissible symbol it follows that

$$T_J F = \sum_{n=0}^{\infty} \sum_{k=1}^{2n+1} ((\Phi_J)^{\wedge}(n))^2 F^{\wedge}(n,k) H_{n,k}. \tag{10}$$

But this implies that

$$\|T_J\| = \sup_{\substack{G \in \mathcal{H} \\ \|G\|_{\mathcal{H}}=1}} \|T_J G\|_{\mathcal{H}} \tag{11}$$

$$= \sup_{\substack{G \in \mathcal{H} \\ \|G\|_{\mathcal{H}}=1}} \left(\sum_{n=0}^{\infty} \sum_{k=1}^{2n+1} ((\Phi_J)^{\wedge}(n))^4 (G^{\wedge}(n,k))^2 \right)^{\frac{1}{2}}$$

$$\leq \sup_{\substack{G \in \mathcal{H} \\ \|G\|_{\mathcal{H}}=1}} \sup_{n \in \mathbb{N}_0} ((\Phi_J)^{\wedge}(n))^2 \left(\sum_{n=0}^{\infty} \sum_{k=1}^{2n+1} (G^{\wedge}(n,k))^2 \right)^{\frac{1}{2}}$$

$$= \sup_{n \in \mathbb{N}_0} ((\Phi_J)^{\wedge}(n))^2 < \infty$$

for every $J \in \mathbb{Z}$, since $\{(\Phi_J)^{\wedge}(n)\}_{n=0,1,\dots}$ is $\mathcal{H}$-admissible. Now, from Parseval's identity, we obtain

$$\lim_{J \to \infty} \|T_J F - F\|_{\mathcal{H}}^2 \tag{12}$$

$$= \lim_{J \to \infty} \sum_{n=0}^{\infty} \sum_{k=1}^{2n+1} (1 - ((\Phi_J)^{\wedge}(n))^2)^2 (F^{\wedge}(n,k))^2.$$

From the conditions (i), (ii), and (iv) of Definition 1.4 we are able to deduce that $((\Phi_J)^{\wedge}(n))^2 \leq 1$ for $n \in \mathbb{N}_0$. But this shows us that

$$0 \leq (1 - ((\Phi_J)^{\wedge}(n))^2)^2 \leq 1 \tag{13}$$

is valid for all $n \in \mathbb{N}_0$. Therefore the limit and the infinite sum in (12) may be interchanged. By applying (i) and (iv) we finally arrive at the desired result. $\square$

Note that condition (iii) has not been used yet. This condition, however, is needed as assumption for defining $\mathcal{H}$-wavelets and establishing a multiresolution analysis.

From arguments of potential theory (cf. [2]) it is known that Theorem 1.5 can be formulated in uniform convergence for every subset $K \subset \overline{\Omega_R^{ext}}$ with $dist(K, \Omega_R) \geq \rho > 0$. This leads us to the following corollary:

Corollary 1.6. *Under the assumptions of Theorem 1.5*

$$\lim_{J \to \infty} \sup_{x \in K} |F(x) - F_J(x)| = 0$$

holds for all $K \subset \overline{\Omega_R^{ext}}$ satisfying $dist(K, \Omega_R) \geq \rho > 0$.

According to our construction, for any $F \in \mathcal{H}$, each $T_J F$ defined by (9) provides an approximation of F at scale J. In terms of filtering $\Phi_J *_{\mathcal{H}} \Phi_J$ may be interpreted as a *low-pass filter*. T_J is the convolution operator of this low-pass filter. Accordingly we understand the *scale space* $\mathcal{V}_J$ to be the image of $\mathcal{H}$ under the operator T_J:

$$\mathcal{V}_J = T_J(\mathcal{H}) = \{(\Phi_J *_{\mathcal{H}} \Phi_J) *_{\mathcal{H}} F \mid F \in \mathcal{H}\}. \tag{14}$$

As an immediate consequence we obtain the following result.

Theorem 1.7. *The scale spaces satisfy the following properties:*

$$(i) \quad \{H_{0,1}\} \subset \mathcal{V}_J \subset \mathcal{V}_{J'} \subset \mathcal{H}, \qquad J \leq J', \tag{15}$$

$$(ii) \quad \bigcap_{J=-\infty}^{\infty} \mathcal{V}_J = \{H_{0,1}\}, \tag{16}$$

$$(iii) \quad \overline{\bigcup_{J=-\infty}^{\infty} \mathcal{V}_J}^{\|\cdot\|_{\mathcal{H}}} = \mathcal{H}, \tag{17}$$

$$(iv) \quad \text{if } F_J \in \mathcal{V}_J \text{ then } D_{-1} F_J \in \mathcal{V}_{J-1}, \qquad J \in \mathbb{Z}. \tag{18}$$

Proof. From the conditions (ii) and (iv) of Definition 1.4 we easily get the validity of the first assertion of Theorem 1.7. The identity (16) follows directly from the conditions (iii) and (iv) of Definition 1.4. The formula (17) is a consequence of Theorem 1.5, while (18) follows immediately from the definition of the shifting operator D_J. $\qquad\qquad\square$

If a collection of subspaces of $\mathcal{H}$ satisfies the conditions of Theorem 1.7 we call it a *multiresolution analysis* (MRA).

The definition of the scaling functions now allows us to introduce $\mathcal{H}$-wavelets. Basic tool again is the concept of $\mathcal{H}$-kernels.

1.3 $\mathcal{H}$-wavelets

We start with the definition of wavelets by aid of a "refinement (scaling) equation".

Definition 1.8.   Let $\{(\Phi_J)^\wedge(n)\}_{n=0,1,\dots}$, $J \in \mathbb{Z}$, be the generating symbol of an $\mathcal{H}$-scaling function as defined by Definition 1.4. Then the generating symbol $\{(\Psi_j)^\wedge(n)\}_{n=0,1,\dots}$, $j \in \mathbb{Z}$, of the associated $\mathcal{H}$-wavelet is defined via the "refinement equation"

$$(\Psi_j)^\wedge(n) = (((\Phi_{j+1})^\wedge(n))^2 - ((\Phi_j)^\wedge(n))^2)^{\frac{1}{2}}. \tag{19}$$

The family $\{\Psi_j\}$, $j \in \mathbb{Z}$, of $\mathcal{H}$-kernels given by

$$\Psi_j(x,y) = \sum_{n=0}^{\infty} \sum_{k=1}^{2n+1} (\Psi_j)^\wedge(n) H_{n,k}(x) H_{n,k}(y), \qquad x,y \in \overline{\Omega_R^{ext}}, \tag{20}$$

is called $\mathcal{H}$-*wavelet associated to the* $\mathcal{H}$-*scaling function* $\{\Phi_J\}$, $J \in \mathbb{Z}$. The corresponding mother wavelet is denoted by Ψ_0.

Clearly we are able to define a dilation and a shifting operator in the same way as we did before. For this reason any wavelet can be interpreted as a dilated and shifted copy of the corresponding mother wavelet like $\Psi_j(x,\cdot) = S_x D_j \Psi_0(\cdot,\cdot)$. We can easily derive from (19) that

$$((\Phi_{J+1})^\wedge(n))^2 = \sum_{j=-\infty}^{J} ((\Psi_j)^\wedge(n))^2 \tag{21}$$

$$= ((\Phi_0)^\wedge(n))^2 + \sum_{j=0}^{J} ((\Psi_j)^\wedge(n))^2. \tag{22}$$

Similar to the definition of the operator T_J we are now led to convolution operators $R_j : \mathcal{H} \to \mathcal{H}$, given by

$$R_j F = (\Psi_j *_{\mathcal{H}} \Psi_j) *_{\mathcal{H}} F, \qquad F \in \mathcal{H}. \tag{23}$$

Thus the identity

$$\Phi_{J+1} *_{\mathcal{H}} \Phi_{J+1} = \sum_{j=-\infty}^{J} \Psi_j *_{\mathcal{H}} \Psi_j \tag{24}$$

$$= \Phi_0 *_{\mathcal{H}} \Phi_0 + \sum_{j=0}^{J} \Psi_j *_{\mathcal{H}} \Psi_j$$

can be written in operator formulation as follows:

$$T_{J+1} = \sum_{j=-\infty}^{J} R_j = T_0 + \sum_{j=0}^{J} R_j. \tag{25}$$

The convolution operators R_j describe the "detail information" of F at scale j. In terms of filtering, $\Psi_j *_{\mathcal{H}} \Psi_j$, $j \in \mathbb{Z}$, may be interpreted as a *band-pass filter*. This fact immediately gives rise to introduce the *detail spaces* as follows:

$$\mathcal{W}_j = R_j(\mathcal{H}) = \{(\Psi_j *_{\mathcal{H}} \Psi_j) *_{\mathcal{H}} F \mid F \in \mathcal{H}\} . \tag{26}$$

$\mathcal{W}_J$ contains the "detail information" needed to go from an approximation at level J to an approximation at level $J + 1$. Hence we get

$$\sum_{j=-\infty}^{J} \mathcal{W}_j = \mathcal{V}_0 + \sum_{j=0}^{J} \mathcal{W}_j = \mathcal{V}_{J+1} \tag{27}$$

and

$$\mathcal{V}_J + \mathcal{W}_J = \mathcal{V}_{J+1}, \qquad J \in \mathbb{Z}. \tag{28}$$

It should be noted that, in general, the sum in (27) is neither direct nor orthogonal. Later on we mention an example leading to an orthogonal multiresolution.

Summarizing our results we are led to the following conclusion: Starting with $T_0 F$, $F \in \mathcal{H}$ we find (in connection with (25))

$$T_{J+1} F = T_0 F + \sum_{j=0}^{J} R_j F \tag{29}$$

for any $J \in \mathbb{Z}$. In other words, the partial "reconstruction" $R_j F$ is nothing else than the "difference of two smoothings" at two consecutive scales

$$R_j F = T_{j+1} F - T_j F. \tag{30}$$

Definition 1.9. The *wavelet transform WT at scale $j \in \mathbb{Z}$ and position $x \in \overline{\Omega_R^{ext}}$ is given by*

$$WT(F)(j;x) = (\Psi_j(x, \cdot), F)_{\mathcal{H}}, \qquad F \in \mathcal{H}. \tag{31}$$

Combining (31) and (24) we can formulate the main result of wavelet theory as follows.

Theorem 1.10. *Let $\{(\Phi_J)^\wedge(n)\}_{n=0,1,\ldots}$, $J \in \mathbb{Z}$, be the generating symbol of an $\mathcal{H}$-scaling function. Suppose that $\{(\Psi_j)^\wedge(n)\}_{n=0,1,\ldots}$, $j \in \mathbb{Z}$, is the generating symbol of the corresponding $\mathcal{H}$-wavelet. Furthermore, let F be of class $\mathcal{H}$. Then*

$$F_J = (\Phi_0 *_{\mathcal{H}} \Phi_0) *_{\mathcal{H}} F + \sum_{j=0}^{J} (\Psi_j *_{\mathcal{H}} WT(F)(j, \cdot)) \tag{32}$$

is the J-level approximation of F satisfying

$$\lim_{J \to \infty} \|F_J - F\|_{\mathcal{H}} = 0. \tag{33}$$

From Corollary 1.6 we obtain the uniform convergence in every subset $K \subset \overline{\Omega_R^{ext}}$ with positive distance to the sphere Ω_R.

Corollary 1.11. *Under the assumptions of Theorem 1.10*

$$\lim_{J \to \infty} \sup_{x \in K} |(\Phi_0 *_{\mathcal{H}} \Phi_0) *_{\mathcal{H}} F + \sum_{j=0}^{J} (\Psi_j *_{\mathcal{H}} \Psi_j) *_{\mathcal{H}} F| = 0$$

holds for all $K \subset \overline{\Omega_R^{ext}}$ with $dist(K, \Omega_R) \geq \rho > 0$.

The limit relation (33) shows the essential characteristic of wavelets. We change the approximated solution from F_J to F_{J+1} by adding the so-called detail information of level J as the difference of two smoothings of two consecutive scales J and $J+1$ and what is more important, we are able to guarantee $\lim_{J\to\infty} F_J = F$ in the sense of the $\|\cdot\|_{\mathcal{H}}$-topology and the topology of locally uniform convergence in $\overline{\Omega_R^{ext}}$ provided that $F \in \mathcal{H}$.

The following scheme illustrates the essential steps of our wavelet approach in the framework of a separable functional Hilbert space.

$$
\begin{array}{ccccccccc}
T_0(G) & & T_1(G)\ldots & & & T_j(G) & & T_{j+1}(G)\ldots & \overset{j\to\infty}{\to} F \\
\cup & & \cup & & & \cup & & \cup & \cup \\
\mathcal{V}_0 & \subset & \mathcal{V}_1 & \ldots & \subset & \mathcal{V}_j & \subset & \mathcal{V}_{j+1} \ldots & = \mathcal{H} \\
\mathcal{V}_0 & + & \mathcal{W}_0 & + \ldots + & \mathcal{W}_{j-1} & + & \mathcal{W}_j & + \quad \ldots & = \mathcal{H} \\
\cap & & \cap & & \cap & & \cap & & \cap \\
T_0(G)+R_0(G) & + & & \ldots + R_{j-1}(G) & + & R_j(G) & + & \ldots & = F
\end{array}
$$

2 Fast Multiscale Evaluation

Until now efforts have been made to establish the basis and decorrelation properties of harmonic wavelets. Next we come to the third feature of wavelet approximation, viz. fast computation, which will be realized here in form of a pyramid scheme for bandlimited wavelets.

In order to deal with finite-dimensional scale spaces we assume that $\{\Phi_j(\cdot,\cdot)\}_{j\in\mathbb{Z}}$ and $\{\Psi_j(\cdot,\cdot)\}_{j\in\mathbb{Z}}$ are families of bandlimited kernels such that

$$
\Phi_j(x,\cdot) \in \mathcal{H}_{0,\ldots,2^j-1} = span_{\substack{n=0,\ldots,2^j-1 \\ k=1,\ldots,2n+1}} \{H_{n,k}\}
$$

and

$$
\Psi_j(x,\cdot) \in \mathcal{H}_{0,\ldots,2^{j+1}-1} = span_{\substack{n=0,\ldots,2^{j+1}-1 \\ k=1,\ldots,2n+1}} \{H_{n,k}\}
$$

holds for all $x \in \overline{\Omega_R^{ext}}$. Consequently, the scale spaces and the detail spaces, respectively, fulfill the relations

$$
\mathcal{V}_j = \mathcal{H}_{0,\ldots,2^j-1}, \qquad \mathcal{W}_j \subset \mathcal{H}_{0,\ldots,2^{j+1}-1}.
$$

Simple examples are as follows:

(a) orthogonal (Shannon)

$$(\Phi_j)^\wedge(n) = \begin{cases} 1 & for \quad n = 0,\dots,M_j \\ 0 & for \quad n \geq M_j + 1 \end{cases},$$

(b) non-orthogonal (CP)

$$(\Phi_j)^\wedge(n) = \begin{cases} (1 - 2^{-j}n)^2(1 + 2^{-j+1}n) & for \quad n = 0,\dots,M_j \\ 0 & for \quad n \geq M_j + 1 \end{cases}$$

with

$$M_j = \begin{cases} 0 & for \quad j \in \mathbb{Z}, j < 0 \\ 2^j - 1 & for \quad j \in \mathbb{Z}, j \geq 0 \end{cases}.$$

Note that the case (a) (i.e. the Shannon scaling function and wavelet) leads to an orthogonal multiresolution analysis, i.e. the detail and scale spaces satisfy $\mathcal{V}_{j+1} = \mathcal{V}_j \oplus \mathcal{W}_j$, $\mathcal{W}_j \perp \mathcal{W}_k$, $k \neq j$, $j \geq 0$. In the case (b) (i.e. the $\underline{C}$ubic $\underline{P}$olynomial scaling function and wavelet) the scale and detail spaces are still finite-dimensional, but the detail spaces are no longer orthogonal. Further choices for $(\Phi_j)^\wedge(n)$ can be found in [4, 5].

The key ideas of our bandlimited wavelet discretization are based on the following observations:

(1) For $j = 0,\dots,J$, there are coefficients $w_l^{N_j} \in \mathbb{R}$ and points $y_l^{N_j} \in \Omega_R$ (cf. [5]) of exact integration formulae in $\mathcal{H}_{0,\dots,2^{j+2}-2}$, $N_j \geq (2^{j+2} - 1)^2$, such that

$$(\Gamma_{\mathcal{H}_{0,\dots,2^{j+1}-1}}(\cdot,\cdot), P)_{\mathcal{H}} = \sum_{l=1}^{N_j} w_l^{N_j} \Gamma_{\mathcal{H}_{0,\dots,2^{j+1}-1}}(\cdot, y_l^{N_j}) P(y_l^{N_j}) \tag{34}$$

holds for all $P \in \mathcal{H}_{0,\dots,2^{j+1}-1}$, where

$$\Gamma_{\mathcal{H}_{0,\dots,2^{j+1}-1}}(x,y) = \sum_{n=0}^{2^{j+1}-1} \sum_{k=1}^{2n+1} H_{n,k}(x) H_{n,k}(y), \quad (x,y) \in \overline{\Omega_R^{ext}} \times \overline{\Omega_R^{ext}}$$

is the uniquely determined reproducing kernel in $\mathcal{H}_{0,\dots,2^{j+1}-1}$ with respect to the $\mathcal{H}$-topology. The coefficients $w_l^{N_j}$ may be calculated from the linear equations

$$\sum_{l=1}^{N_j} w_l^{N_j} = 1, \tag{35}$$

$$\sum_{l=1}^{N_j} w_l^{N_j} H_{n,k}(y_l^{N_j}) = 0, \quad n = 1,\dots,2^{j+2} - 2, \; k = 1,\dots,2n+1 \tag{36}$$

in an a-priori step and stored elsewhere.

(2) For some suitably large J, the scale space $\mathcal{V}_{J+1}$ is 'sufficiently close' to $\mathcal{H}$. Consequently, for each $F \in \mathcal{H}$, there exists a bandlimited function of class $\mathcal{V}_{J+1}$ such that the error between F and $(\Phi_{J+1} *_{\mathcal{H}} \Phi_{J+1}) *_{\mathcal{H}} F$ (understood in the $\|\cdot\|_{\mathcal{H}}$–topology) is negligible. This is the reason why the input data are assumed to be data of the function $(\Phi_{J+1} *_{\mathcal{H}} \Phi_{J+1}) *_{\mathcal{H}} F \in \mathcal{V}_{J+1}$ rather than of $F \in \mathcal{H}$ for the remainder of this section. In other words, the geopotential data $F(y_l^{N_J}) = \alpha_l^{N_J}$, $l = 1,\dots,N_J$ are supposed to be known from a potential $F \in \mathcal{V}_{J+1}$.

What we are going to realize is the following pyramid scheme: Starting from a sufficiently large J, there exist vectors $a^{N_j} \in \mathbb{R}^{N_j}$, $j = 0,\dots,J$ (being, of course, dependent on the function $F \in \mathcal{H}$ under consideration) such that the following statements hold true:

(i) For $j = 0,\dots,J$, all wavelet transforms can be calculated via the formulae

$$\Psi_j *_{\mathcal{H}} F = (WT)(F)(j;\cdot) = \sum_{i=1}^{N_j} a_i^{N_j} \Psi_j(\cdot, y_i^{N_j}).$$

(ii) The vectors $a^j \in \mathbb{R}^{N_j}$, $j = 0,\dots,J-1$ are obtainable from $a^{j+1} \in \mathbb{R}^{N_{j+1}}$ by recursion.

(iii) The vectors satisfy, in addition,

$$(\Phi_{j+1} *_{\mathcal{H}} \Phi_{j+1}) *_{\mathcal{H}} F = \sum_{i=1}^{N_j} a_i^{N_j}(\Phi_{j+1} *_{\mathcal{H}} \Phi_{j+1})(\cdot, y_i^{N_j}),$$

$$(\Psi_j *_{\mathcal{H}} \Psi_j) *_{\mathcal{H}} F = \sum_{i=1}^{N_j} a_i^{N_j}(\Psi_j *_{\mathcal{H}} \Psi_j)(\cdot, y_i^{N_j}).$$

Our considerations are divided into two parts, viz. the initial step concerning the scale level J and the pyramid step establishing the recursion relation.

2.1   The Initial Step

Observing the exact integration formulae (34) and the reproducing property of $\Gamma_{\mathcal{H}_{0,\dots,2^{J+1}-1}}$ in $\mathcal{V}_{J+1} = \mathcal{H}_{0,\dots,2^{J+1}-1}$ we obtain for $F \in \mathcal{V}_{J+1}$

$$\Gamma_{\mathcal{H}_{0,\dots,2^{J+1}-1}} *_{\mathcal{H}} F = \sum_{l=1}^{N_J} w_l^{N_J} F(y_l^{N_J}) \Gamma_{\mathcal{H}_{0,\dots,2^{J+1}-1}}(\cdot, y_l^{N_J}). \tag{37}$$

Letting

$$a_l^{N_J} = w_l^{N_J} F(y_l^{N_J}) = w_l^{N_J} \alpha_l^{N_J} \tag{38}$$

for $l = 1, \dots, N_J$ we find

$$\Gamma_{\mathcal{H}_{0,\dots,2^{J+1}-1}} *_{\mathcal{H}} F = \sum_{l=1}^{N_J} a_l^{N_J} \Gamma_{\mathcal{H}_{0,\dots,2^{J+1}-1}}(\cdot, y_l^{N_J}).$$

Note that the coefficients $a_l^{N_J}$ are dependent on $F \in \mathcal{V}_{J+1}$.

We now prove an auxiliary result, which is of basic significance for our considerations.

Lemma 2.1. *Let F be of class $\mathcal{V}_{J+1}$. Suppose that Γ is an $\mathcal{H}$-kernel of the form (1) such that $\Gamma^{\wedge}(n) = 0$ for all $n \geq 2^{J+1} - 1$. Then the coefficients $a_i^{N_J} = w_i^{N_J} F(y_i^{N_J})$, $i = 1, \dots, N_J$ satisfy the equation*

$$\Gamma *_{\mathcal{H}} F = \sum_{i=1}^{N_J} a_i^{N_J} \Gamma(\cdot, y_i^{N_J}).$$

Proof. $\Gamma_{\mathcal{H}_{0,\dots,2^{J+1}-1}}$ expressed in terms of the $\mathcal{H}$–orthonormal system $\{H_{n,k}\}_{n=0,1,\dots}$ reads as follows:

$$\Gamma_{\mathcal{H}_{0,\dots,2^{J+1}-1}}(x, y) = \sum_{n=0}^{2^{J+1}-1} \sum_{k=1}^{2n+1} H_{n,k}(x) H_{n,k}(y) \tag{39}$$

for all $(x,y) \in \overline{\Omega_R^{ext}} \times \overline{\Omega_R^{ext}}$. On the one hand this implies

$$\Gamma_{\mathcal{H}_{0,\dots,2^{J+1}-1}} *_{\mathcal{H}} F = \sum_{n=0}^{2^{J+1}-1} \sum_{k=1}^{2n+1} F^{\wedge}(n,k) H_{n,k}.$$

On the other hand (observe that $F^{\wedge}(n,k) = F *_{\mathcal{H}} H_{n,k}$) we have

$$
\begin{aligned}
\sum_{i=1}^{N_J} a_i^{N_J} \Gamma_{\mathcal{H}_{0,\dots,2^{J+1}-1}}(\cdot, y_i^{N_J}) &= \sum_{n=0}^{2^{J+1}-1} \sum_{k=1}^{2n+1} \sum_{i=1}^{N_J} a_i^{N_J} H_{n,k}(y_i^{N_J}) H_{n,k} \quad (40) \\
&= \sum_{n=0}^{2^{J+1}-1} \sum_{k=1}^{2n+1} F^{\wedge}(n,k) H_{n,k}.
\end{aligned}
$$

By comparing the coefficients we therefore obtain

$$F^{\wedge}(n,k) = \sum_{i=1}^{N_J} a_i^{N_J} H_{n,k}(y_i^{N_J})$$

for all $n = 0,\dots,2^{J+1} - 1$, $k = 1,\dots,2n+1$. But this shows us that

$$
\begin{aligned}
\Gamma *_{\mathcal{H}} F &= \sum_{n=0}^{2^{J+1}-1} \sum_{k=1}^{2n+1} \Gamma^{\wedge}(n) F^{\wedge}(n,k) H_{n,k} \\
&= \sum_{i=1}^{N_J} \sum_{n=0}^{2^{J+1}-1} \sum_{k=1}^{2n+1} a_i^{N_J} \Gamma^{\wedge}(n) H_{n,k}(y_i^{N_J}) H_{n,k} \\
&= \sum_{i=1}^{N_J} a_i^{N_J} \Gamma(\cdot, y_i^{N_J}).
\end{aligned}
$$

This is the desired result. $\qquad\square$

As special cases we obtain from Lemma 2.1 the following identities:

$$\Phi_{J+1} *_{\mathcal{H}} F = \sum_{i=1}^{N_J} a_i^{N_J} \Phi_{J+1}(\cdot, y_i^{N_J}),$$

$$\Phi_{J+1} *_{\mathcal{H}} \Phi_{J+1} *_{\mathcal{H}} F = \sum_{i=1}^{N_J} a_i^{N_J} (\Phi_{J+1} *_{\mathcal{H}} \Phi_{J+1})(\cdot, y_i^{N_J}),$$

and

$$\Psi_J *_{\mathcal{H}} F = \sum_{i=1}^{N_J} a_i^{N_J} \Psi_J(\cdot, y_i^{N_J}),$$

$$\Psi_J *_{\mathcal{H}} \Psi_J *_{\mathcal{H}} F = \sum_{i=1}^{N_J} a_i^{N_J} (\Psi_J *_{\mathcal{H}} \Psi_J)(\cdot, y_i^{N_J}).$$

2.2   The Pyramid Step

An essential tool for the pyramid step is the following lemma.

Lemma 2.2. *Let F be of class $\mathcal{V}_{J+1}$. Suppose that Γ is a kernel with $\Gamma^{\wedge}(n) = 0$ for all $n > 2^J - 1$. Then the vector $a^{N_{J-1}}$, $a^{N_{J-1}} = (a_1^{N_{J-1}}, \dots, a_{N_{J-1}}^{N_{J-1}})^T$ given by*

$$a_i^{N_{J-1}} = w_i^{N_{J-1}} (\Gamma_{\mathcal{H}_{0,\dots,2^J-1}} *_{\mathcal{H}} F)(y_i^{N_{J-1}}), \qquad i = 1, \dots, N_{J-1}$$

satisfies the equations

$$\Gamma *_{\mathcal{H}} F = \sum_{i=1}^{N_{J-1}} a_i^{N_{J-1}} \Gamma(\cdot, y_i^{N_{J-1}}).$$

Proof. Since $\Gamma_{\mathcal{H}_{0,\dots,2^J-1}}(\cdot,\cdot)$ is the (uniquely determined) reproducing kernel of $\mathcal{V}_J = \mathcal{H}_{0,\dots,2^J-1}$ (with respect to the $\mathcal{H}$–topology), we readily get

$$\Gamma *_{\mathcal{H}} F = \Gamma *_{\mathcal{H}} \Gamma_{\mathcal{H}_{0,\dots,2^J-1}} *_{\mathcal{H}} F \qquad (41)$$

$$= \sum_{i=1}^{N_{J-1}} w_i^{N_{J-1}} (\Gamma_{\mathcal{H}_{0,\dots,2^J-1}} *_{\mathcal{H}} F)(y_i^{N_{J-1}}) \Gamma(\cdot, y_i^{N_{J-1}}).$$

But this implies Lemma 2.2. $\qquad\square$

Looking at our foregoing results we notice that there are two ways of discretizing an $\mathcal{H}$–convolution $\Gamma *_{\mathcal{H}} F$ with Γ satisfying the assumptions of Lemma 2.2. On the one hand we obtain from Lemma 2.1

$$\Gamma *_{\mathcal{H}} F = \sum_{i=1}^{N_J} a_i^{N_J} \Gamma(\cdot, y_i^{N_J}) \qquad (42)$$

with coefficients $a_1^{N_J}, \dots, a_{N_J}^{N_J}$ given by

$$a_i^{N_J} = w_i^{N_J} \alpha_i^{N_J}, \qquad i = 1, \dots, N_J. \qquad (43)$$

It is remarkable that the coefficients are independent of the choice of the kernel Γ. As particularly important case we mention

$$\Gamma_{\mathcal{H}_{0,\dots,2^J-1}} *_{\mathcal{H}} F = \sum_{i=1}^{N_J} a_i^{N_J} \Gamma_{\mathcal{H}_{0,\dots,2^J-1}}(\cdot, y_i^{N_J}). \qquad (44)$$

On the other hand we are able to deduce from Lemma 2.2 that

$$\Gamma *_{\mathcal{H}} F = \sum_{i=1}^{N_{J-1}} a_i^{N_{J-1}} \Gamma(\cdot, y_i^{N_{J-1}}) \qquad (45)$$

with coefficients $a_1^{N_{J-1}}, \dots, a_{N_{J-1}}^{N_{J-1}}$ given by

$$a_i^{N_{J-1}} = w_i^{N_{J-1}} (\Gamma_{\mathcal{H}_{0,\dots,2^J-1}} *_{\mathcal{H}} F)(y_i^{N_{J-1}}), \quad i = 1, \dots, N_{J-1}. \qquad (46)$$

Inserting (44) into (46) we find

$$a_i^{N_{J-1}} = w_i^{N_{J-1}} \sum_{l=1}^{N_J} a_l^{N_J} \Gamma_{\mathcal{H}_{0,\dots,2^J-1}}(y_i^{N_{J-1}}, y_l^{N_J}) \qquad (47)$$

for $i = 1, \ldots, N_{J-1}$. In other words, the coefficients $a_i^{N_{J-1}}$ can be calculated recursively. Moreover, the coefficients $a_i^{N_{J-1}}$ are independent of the special choice of the kernel Γ. This finally leads us to the following discretization of the $\mathcal{H}$–convolutions as follows:

$$\Phi_J *_{\mathcal{H}} F = \sum_{i=1}^{N_{J-1}} a_i^{N_{J-1}} \Phi_J(\cdot, y_i^{N_{J-1}}),$$

$$(\Phi_J *_{\mathcal{H}} \Phi_J) *_{\mathcal{H}} F = \sum_{i=1}^{N_{J-1}} a_i^{N_{J-1}} (\Phi_J *_{\mathcal{H}} \Phi_J)(\cdot, y_i^{N_{J-1}}),$$

and

$$\Psi_{J-1} *_{\mathcal{H}} F = \sum_{i=1}^{N_{J-1}} a_i^{N_{J-1}} \Psi_{J-1}(\cdot, y_i^{N_{J-1}}),$$

$$(\Psi_{J-1} *_{\mathcal{H}} \Psi_{J-1}) *_{\mathcal{H}} F = \sum_{i=1}^{N_{J-1}} a_i^{N_{J-1}} (\Psi_{J-1} *_{\mathcal{H}} \Psi_{J-1})(\cdot, y_i^{N_{J-1}}).$$

In conclusion, we end up with the following pyramid scheme for the decomposition of a function $F \in \mathcal{V}_{J+1}$:

$$
\begin{array}{ccccccc}
F \longrightarrow & a^{N_J} & \longrightarrow & a^{N_{J-1}} & \longrightarrow \cdots \longrightarrow & a^{N_0} \\
& \downarrow & & \downarrow & & \downarrow \\
& (WT)(F)(J;\cdot) & & (WT)(F)(J-1;\cdot) & & (WT)(F)(0;\cdot)
\end{array}
$$

The reconstruction of the wavelet coefficients can be performed as described before via the formula

$$
\begin{aligned}
R_j(F) &= \Psi_j *_{\mathcal{H}} (WT)(F)(j;\cdot) \\
&= \sum_{i=1}^{N_j} w_i^{N_j} (WT)(F)(j; y_i^{N_j}) \Psi_j(\cdot, y_i^{N_j}).
\end{aligned}
$$

This leads us to the following scheme:

$$
\begin{array}{ccccccc}
(WT)(F)(0; y_i^{N_0}) & & (WT)(F)(1; y_i^{N_1}) & & (WT)(F)(2; y_i^{N_2}) \\
\downarrow & & \downarrow & & \downarrow \\
R_0(F) & & R_1(F) & & R_2(F) \\
& \searrow & & \searrow & & \searrow \\
T_0(F) \longrightarrow + \longrightarrow & & T_1(F) \longrightarrow + \longrightarrow & & T_2(F) \longrightarrow & \cdots
\end{array}
$$

According to our approach the wavelet transform $(WT)(F)(j;\cdot)$ is given by the coefficients $a_1^{N_j},\ldots,a_{N_j}^{N_j}$. This also enables us to reconstruct the function only by use of the numbers $a_i^{N_j}$, rather than calculating the wavelet coefficients of F:

$$R_j(F) = \sum_{i=1}^{N_j} a_k^{N_j}(\Psi_j *_{\mathcal{H}} \Psi_j)(\cdot,y_i^{N_j}).$$

Thus the decomposition and reconstruction, respectively, can be simplified as follows:

$$F \longrightarrow a^{N_J} \longrightarrow a^{N_J-1} \longrightarrow \ldots \longrightarrow a^{N_0}$$

and

$$
\begin{array}{ccccccccc}
a^{N_0} & & & & a^{N_1} & & & & a^{N_2} \\
\downarrow & & & & \downarrow & & & & \downarrow \\
R_0(F) & & & & R_1(F) & & & & R_2(F) \\
& \searrow & & & & \searrow & & & & \searrow \\
T_0(F) & \longrightarrow & + & \longrightarrow & T_1(F) & \longrightarrow & + & \longrightarrow & T_2(F) & \longrightarrow & + & \longrightarrow \ldots
\end{array}
$$

That means the reconstruction of the function is not performed with $\Psi_j(\cdot,\cdot)$. Instead we have used the $\mathcal{H}$-convolution $\Psi_j *_{\mathcal{H}} \Psi_j$. Of particular significance is that the vectors a^{N_j} do <u>not</u> depend on the special choice of the bandlimited scaling function. As a matter of fact, we are able to reconstruct the function with respect to different types of wavelets just by use of the vectors a^{N_j}.

Remark: The critical point of our pyramid scheme is the determination of the coefficients $w_l^{N_j}$, $j = 0,\ldots,J$, from the linear equations (35) and (36). It should be mentioned that the solution of the linear systems can be avoided completely if we place the knots for each detail step $j = 0,\ldots,J$ on a special longitude-latitude grid on the sphere Ω_R. The corresponding set of weights is explicitly available without solving any linear system from results in [1].

3   Gravitational Field Determination

In the foregoing we have seen that harmonic wavelets are 'building blocks' that enable fast decorrelation of geopotential data. Now we discuss in more detail the concept of multiresolution analysis. The multiresolution analysis 'looks at' the earth's gravitational potential through a microscope, whose resolution gets finer and finer. Thus it associates to the gravitational potential a sequence of smoothed versions, labelled by the scale parameter. This aspect is illustrated by the figures below for the NASA EGM96-model. The computations have been performed on the basis of the CP-wavelets. The point systems and the weights of the exact integration formulae in (34) have been chosen as described in [1].

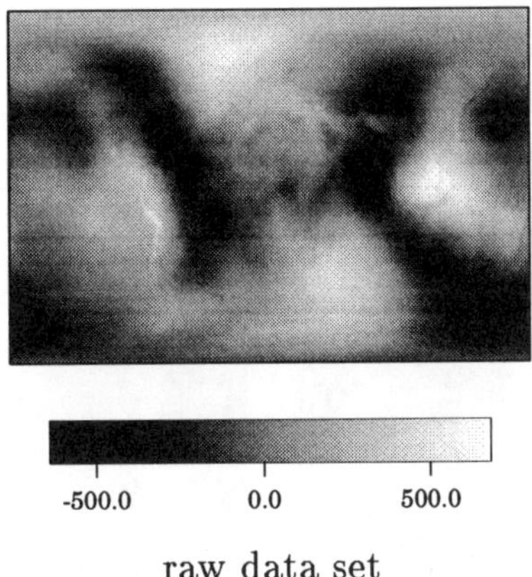

raw data set

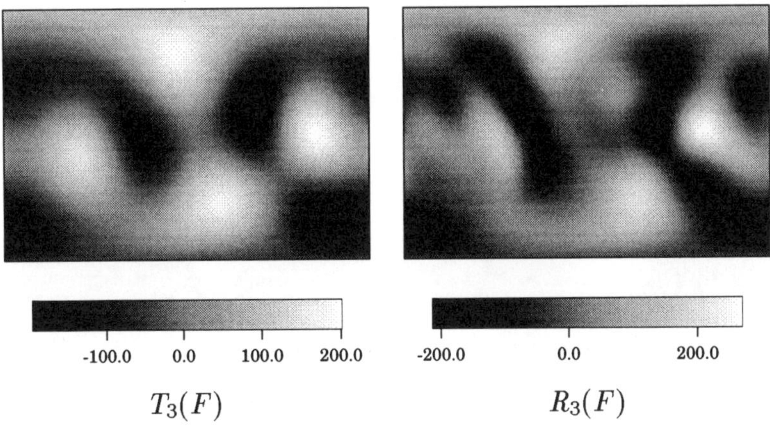

$T_3(F)$ $\qquad\qquad\qquad\qquad R_3(F)$

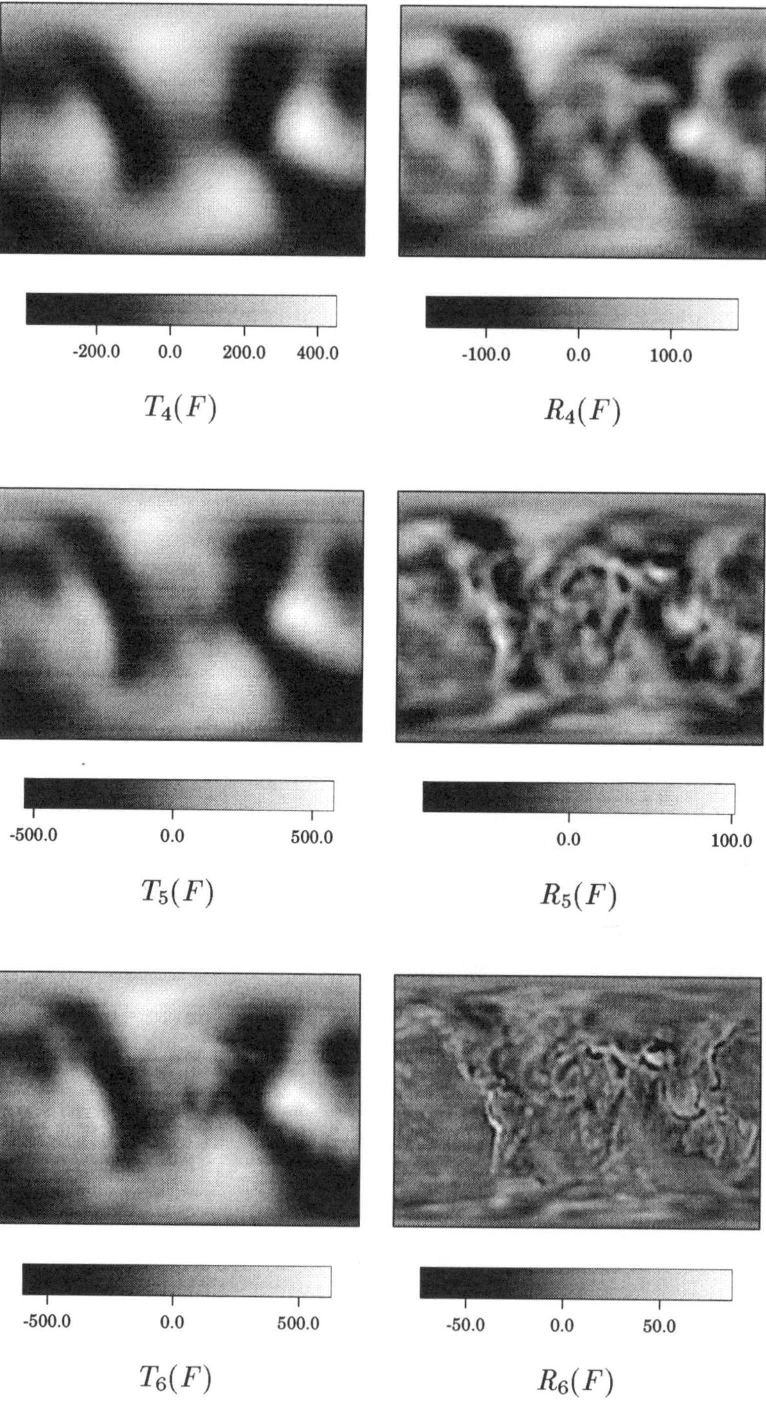

$$T_4(F) \qquad\qquad R_4(F)$$

$$T_5(F) \qquad\qquad R_5(F)$$

$$T_6(F) \qquad\qquad R_6(F)$$

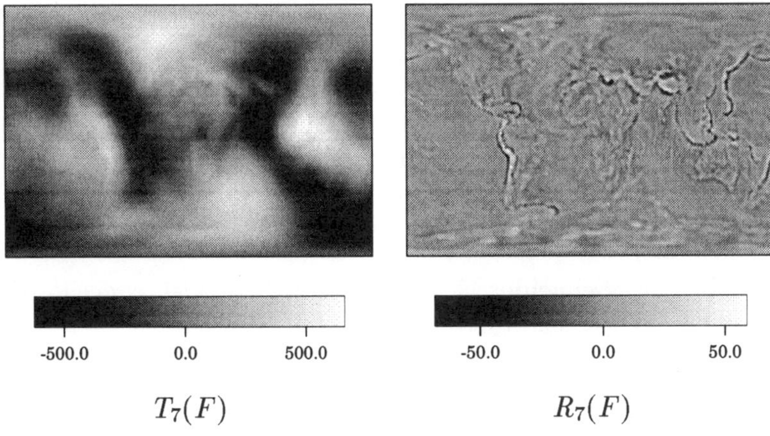

$T_7(F)$ $\qquad\qquad\qquad$ $R_7(F)$

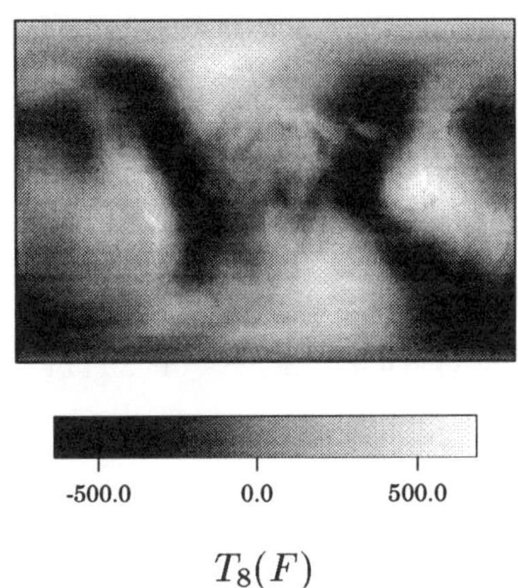

$T_8(F)$

References

[1] J. R. Driscoll, D. M. Healy: *Computing Fourier transforms and convolutions on the 2-sphere*, Adv. Appl. Math. **15** (1994), 202–250.

[2] W. Freeden: *On the approximation of external gravitational potential with closed systems of (trial) functions*, Bull. Geod. **54** (1980), 1–20.

[3] W. Freeden: *A spline interpolation method for solving boundary value problems of potential theory from discretely given data*, Numer. Meth. Part. Diff. Equations **3** (1987), 375–398.

[4] W. Freeden, T. Gervens, M. Schreiner: *Constructive Approximation on the Sphere (With Applications to Geomathematics)*. Oxford Science Publications, Clarendon, 1998.

[5] W. Freeden, F. Schneider: *An integrated wavelet concept of physical geodesy*, J. Geodesy **72** (1998), 259–281.

[6] F. Narcowich, J. Ward: *Nonstationary wavelets on the m–sphere for scattered data*, Appl. Comp. Harm. Anal. **3** (1996), 324–336.

[7] D. Potts, G. Steidl, M. Tasche: *Kernels of spherical harmonics and spherical frames*, in: *Advanced Topics in Multivariate Approximation*, F. Fontanella et al. (eds.), 287–301 World Scientific, Singapore 1996

[8] M. Schreiner: *A pyramid scheme for spherical wavelets*, AGTM Report No. 170, University of Kaiserslautern, 1997.

Address:

WILLI FREEDEN, OLIVER GLOCKNER, ROLF LITZENBERGER
Geomathematics Group
University of Kaiserslautern
P.O. Box 3049
D–67653 Kaiserslautern
Germany

On a Family of Orthogonal Wavelets on the Quincunx Grid: Open Regularity Questions

Achim Gottscheber and Gabriele Steidl

Dedicated to Professor M. Reimer on the occasion of his 65th birthday

Abstract

In this paper, we introduce a new family of nonseparable orthogonal wavelets on the quincunx grid arising from Butterworth wavelets with an odd number of vanishing moments. Our wavelets are closely related to bireciprocal wave digital filters. Unfortunately, the determination of the regularity of our wavelets is still open.

Introduction

Wavelet techniques have attained attention during the last years. Many applications, such as image processing, make use of bivariate wavelet bases. Most of these bivariate wavelets are simply tensor products of univariate wavelets which have a number of drawbacks, e.g. the "preferred directions effect" due to their separabilty. Only a few nonseparable bivariate bases have been constructed. See [10, 14] for the design of orthogonal, not differentiable wavelets, [3, 11] for a biorthogonal approach and the new paper [1] for the first construction of orthogonal nonseparable wavelets with compact support and arbitrary high smoothness.

In this paper, we propose a new family of nonseparable orthogonal wavelets on the quincunx grid. Note that the quincunx grid is a very popular choice since the corresponding multiresolution involves only one scaling function and one wavelet [10, 14, 3]. Our construction follows in a simple way from the design of univariate Butterworth wavelets of odd degree from the point of view of wave digital filters. It has the advantage that the approach can be extended to higher dimensions and to four–channel perfect reconstruction filter banks [7]. Furthermore, our design gives rise to an efficient implementation in the frequency domain by using the polyphase decomposition and to an efficient sampling of the continuous function at the starting point of the wavelet analysis [7].

Advances in Multivariate Approximation; W. Haußmann, K. Jetter and M. Reimer (eds.)
Mathematical Research, Vol. 107, pp. 175–184, ISBN 3-527-40236-5
© WILEY–VCH, Berlin 1999

Problem. Unfortunately, our wavelets are not of compact support and the determination of their smoothness is still open.

1   Univariate Butterworth Wavelets

We consider the scaling function $\varphi_N \in L_2(\mathbb{R})$ satisfying the refinement relation

$$\varphi_N(x) \;=\; 2 \sum_{k \in Z} h_k\, \varphi_N(2x - k)\,,$$

$$\hat{\varphi}_N(v) \;=\; H_N(e^{-iv/2})\hat{\varphi}_N(\tfrac{v}{2})\,, \qquad H_N(z) = \sum_{k \in Z} h_k\, z^k\,,$$

where H_N denotes the *N-th order Butterworth filter* [15] arising from

$$|H_N(z)|^2 \;=\; \frac{(z+1)^{2N}}{(z+1)^{2N} + (-1)^N (z-1)^{2N}}$$

$$\;=\; \frac{(\cos \tfrac{v}{2})^{2N}}{(\cos \tfrac{v}{2})^{2N} + (\sin \tfrac{v}{2})^{2N}} \qquad (z := e^{-iv})\,. \tag{1.1}$$

For odd N $(1 < N < \infty)$, the poles of (1.1) are given by 0 and $\pm i \cot \frac{k\pi}{2N}$ $(k = 1, ..., N-1)$. From (1.1) we can obtain different filters H_N. As an example, let us choose the filter having the poles $\pm i \cot \frac{k\pi}{2N}$ $(k = 1, ..., (N-1)/2)$. With $\alpha_k := (\cot \frac{k\pi}{2N})^2$, $k = 1, ..., (N-1)/2$, our filter can be written as

$$H_N(z) = C_N \frac{(z+1)^N}{(z^2 + \alpha_1)...(z^2 + \alpha_{\frac{N-1}{2}})} = C_N \frac{(z+1)^N}{U_N(z^2)}\,, \tag{1.2}$$

where the constant $C_N := (1 + \alpha_1)...(1 + \alpha_{\frac{N-1}{2}})/2^N$ is chosen such that $H(1) = 1$.

For $N = 1$, we obtain $H_1(e^{-iv}) = (1 + e^{-iv})/2$ and φ_1 is the characteristic function on $[0,1)$.

For $N = \infty$, we see that

$$H_\infty(e^{iv}) = \begin{cases} 1 & v \in [\tfrac{\pi}{2}, \tfrac{\pi}{2})\,, \\ 0 & \text{otherwise} \end{cases} \tag{1.3}$$

is the ideal lowpass filter which corresponds to the sinc–function φ_∞.

Remark. The above choice of the filter is of particular interest for practical computations. It is easy to check that H_N (N odd) possesses the following *polyphase decomposition*

$$H_N(z) = \frac{1}{2}(E_N(z^2) + zF_N(z^2)) \tag{1.4}$$

with the *allpass filters* E_N, F_N given by

$$
\begin{aligned}
E_N(z^2) &= H_N(z) + H_N(-z) \\
&= \frac{(\alpha_1 z^2 + 1)(\alpha_3 z^2 + 1)...(\alpha_M z^2 + 1)}{(z^2 + \alpha_1)(z^2 + \alpha_3)...(z^2 + \alpha_M)},
\end{aligned}
$$

and

$$
\begin{aligned}
F_N(z^2) &= \frac{1}{z}(H_N(z) - H_N(-z)) \\[2mm]
&= \frac{(\alpha_2 z^2 + 1)(\alpha_4 z^2 + 1)...(\alpha_L z^2 + 1)}{(z^2 + \alpha_2)(z^2 + \alpha_4)...(z^2 + \alpha_L)}
\end{aligned}
$$

with

$$
M := \left\{ \begin{array}{ll} (N-3)/2 & \text{for } N \equiv 1 \bmod 4 \\ (N-1)/2 & \text{for } N \equiv 3 \bmod 4 \end{array} \right. ,\quad
L := \left\{ \begin{array}{ll} (N-1)/2 & \text{for } N \equiv 1 \bmod 4 \\ (N-3)/2 & \text{for } N \equiv 3 \bmod 4 \end{array} \right. .
$$

Symbols H_N with polyphase decomposition (1.4), where E and F are allpass filters can be obtained in a more general way from bireciprocal wave digital filters (WDF). For a consideration of H_N from the point of view of WDF, we refer to [5, 6, 7].

Scaling functions and wavelets related to Butterworth filters were firstly considered in [13]. Since H_N has a finite number of poles, the coefficients h_k and the scaling function φ_N are of *exponential decay*. Further, -1 is a N-fold zero of H_N such that φ_N satisfies a *Strang–Fix condition of order N* and polynomials of degree $< N$ can be locally represented as linear combinations of integer translates of φ_N. By (1.1), we see that H_N satisfies the *orthogonality relation*

$$|H_N(e^{iv})|^2 + |H_N(e^{i(v+\pi)})|^2 = 1 \qquad (v \in [-\pi, \pi)). \tag{1.5}$$

Since moreover $|H_N(e^{iv})| > 0$ for all $v \in (-\pi, \pi)$, Cohen's criterion [2, p. 39] implies that the integer translates of φ_N are orthonormal scaling functions. By [4, 12], the scaling functions φ_N become arbitrary smooth with increasing N.

The usual way to construct corresponding orthogonal wavelets consists in the *CQF-setting*

$$\psi_N(x) \;=\; 2\sum_{k \in Z} g_k\, \varphi_N(2x - k),$$

$$\hat{\psi}_N(v) \;=\; e^{iv/2} H_N(-e^{iv/2})\, \hat{\varphi}_N(\tfrac{v}{2}), \quad G(z) = z^{-1} H(-z^{-1}).$$

Clearly, other choices of the wavelet filter are possible. See [13] for the characterization of all orthogonal filters with perfect reconstruction property. For our purposes, we define the wavelet filter by the *WDF-setting*

$$G_N(z) := H_N(-z) \quad (z := e^{-iv}), \tag{1.6}$$

i.e.

$$\hat{\psi}_N(v) = H_N(-e^{-iv/2})\, \hat{\varphi}_N(\tfrac{v}{2}).$$

Indeed, by (1.2),

$$
\begin{aligned}
H_N(z)G_N(z^{-1}) + H_N(-z)G_N(-z^{-1}) &= H_N(z)H_N(-z^{-1}) + H_N(-z)H_N(z^{-1}) \\
&= \frac{C_N^2\left((z - z^{-1})^N + (z^{-1} - z)^N\right)}{U_N(z^2)U_N(z^{-2})} \\
&= 0, \tag{1.7}
\end{aligned}
$$

such that $(\varphi_N, \psi_N(\cdot - k))_{L_2} = 0$ $(k \in \mathbb{Z})$ and ψ_N is an orthogonal wavelet. Note that (1.7) follows also immediately by (1.4) since $E(z^{-1}) = 1/E(z)$ and $F(z^{-1}) = 1/F(z)$. However, neither for Butterworth filters of even order nor for Daubechies filters, (1.6) determines an orthonormal wavelet.

2   Butterworth Wavelets on the Quincunx Grid

Let $\mathbf{A}$ denote one of the matrices $\mathbf{S} := \begin{pmatrix} 1 & 1 \\ 1 & -1 \end{pmatrix}$ or $\mathbf{R} := \begin{pmatrix} 1 & -1 \\ 1 & 1 \end{pmatrix}$ related to the quincunx grid $\mathbf{A}\mathbb{Z}^2$ and let $\mathbf{B} := \mathbf{A}^T$. Note that $\mathbf{S}^2 = 2\mathbf{I}$ and that $\mathbf{R}^4 = -4\mathbf{I}$ with the 2×2 identitiy matrix $\mathbf{I}$. Since $|\det \mathbf{A}| = 2$, the representation system of $\mathbf{A}\mathbb{Z}^2$ in $\mathbb{Z}^2$ consists only of two representatives, for example $\{(0,0)^T, (1,0)^T\}$ and we have as in the univariate case only one wavelet Ψ such that $\{|\det \mathbf{A}|^{j/2}\, \Psi(\mathbf{A}^j x - k) : k \in \mathbb{Z}^2; j \in \mathbb{Z}\}$ is an orthonormal basis of $L_2(\mathbb{R}^2)$. Therefore, the quincunx grid is a popular choice for the construction of two–dimensional wavelets [1, 3, 10, 14, 16]. We consider scaling functions Φ defined by the refinement equation

$$\Phi(\mathbf{x}) = 2 \sum_{k \in \mathbf{Z}^2} h_k \, \Phi(\mathbf{A}x + k),$$

$$\hat{\Phi}(\omega) = H(e^{-i\mathbf{B}^{-1}\omega})\Phi(\mathbf{B}^{-1}\omega), \qquad H(e^{-i\mathbf{B}^{-1}\omega}) := \sum_{k \in \mathbf{Z}^2} h_k \, e^{-ik\mathbf{B}^{-1}\omega}$$

with the symbol

$$
\begin{aligned}
H(e^{i\omega}) \;:=\; & H_N(e^{i\frac{\omega_1+\omega_2}{2}})H_N(e^{i\frac{\omega_1-\omega_2}{2}}) + G_N(e^{i\frac{\omega_1+\omega_2}{2}})G_N(e^{i\frac{\omega_1-\omega_2}{2}}) \qquad (2.1) \\
\;=\; & H_N(e^{i\frac{\omega_1+\omega_2}{2}})H_N(e^{i\frac{\omega_1-\omega_2}{2}}) + H_N(-e^{i\frac{\omega_1+\omega_2}{2}})H_N(-e^{i\frac{\omega_1-\omega_2}{2}}).
\end{aligned}
$$

In the following, we always denote by H the bivariate mask (2.1), by H_N ($N \in \mathbb{N}$ odd) the univariate Butterworth filter (1.2) and by Φ and φ_N the corresponding scaling functions, respectively. Clearly, we have by definition of H_N that $H(1) = 1$. Again, our scaling functions Φ have exponential decay.

The idea for the construction of (2.1) arises from the consideration of the sinc–wavelets, i.e. from the special case $N = \infty$ [8]. By (1.3) and since

$$H(e^{i\mathbf{B}\omega}) := H_N(e^{i\omega_1})H_N(e^{i\omega_2}) + H_N(-e^{i\omega_1})H_N(-e^{i\omega_2}), \qquad (2.2)$$

the filters $H(e^{-i\omega})$ and $H(e^{-i\mathbf{B}\omega})$ behave for $N = \infty$ as in Figure 1 and by

$$\hat{\Phi}(\omega) = \prod_{j=1}^{\infty} H(e^{-i\mathbf{B}^{-j}\omega}) \qquad (2.3)$$

it is easy to check that $\hat{\Phi}(\omega) = \hat{\varphi}_\infty(\omega_1)\hat{\varphi}_\infty(\omega_2)$.

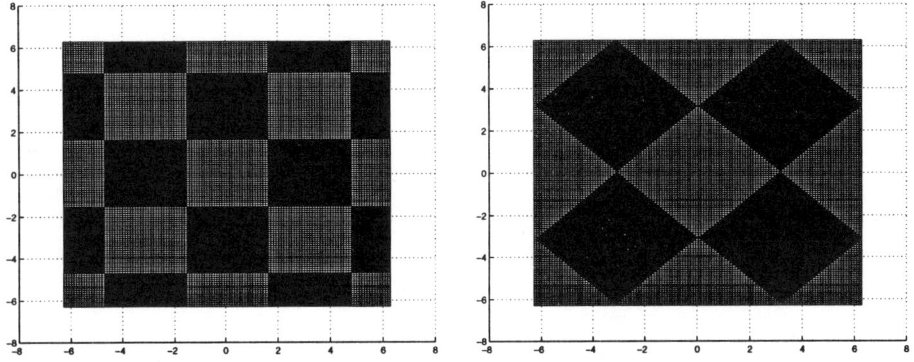

Figure 1. $|H(e^{i\mathbf{B}\omega})|$ (left) and $|H(e^{i\omega})|$ (right) for $N = \infty$ (view from above).

For the "Haar–case" $N = 1$, definition (2.1) results in $H(e^{i\omega}) = H_1(e^{i\omega_1})$. "Induced" scaling functions of the above type were introduced in [3]. If $\mathbf{A} = \mathbf{S}$, then we obtain as corresponding scaling function $\Phi(x) = \varphi_1(x_2)\varphi_1(x_1 - x_2)$. For $\mathbf{A} = \mathbf{R}$, the scaling function is the characteristic function of the "twin dragon".

Similarly, we define our wavelets by

$$\Psi(x) \;=\; 2 \sum_{k \in Z^2} g_k\, \Phi(\mathbf{A}x - k)\,,$$

$$\hat{\Psi}(\omega) \;=\; G(e^{-i\mathbf{B}^{-1}\omega})\,\Phi(\mathbf{B}^{-1}\omega)$$

with

$$
\begin{aligned}
G(e^{i\omega}) \;&:=\; G_N(e^{i\frac{\omega_1+\omega_2}{2}})H_N(e^{i\frac{\omega_1-\omega_2}{2}}) + H_N(e^{i\frac{\omega_1+\omega_2}{2}})G_N(e^{i\frac{\omega_1-\omega_2}{2}}) \qquad (2.4)\\
\;&=\; H_N(-e^{i\frac{\omega_1+\omega_2}{2}})H_N(e^{i\frac{\omega_1-\omega_2}{2}}) + H_N(e^{i\frac{\omega_1+\omega_2}{2}})H_N(-e^{i\frac{\omega_1-\omega_2}{2}})\,.
\end{aligned}
$$

Theorem 2.1.   *For all $\omega \in [-\pi,\pi)$, the filters (2.1) and (2.4) satisfy the orthogonality relations*

i) $|H(e^{i\omega})|^2 + |H(e^{i(\omega+(\pi,\pi)^T)})|^2 = 1$.

ii) $H(e^{i\omega})G(e^{-i\omega}) + H(e^{i(\omega+(\pi,\pi)^T)})G(e^{-i(\omega+(\pi,\pi)^T)}) = 0$.

Proof. 1. Since $\det \mathbf{B} \neq 0$, equation i) is equivalent to

$$|H(e^{i\mathbf{B}\omega})|^2 + |H(e^{i\mathbf{B}(\omega+(\pi,0)^T)})|^2 = 1\,.$$

By (2.2), the left–hand side can be rewritten as

$$
\begin{aligned}
|H(e^{i\mathbf{B}\omega})|^2 + |H(e^{i\mathbf{B}(\omega+(\pi,0)^T)})|^2 \;=\; & (|H_N(e^{i\omega_1})|^2 + |H_N(-e^{i\omega_1})|^2)\\
& (|H_N(e^{i\omega_2})|^2 + |H_N(-e^{i\omega_2})|^2)\\
+ \; & (H_N(e^{i\omega_1})H_N(-e^{-i\omega_1}) + H_N(-e^{i\omega_1})H_N(e^{-i\omega_1}))\\
& (H_N(e^{i\omega_2})H_N(-e^{-i\omega_2}) + H_N(-e^{i\omega_2})H_N(e^{-i\omega_2}))
\end{aligned}
$$

Since by (1.5) and (1.7), the first product on the right–hand side is equal to 1 and the second product is equal to 0, we have proved assertion i).

2. Similarly, we obtain by the definitions (2.1) and (2.4) and by (1.7) that

$$
\begin{aligned}
H(e^{i\mathbf{B}\omega})G(e^{-i\mathbf{B}\omega}) \quad + \quad & H(e^{i\mathbf{B}(\omega+(\pi,0)^T)})G(e^{-i\mathbf{B}(\omega+(\pi,0)^T)}) = \\
& (|H_N(e^{i\omega_1})|^2 + |H_N(-e^{i\omega_1})|^2) \\
& (H_N(e^{i\omega_2})H_N(-e^{-i\omega_2}) + H_N(-e^{i\omega_2})H_N(e^{-i\omega_2})) \\
+ \quad & (|H_N(e^{i\omega_2})|^2 + |H_N(-e^{i\omega_2})|^2) \\
& (H_N(e^{i\omega_1})H_N(-e^{-i\omega_1}) + H_N(-e^{i\omega_1})H_N(e^{-i\omega_1})) \\
= \quad & 0. \qquad\qquad\qquad\qquad\qquad\qquad\qquad\qquad\qquad \square
\end{aligned}
$$

Moreover, it is easy to check the Cohen criterion (see [2, p. 212 and p. 213, Figure B.2]) such that (2.3) converges in $L_2(\mathbb{R}^2)$. Therefore, we deal indeed with orthogonal scaling functions and corresponding orthogonal wavelets.

By definition of H and since -1 is an N–fold zero of H_N, we obtain immediately that

$$
\frac{\partial^{n_1}}{\partial\omega_1^{n_1}}\frac{\partial^{n_2}}{\partial\omega_2^{n_2}}H(e^{i\omega})\Big|_{\omega=(\pi,\pi)^T} = 0 \qquad \text{for } 0 \le n_1 + n_2 \le N - 1.
$$

Thus, bivariate polynomials of degree $< N$ can be locally represented as linear combinations of $\phi(\cdot - \mathbf{k})$ ($\mathbf{k} \in \mathbb{Z}^2$), see [9].

Problem. Except for $N = 1, \infty$, the smoothness of our wavelets is still an open question.

It may be that the following form of the filters can be useful. By definition, we can write our filters in the form

$$
\begin{aligned}
|H(e^{i\mathbf{B}\omega})|^2 \quad = \quad & |H_N(e^{i\omega_1})|^2|H_N(e^{i\omega_2})|^2 + |H_N(-e^{i\omega_1})|^2|H_N(-e^{i\omega_2})|^2 \\
+ \quad & H_N(e^{i\omega_1})H_N(e^{i\omega_2})H_N(-e^{-i\omega_1})H_N(-e^{-i\omega_2}) \\
+ \quad & H_N(e^{-i\omega_1})H_N(e^{-i\omega_2})H_N(-e^{i\omega_1})H_N(-e^{i\omega_2})
\end{aligned}
$$

and further by (1.7) and (1.1) as

$$
\begin{aligned}
|H(e^{i\mathbf{B}\omega})|^2 \quad = \quad & \frac{(\cos\frac{\omega_1}{2}\cos\frac{\omega_2}{2})^{2N} + (\sin\frac{\omega_1}{2}\sin\frac{\omega_2}{2})^{2N}}{((\cos\frac{\omega_1}{2})^{2N} + (\sin\frac{\omega_1}{2})^{2N})((\cos\frac{\omega_2}{2})^{2N} + (\sin\frac{\omega_2}{2})^{2N})} \\
+ \quad & 2H_N(e^{i\omega_1})H_N(-e^{-i\omega_1})H_N(e^{i\omega_2})H_N(-e^{-i\omega_2}).
\end{aligned}
$$

By (1.2), we verify that

$$
H_N(z)H_N(-z^{-1}) = \frac{C_N^2(z-z^{-1})^N}{U_N(z^2)U_N(z^{-2})} = \frac{i^N(\sin\frac{v}{2}\cos\frac{v}{2})^N}{(\cos\frac{v}{2})^{2N} + (\sin\frac{v}{2})^{2N}} \qquad (z := e^{iv}).
$$

Consequently, we obtain the following simple form of our symbols

$$|H(e^{i\mathbf{B}\omega})|^2 = \frac{((\cos\frac{\omega_1}{2}\cos\frac{\omega_2}{2})^N - (\sin\frac{\omega_1}{2}\sin\frac{\omega_2}{2})^N)^2}{((\cos\frac{\omega_1}{2})^{2N} + (\sin\frac{\omega_1}{2})^{2N})((\cos\frac{\omega_2}{2})^{2N} + (\sin\frac{\omega_2}{2})^{2N})}.$$

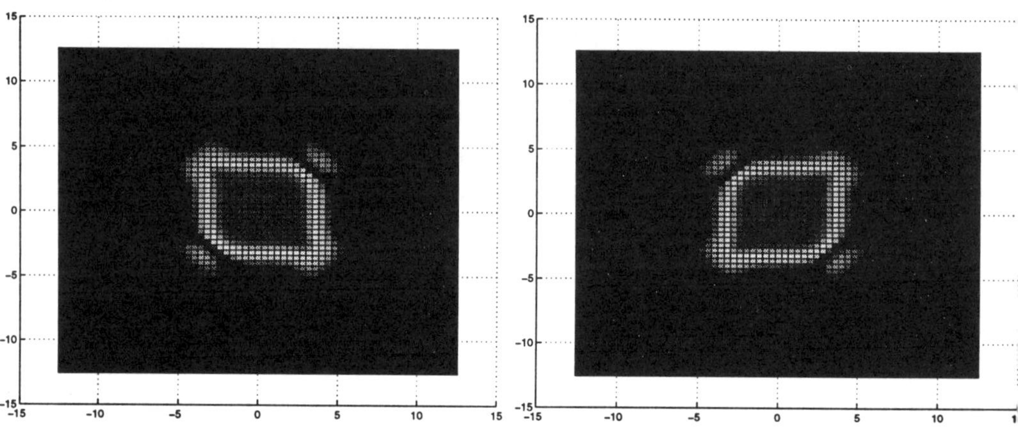

Figure 2. $|\hat{\mathbf{\Phi}}(\omega)|$ with $\mathbf{A} = \mathbf{S}$ (left) and $\mathbf{A} = \mathbf{R}$ (right) for $N = 3$ (view from above).

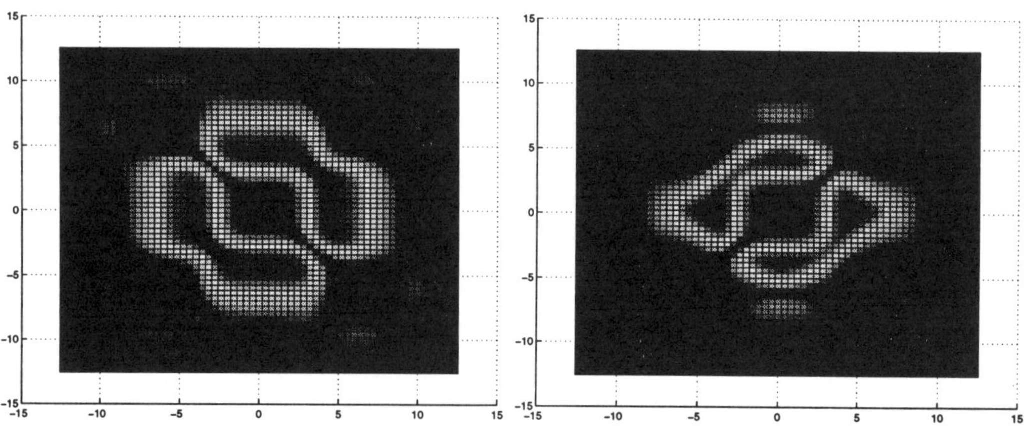

Figure 3. $|\hat{\mathbf{\Psi}}(\omega)|$ with $\mathbf{A} = \mathbf{S}$ (left) and $\mathbf{A} = \mathbf{R}$ (right) for $N = 3$ (view from above).

References

[1] E. Belogay, Y. Wang: *Arbitrarily smooth orthogonal nonseparable wavelets in* $\mathbf{R}^2$, Preprint 1997.

[2] A. Cohen, R. D. Ryan: *Wavelets and Multiscale Signal Processing*, Chapman and Hall, London 1995.

[3] A. Cohen, I. Daubechies: *Nonseparable bidimensional wavelet bases*, Mat. Iberoamericana **9** (1993), 51–137.

[4] A. Cohen, I. Daubechies: *A new technique to estimate the regularity of refineable functions*, Rev. Mat. Iberoamericana **12** (1996), 527–591.

[5] A. Fettweis: *Wave digital filters: theory and practice*, Proc. IEEE **74**, No.2 (1986), 270–324.

[6] L. Gaszi: *Explicit formulae for lattice wave digital filters*, IEEE Trans. CAS **32** (1985), 66–88.

[7] A. Gottscheber: *Multi–dimensional Wave Digital Filters and Wavelets*, Doctoral Thesis, University of Mannheim 1998.

[8] A. Gottscheber, A. Nishihara: *Passive two–dimensional wave digital filters used in a multirate system having perfect reconstruction*, IEICE Trans. on Fundamentals, **80-A**, No. 1 (1997), 133–139.

[9] K. Gröchenig, A. Cohen, L. Villemoes: *Regularity of multivariate refineable functions*, Preprint 1996.

[10] K. Gröchenig, W. R. Madych: *Multiresolution analysis, Haar bases and self–similar rilings of* $\mathbf{R}^n$, IEEE Trans. Inform. Theory **38** (1992), 556–568.

[11] W. He, M. J. Lai: *Construction of bivariate compactly supported box spline wavelets with arbitrary high regularity*, Preprint 1997.

[12] L. Hervé: *Construction et régularité des fonctions d'échelle*, SIAM J. Math. Anal. **25** (1995), 1361–1385.

[13] C. Herley, M. Vetterli: *Wavelets and recursive filter banks*, IEEE Trans. on Signal Proc. **41**, No. 8 (1993), 2536–2556.

[14]   J. Kovačevic, M. Vetterli: *Nonseperable multidimensional perfect reconstruction filter banks and wavelet bases for* $\mathrm{I\!R}^n$, IEEE Trans. Inform. Theory **38** (1992), 533–555.

[15]   A. Oppenheim, R. Schafer: *Digital Signal Processing,* Prentice Hall, New York 1975.

[16]   L. Villemoes: *Continuity of nonseperable quincunx wavelets,* Appl. Comput. Harmon. Anal. **1** (1994), 180–187.

Addresses:

ACHIM GOTTSCHEBER
Forschungszentrum Karlsruhe
Hermann-von–Helmholtz–Platz 1
76344 Eggenstein–Leopoldshafen
Germany

GABRIELE STEIDL
Department of Mathematics
University of Mannheim
D–68131 Mannheim
Germany

Average Case Analysis of Numerical Integration

Peter J. Grabner*†, Pierre Liardet‡ and Robert F. Tichy* **

Dedicated to Professor Manfred Reimer on the occasion of his 65th birthday

Abstract

We extend the average case analysis of numerical integration in the sense of Traub and Woźniakowski to homogeneous spaces. Various examples are described and the connections to minimal energy point sets are outlined.

† This author was supported by the Austrian Academy of Sciences

* These authors were supported by the START–project Y96–MAT of the Austrian Science Fund

‡ This author was supported by MENRT–CNRS

** This author was supported by the Austrian Science Fund project Nr. P12005–MAT

1   Introduction

Persi Diaconis [5] considered numerical integration of real valued functions on the unit interval $[0, 1]$ from a Bayesian point of view. He surveyed on methods how to get information on the integral, provided that the prior distribution of the functions is known (for instance Brownian motion). He also mentioned several applications such as random surfaces (as used in tomography) and statistical geodesy. In particular, Lauritzen [20, 21, 22] used such methods and ideas to estimate the gravity potential of the earth from the knowledge of a certain number of measurements.

From a different point of view numerical integration of functions on the s–dimensional unit cube $[0, 1]^s$ was analyzed by Traub, Wasilkowski, and Woźniakowski (cf. [33, 32, 34]). These authors approximated the integral by the arithmetic mean of N function values. They equipped the space of continuous functions on $[0, 1]^s$ with Wiener measure and investigated the average case error in the sense of quadratic mean. Furthermore, they proved that the average case complexity is optimal, provided that the integration points have L^2–discrepancy of optimal order of magnitude.

Advances in Multivariate Approximation; W. Haußmann, K. Jetter and M. Reimer (eds.)
Mathematical Research, Vol. 107, pp. 185–200, ISBN 3-527-40236-5
© WILEY-VCH, Berlin 1999

We extend these ideas to functions defined on homogeneous spaces X. In Section 2 we construct a Gaussian stochastic process on X by prescribing the covariance function. Of course, a very interesting special case is the sphere, and in this case such processes have been investigated in [2, 3]. For similar processes in the more general setting of hypergroups, we refer to [1].

In Section 3 we outline several examples and related concepts of discrepancy. We start with a discontinuous process defined via the Walsh orthogonal system. Then we introduce the diaphony, the L^2-discrepancy, and various kinds of geodesic discrepancies and relate them to suitably defined Gaussian processes. For a more detailed study of different notions of discrepancies on spheres, we refer to our earlier paper [11].

In the final Section 4 we outline an extension of the Traub–Woźniakowski approach to numerical integration of functions on homogeneous spaces and its connection to minimal energy point sets.

2 Construction of the Process

Let G be a compact group acting transitively and continuously on a metric space X and μ_0 the normalized Haar measure on G. Then X can be topologically identified with the quotient G/K of G modulo the stabilizer $K \subseteq G$ of a point. A natural probability measure on $X \cong G/K$ is given by $\mu(M) = \mu_0(\{g \in G \mid gK \in M\})$ (cf. [15, 6, 27]).

For any $\rho \in G^*$ (the dual of G modulo equivalence of representations) let

$$m_{ij}^{(\rho)}(g) = (U^{(\rho)}(g)a_j, a_i), \quad i,j = 1,\ldots,n_\rho$$

be the coordinate functions of the irreducible unitary representation $U^{(\rho)}$ in the finite dimensional Hilbert space generated by an orthonormal system $a_1,\ldots,a_{n_\rho}$. We will denote the trivial representation by ρ_0. This system can be chosen such that for a certain $d_\rho \leq n_\rho$

$$\mathrm{Sp}(\{m_{ij}^{(\rho)} \mid i,j = 1,\ldots,n_\rho\}) \cap L^2(G/K) = \mathrm{Sp}(\{m_{ij}^{(\rho)} \mid 1 \leq i \leq n_\rho, 1 \leq j \leq d_\rho\}),$$

where $\mathrm{Sp}(S)$ denotes the linear space generated by S. By the theorem of Peter–Weyl and well–known algebraic tools it can be deduced that for $J_\rho = \mathrm{Sp}(\{m_{ij}^{(\rho)} \mid i = 1,\ldots,n_\rho, j = 1,\ldots,m_\rho\})$

$$L^2(G/K) = \widehat{\bigoplus_{\rho \in G^*}} J_\rho \quad \text{(Hilbert orthogonal sum)}.$$

By reordering orthonormal bases of the J_ρ's we obtain a complete orthonormal system $\psi_{n,\rho}(x)$ $(\rho \in G^*, n = 1,\ldots,d_\rho = \dim(J_\rho))$ of continuous real functions.

Define now a symmetric kernel function by

$$K(x,y) = \sum_{\rho \in G^*} a_\rho \sum_{n=1}^{m_\rho} \psi_{n,\rho}(x)\psi_{n,\rho}(y) \tag{2.1}$$

with strictly positive coefficients a_ρ and

$$\sum_{\rho \in G^*} a_\rho m_\rho = 1.$$

Furthermore, $K(x,y) = 1$, if and only if $x = y$, the series (2.1) is uniformly convergent and the kernel function is positive definite. The term "positive definite" refers to the fact that

$$\sum_{i,j=1}^{M} K(x_i, x_j)u_i u_j \geq 0$$

for all choices of $x_i \in X$ and $u_i \in \mathbb{R}$, $i = 1,\ldots,M$. This is the generalization of the usual concept of positive definite functions to harmonic spaces (cf. [15], [30]). If K acts transitively on every geodesic sphere around the point stable under K, then the kernel can be written as a function of the distance $d(x,y)$, which would give the notion of positive definiteness studied in [29] and [30]. This will always be the case in our examples.

Remark 1. *The function $d(x,y) = \sqrt{1 - K(x,y)}$ defines a G-invariant metric on X. Conversely, for an arbitrarily given metric d the function $1 - d(x,y)^2$ can be considered, and it is a natural question, whether it is positively definite. The identity*

$$\begin{vmatrix} 1 & 1 - d(x,y)^2 & 1 - d(x,z)^2 \\ 1 - d(x,y)^2 & 1 & 1 - d(y,z)^2 \\ 1 - d(x,z)^2 & 1 - d(y,z)^2 & 1 \end{vmatrix}$$

$$= (d(x,y) + d(y,z) - d(x,z))(d(x,z) + d(z,y) - d(x,y)) \times$$
$$(d(x,z) + d(x,y) - d(y,z))(d(x,y) + d(x,z) + d(y,z))$$
$$-2d(x,y)^2 d(x,z)^2 d(y,z)^2$$

shows that in the case of a geodesic metric (take $d(x,z) = d(x,y) + d(y,z)$) there is no positive definiteness of the function $1 - d(x,y)^2$.

Definition 1. *Let $A_{n,\rho}$, $\rho \in G^*$, $n = 1, \ldots, m_\rho$, be independent normal random variables with mean 0 and variance a_ρ. Define the process*

$$Y(x) = \sum_{\rho \in G^*} \sum_{n=1}^{m_\rho} A_{\rho,n} \psi_{n,\rho}(x). \qquad (2.2)$$

Equip now the space of continuous functions $C(X)$ with the Lévy measure λ defined by this process.

Remark 2. *The series (2.2) is almost everywhere uniformly convergent by Kolmogorov's three series theorem.*

Remark 3. *The process $Y(x)$ could be defined alternatively as a Gaussian process with mean $\mathbb{E}Y(x) = 0$ and covariance function $\mathbb{E}Y(x)Y(y) = K(x,y)$. This equivalent definition would avoid Fourier series.*

We are now ready to compute the integral of the square of the integration error with respect to λ.

Proposition 1.

$$\int_{C(X)} \left(\frac{1}{N} \sum_{n=1}^{N} y(x_n) - \int_X y(x)\,d\mu(x) \right)^2 d\lambda(y) = \qquad (2.3)$$

$$\sum_{\rho \in G^* \setminus \{\rho_0\}} a_\rho \sum_{m=1}^{m_\rho} \left(\frac{1}{N} \sum_{n=1}^{N} \psi_{m,\rho}(x_n) \right)^2. \qquad (2.4)$$

Proof. Observe first that $\int_X Y(x)\,d\mu(x) = A_{\rho_0,1}$. Thus we can rewrite the first integral as

$$\int_{C(X)} \left(\frac{1}{N} \sum_{n=1}^{N} \sum_{\rho \neq \rho_0} \sum_{m=1}^{m_\rho} A_{m,\rho} \psi_{m,\rho}(x_n) \right)^2 \lambda(dY).$$

Interchanging the orders of summation and using the independence of the $A_{m,\rho}$'s gives the result. $\qquad\qquad\square$

3   Examples and Geodesic L^2–Discrepancy

We will now present a list of examples which will show that several concepts of diaphony and L^2–discrepancy fit into the general approach decribed above.

3.1   Example 1: Dyadic Diaphony

In [14] a concept of diaphony based on the orthonormal system of Walsh functions on the unit cube $[0,1)^s$ is introduced. As general references for Walsh functions we refer to [10] and [28].

For $x = \sum_{k=0}^{\infty} \frac{\varepsilon_k}{2^{k+1}}$ given by its binary digital expansion (in case of ambiguity we take the finite version of the digital expansion) and $m = \sum_{k=0}^{K} \delta_k 2^k$ we define the m–th Walsh function by

$$w_m(x) = (-1)^{\sum_{k=0}^{K} \varepsilon_k \delta_k}. \tag{3.1}$$

For $\mathbf{x} \in [0,1)^s$ and $\mathbf{m} \in \mathbf{N}^s$ we define

$$w_{\mathbf{m}}(\mathbf{x}) = \prod_{j=1}^{s} w_{m_j}(x_j).$$

The dyadic diaphony of a sequence $\omega = (\mathbf{x}_1, \mathbf{x}_2, \ldots)$ is then given by

$$F_N(\omega) = \left(\frac{1}{3^s - 1} \sum_{\mathbf{m} \neq 0} \rho(\mathbf{m}) |S_N(w_{\mathbf{m}}, \omega)|^2 \right)^{\frac{1}{2}}, \tag{3.2}$$

where $\rho(\mathbf{m}) = \prod_{j=1}^{s} \rho(m_j)$ with

$$\rho(m) = \begin{cases} 2^{-2g} & \text{for } 2^g \leq m < 2^{g+1}, \quad g \in \mathbf{N} \\ 1 & \text{for } m = 0 \end{cases} \tag{3.3}$$

and

$$S_N(w_{\mathbf{m}}, \omega) = \frac{1}{N} \sum_{n=1}^{N} w_{\mathbf{m}}(\mathbf{x}_n)$$

denotes the Weyl sum with respect to the Walsh functions.

At first sight this concept does not fit into the scheme introduced above, since the Walsh functions are not the group characters on $[0,1)^s$. But there is the bijection

$$\xi : [0,1) \ni \sum_{k=0}^{\infty} \frac{\varepsilon_k}{2^{k+1}} \mapsto (\varepsilon_0, \varepsilon_1, \ldots) \in \mathbf{F}_2^{\mathbf{N}}$$

between $[0,1)$ and a subset of measure 1 of the group $\mathbf{F}_2^{\mathbf{N}}$ (the group of infinite sequences of $0,1$ equipped with the product topology and the Haar measure).

This map is also continuous except for the dyadic rational points. Clearly the component–wise map between s–tuples enjoys the same properties. $[0,1)$ also inherits an addition law from $\mathbb{F}_2^{\mathbb{N}}$, which is just $x\dotplus y = \xi^{-1}(\xi(x)+\xi(y))$. Again, this addition will be performed component–wise on the s–tuples.

Since the function

$$\varphi(x) = \begin{cases} 3 - 3 \cdot 2^{1+\lfloor \log_2 x \rfloor} & \text{for } x \in (0,1) \\ 3 & \text{for } x = 0 \end{cases} \tag{3.4}$$

has the Walsh expansion

$$\varphi(x) = \sum_{k=0}^{\infty} \rho(k) w_k(x),$$

we can write (under an additional condition on the values of the $\mathbf{x}_n$)

$$F_N^2(\omega) = \frac{1}{3^s - 1} \frac{1}{N^2} \sum_{n,m=1}^{N} \phi(\mathbf{x}_n \dotplus \mathbf{x}_m) \tag{3.5}$$

with

$$\phi(\mathbf{x}) = \prod_{j=1}^{s} \varphi(x_j) - 1.$$

Notice, that $w_k(x\dotplus y) = w_k(x)w_k(y)$, if either one of x and y is a dyadic rational or $x\dotplus y$ is not a dyadic rational (this describes the condition indicated above).

With the restriction on the choice of the values $\mathbf{x}_n$, formula (3.5) fits into the general theory developed above. We will now derive some properties of the stochastic process behind this type of diaphony. The process is given as the Gaussian process with covariance function

$$\mathbb{E}X(\mathbf{x})X(\mathbf{y}) = \phi(\mathbf{x}\dotplus\mathbf{y}), \tag{3.6}$$

which is equivalent to

$$\mathbb{E}(X(\mathbf{x}) - X(\mathbf{y}))^2 = 2(3^s - 1) - 2\phi(\mathbf{x}\dotplus\mathbf{y}) \leq 2 \cdot 3^s \sum_{j=1}^{s} \varphi(x_j\dotplus y_j). \tag{3.7}$$

Equation (3.7) implies that the process is continuous whenever $\mathbf{x} \to \mathbf{y}$ implies $\mathbf{x}\dotplus\mathbf{y} \to \mathbf{0}$. This is the case exactly when $\mathbf{y}$ has no dyadic rational entry. It remains to consider the case when one of the components of the vector $\mathbf{y}$ is a dyadic rational. Then it follows from the right continuity of the function ϕ that $\mathbf{x}\dot{+}\mathbf{y} \to \mathbf{0}$, if the components of $\mathbf{x}$ with the same indices as the dyadic rational components of $\mathbf{y}$ converge monotonically decreasing to this value. It follows from the fact that the Walsh functions only have jump discontinuities, that the limits in the other $2^s - 1$ orthants exist. For these limits to be the same, the Fourier coefficients would have to satisfy linear relations, which are certainly only satisfied with probability 0. Thus we have proved

Proposition 2. *The Gaussian process $X(\mathbf{x})$, $\mathbf{x} \in [0,1)^s$ given by $\mathbb{E}X(\mathbf{x}) = 0$ and (3.6) gives the dyadic diaphony as defined in [14] and has trajectories which are continuous in every point $\mathbf{x}$ with no dyadic rational component. In all points with at least one dyadic rational component the limits in the different orthants exist but are different with probability 1, if a rational component is approached from two different directions.*

Remark 4. *Similar results could be derived for arbitrary Cantor–type digital expansions and the characters of $\mathbf{A}$–adic numbers as defined in [15]. Especially, there is a generalization of the above results to arbitrary q–adic digital expansions.*

3.2   Example 2: Euclidian L^2–Discrepancy

It was first observed by O. Strauch [31] that the usual L^2–discrepancy on the interval can be written as a Wiener integral

$$\int_0^1 \left(\frac{1}{N}\sum_{n=1}^N \chi_{[0,x)}(x_n) - x\right)^2 dx = 2 \int \left(\frac{1}{N}\sum_{n=1}^N f(x_n) - \int_0^1 f(x)\,dx\right)^2 df,$$

where the second integral is extended over all continuous functions on $[0,1]$ with $f(0) = 0$.

We give an alternative interpretation of L^2–discrepancy which is based on a process with periodic trajectories. The corresponding diaphony can be written as

$$\frac{6}{\pi^2}\sum_{k=1}^\infty \frac{1}{k^2}\left|\frac{1}{N}\sum_{n=1}^N e^{2\pi i k x_n}\right|^2 = \frac{6}{\pi^2}\sum_{k=1}^\infty \frac{1}{k^2}\left(\left|\frac{1}{N}\sum_{n=1}^N \sin(2\pi k x_n)\right|^2 + \left|\frac{1}{N}\sum_{n=1}^N \cos(2\pi k x_n)\right|^2\right).$$

This yields the corresponding kernel function

$$K(x,y) \;=\; \frac{6}{\pi^2}\sum_{k=1}^{\infty}\frac{1}{k^2}\cos(2\pi k(x-y))$$

$$=\; 6\left(\|x-y\|-\frac{1}{2}\right)^2 - \frac{1}{2},$$

where $\|x-y\|$ denotes the chordal distance on the circle $\mathbb{R}/\mathbb{Z}$, i.e. $\|x-y\| = \min_{n\in\mathbb{Z}}|x-y-n|$. This definition of the process is equivalent to

$$\mathbb{E}\left(X(x)-X(y)\right)^2 = 12\|x-y\|\left(1-\|x-y\|\right),$$

which implies that the trajectories are continuous functions.

Remark 5. *Note that the simpler kernel function $K(x,y) = 1 - \alpha\|x-y\|$ (with $0 < \alpha < 4$ for positive definiteness) would give trajectories $X(x)$ with the unwanted property that $X(x) + X(x+\frac{1}{2}) = const$. This is a consequence of*

$$\mathbb{E}\left(X(x)-X(x+\frac{1}{2})-X(y)+X(y+\frac{1}{2})\right)^2 = 2\alpha\left(\|x-y\|+\|x-y+\frac{1}{2}\|-1\right),$$

which vanishes for geometric reasons.

The computations abovecan be immediately generalized to the s–dimensional torus. Here the kernel function corresponding to the usual L^2–discrepancy is given by

$$K_s(\mathbf{x},\mathbf{y}) \;=\; \frac{1}{2^s-1}\left(\prod_{j=1}^{s}\left(6\left(\|x_j-y_j\|-\frac{1}{2}\right)^2+\frac{1}{2}\right)-1\right)$$

$$=\; \frac{1}{2^s-1}\left(\prod_{j=1}^{s}(K(x_j,y_j)+1)-1\right).$$

We once again have continuous periodic trajectories.

Remark 6. *It is also known that the discrepancy function*

$$D_N(x,x_n) = \frac{1}{\sqrt{N}}\left(\sum_{n=1}^{N}\chi_{[0,x)}(x_n)-x\right)$$

behaves like a trajectory of the Brownian bridge as $N \to \infty$. This leads to a new proof of Kolmogorov's theorem (cf. [16]) on the distribution of the discrepancy (cf. [7, 8]). It is possible to use this fact to study the distribution of L^2–discrepancy of sequences (cf. [12]).

3.3   Example 3: Geodesic L^2–Discrepancy

Since X is supposed to be compact and metrizable with G–invariant metric $d : X \times X \to [0, R]$ it is natural to define a discrepancy by

$$
D_N^G(x_n) = \int\limits_X \int\limits_0^R \left(\frac{1}{N} \sum_{n=1}^N \chi_{B(x,r)}(x_n) - \mu(B(x,r)) \right)^2 \, dr \, d\mu(x),
$$

where R denotes the diameter of X. Expanding and integrating term by term yields

$$
D_N^G(x_n) = \frac{1}{N^2} \sum_{m,n=1}^N \int\limits_X \int\limits_0^R \chi_{B(x_m,r)}(x) \chi_{B(x_n,r)}(x) \, dr \, d\mu(x)
$$

$$
- \frac{2}{N} \sum_{n=1}^N \int\limits_X \int\limits_0^R \chi_{B(x_n,r)}(x) \mu(B(x,r)) \, dr \, d\mu(x) + \int\limits_X \int\limits_0^R \mu(B(x,r))^2 \, dr \, d\mu(x)
$$

$$
= \frac{1}{N^2} \sum_{m,n=1}^N \int\limits_{\frac{1}{2}d(x_m,x_n)}^R \mu(B(x_m,r) \cap B(x_n,r)) \, dr - \int\limits_0^R \mu(B(\cdot,r))^2 \, dr. \quad (3.8)
$$

We take the function

$$
K(x,y) = f(d(x,y)) = \int\limits_{\frac{1}{2}d(x,y)}^R \mu(B(x,r) \cap B(y,r)) \, dr - \int_0^R \mu(B(\cdot,r))^2 \, dr \quad (3.9)
$$

as the kernel function.

As a first special case let us consider the circle $\mathbb{R}/\mathbb{Z}$. Then we have by (3.8)

$$
D_N^G(x_n) = \int_0^1 \int_0^{\frac{1}{2}} \left(\frac{1}{N} \sum_{n=1}^N \chi_{(x-r,x+r)}(x_n) - 2r \right)^2 \, dr \, dx
$$

$$
= \frac{1}{N^2} \sum_{m,n=1}^N \left(\int_{\frac{1}{2}\|x_m-x_n\|}^{\frac{1}{2}(1-\|x_m-x_n\|)} (2r - \|x_m - x_n\|) \, dr + \right.
$$

$$
\left. \int_{\frac{1}{2}(1-\|x_m-x_n\|)}^{\frac{1}{2}} (4r - 1) \, dr \right) - \frac{1}{6},
$$

since

$$\frac{1}{2\pi} \int_0^{2\pi} \chi_{(x_m-r,x_m+r)}(x)\chi_{(x_n-r,x_n+r)}(x)\,dx$$

$$= \begin{cases} 0 & \text{if } r \le \frac{1}{2}\|x_m - x_n\| \\[2mm] 2r - \|x_m - x_n\| & \text{if } \frac{1}{2}\|x_m - x_n\| < r \\[1mm] & \qquad \le \frac{1}{2} - \frac{1}{2}\|x_m - x_n\| \\[2mm] 4r - 1 & \text{if } r > \frac{1}{2} - \frac{1}{2}\|x_m - x_n\|. \end{cases}$$

Computing the remaining integrals and collecting terms yields

$$D_N^G(x_n) = \frac{1}{N^2} \sum_{m,n=1}^{N} \left(\frac{1}{6} - \frac{1}{2}\|x_m - x_n\| + \frac{1}{2}\|x_m - x_n\|^2 \right). \tag{3.10}$$

The next special case which will be considered is the sphere S^2. Here we have to compute the kernel function

$$K(\mathbf{x},\mathbf{y}) = \int_{\frac{1}{2}\arccos\langle\mathbf{x},\mathbf{y}\rangle}^{\pi} \mu\left(B(\mathbf{x},\varphi) \cap B(\mathbf{y},\varphi)\right) d\varphi - \frac{1}{4} \int_0^{\pi} (1 - \cos\varphi)^2\, d\varphi. \tag{3.11}$$

In order to compute the normalized surface measure of $B(\mathbf{x},\varphi) \cap B(\mathbf{y},\varphi)$, we express the characteristic function of the cap $B(\mathbf{u},\varphi)$ by its Fourier series (cf. [23, 24])

$$\chi_{B(\mathbf{u},\varphi)}(\mathbf{x}) = \frac{1 - \cos\varphi}{2} + \sum_{n=1}^{\infty} (P_{n+1}(\cos\varphi) - P_{n-1}(\cos\varphi)) \sum_{k=-n}^{n} K_{nk}(\mathbf{u})K_{nk}(\mathbf{x}).$$

P_n denotes the n–th Legendre polynomial. Integrating $\chi_{B(\mathbf{x},\varphi)}(\mathbf{u})\chi_{B(\mathbf{y},\varphi)}(\mathbf{u})$ we obtain

$$\mu\left(B(\mathbf{x},\varphi) \cap B(\mathbf{y},\varphi)\right) = \frac{(1-\cos\varphi)^2}{4} + \sum_{n=1}^{\infty} \frac{(P_{n+1}(\cos\varphi) - P_{n-1}(\cos\varphi))^2}{2n+1} P_n(\langle\mathbf{x},\mathbf{y}\rangle).$$

Inserting this into (3.11) and integrating with respect to φ yields

$$K(\mathbf{x},\mathbf{y}) = \sum_{n=1}^{\infty} a_n P_n\left(\langle\mathbf{x},\mathbf{y}\rangle\right) \tag{3.12}$$

with

$$a_n = 16^{-n}\left(\frac{1}{16}\sum_{m=0}^{n+1}\binom{2m}{m}^2\binom{2n+2-2m}{n+1-m}^2 + 16\sum_{m=0}^{n-1}\binom{2m}{m}^2\binom{2n-2-2m}{n-1-m}^2\right.$$
$$\left. - 2\sum_{m=1}^{n}\binom{2m}{m}\binom{2m-2}{m-1}\binom{2n+2-2m}{n+1-m}\binom{2n-2m}{n-m}\right),$$

where we have used the Fourier expansion of Legendre polynomials (cf. [23], p. 229). Thus we have for the geodesic discrepancy on S^2

$$D_N^G(\mathbf{x}_n) = \frac{1}{N^2}\sum_{i,j=1}^{N} K(\mathbf{x}_i,\mathbf{x}_j),$$

where K is given by the Fourier expansion (3.12).

In the case of the sphere one could obtain a more explicit formula for the L^2–discrepancy by replacing the measure dr by $\sin r\, dr$, which amounts to a change of the metric; this new metric is no more geodesic. This yields as above the Fourier coefficient of the kernel

$$a_n = \frac{2}{(2n+3)(2n-1)},$$

which leads to the explicit expression for the kernel

$$K(\mathbf{x},\mathbf{y}) = \frac{2}{3} - \sqrt{\frac{1-\langle\mathbf{x},\mathbf{y}\rangle}{2}},$$

where we have used [23] p. 238 for $\lambda = \frac{1}{2}$. The expression $\sqrt{\frac{1-\langle\mathbf{x},\mathbf{y}\rangle}{2}}$ is clearly half the Euclidean distance between the points $\mathbf{x}$ and $\mathbf{y}$ in $\mathbb{R}^3$. We obtain for the discrepancy with respect to this new metric

$$\tilde{D}_N^G(\mathbf{x}_n) = \frac{1}{N^2}\sum_{i,j=1}^{N}\left(\frac{2}{3} - \sqrt{\frac{1-\langle\mathbf{x}_i,\mathbf{x}_j\rangle}{2}}\right).$$

4   Average Case Analysis

In this final section we outline Woźniakowski's approach of the average case analysis of numerical integration. In [34] the author considered real valued functions on the s–dimensional unit cube and the classical Wiener measure; for further surveys on this theory we refer to [33, 32, 9].

Here we will extend the basic ideas of this approach to numerical integration of real valued functions on homogeneous spaces. Let f be a continuous real–valued function on the homogeneous space $X \cong G/K$ and set

$$I(f) = \int_X f(x)\,d\mu(x), \tag{4.1}$$

with the natural probability measure μ on X associated to the Haar measure on G. We approximate $I(f)$ by the arithmetic mean

$$I_N(f) = \frac{1}{N}\sum_{n=1}^{N} f(x_n) \tag{4.2}$$

extended over the integration points $x_n \in X$.

Usually integration algorithms with low costs are desirable. In this context the costs would depend on computation time and the approximation error. In the case of the approximation of $I(f)$ by $I_N(f)$ the cost $\mathrm{cost}(I_N)$ is proportional to N, the number of function evaluations. The average error of numerical integration with respect to the Lévy measure λ as defined in section 2 is given by

$$E^{\mathrm{avg}}(I_N) = \left(\int_{C(X)} (I(f) - I_N(f))^2\, \lambda(df) \right)^{\frac{1}{2}}. \tag{4.3}$$

The average case ε–complexity is given as the minimal cost of all algorithms with average error $\leq \varepsilon$:

$$\mathrm{comp}^{\mathrm{avg}}(\varepsilon, \mathrm{C}(\mathrm{X})) = \inf\{\mathrm{cost}(\mathrm{I_N}) \mid \mathrm{E}^{\mathrm{avg}}(\mathrm{I_N}) \leq \varepsilon\}. \tag{4.4}$$

By Proposition 1 we have

$$E^{\mathrm{avg}}(I_N) = F_N^A(x_n), \tag{4.5}$$

where

$$F_N^A(x_n) = \left(\sum_{\rho \in G^* \setminus \{\rho_0\}} a_\rho \sum_{m=1}^{m_\rho} \left(\frac{1}{N} \sum_{n=1}^{N} \psi_{m,\rho}(x_n) \right)^2 \right)^{\frac{1}{2}}$$

is the diaphony of the point set x_n with respect to the coefficients $A = (a_\rho)$. The classical notion of diaphony was introduced by Zinterhof [35] in the case of the s–dimensional unit cube (see also [9]). In this case the coefficients are given by

$$a(\mathbf{h}) = \frac{1}{\prod_{k=1}^{s} \max(1, |h_k|)}.$$

As it is shown in Section 3, in many important cases the diaphony can be expressed in terms of suitably chosen concepts of L^2–discrepancies, such as the classical discrepancy on $[0,1)^s$ (cf. [19, 9]) or the geodesic discrepancy on the sphere.

In order to establish a bound for the average case complexity of numerical integration we need suitable point sets of small diaphony. Note that the diaphony can be interpreted as an energy functional.

Definition 2. *A set of points $z_n \in X$ is called a minimal energy set on X, if the function $F_N^A(x_n)$ attains its minimal value in the point $(x_1, \ldots, x_N) = (z_1, \ldots, z_N)$. The minimal value is denoted by $\Phi(N)$.*

In the classical case of the s–dimensional unit cube $\Phi(N) = \mathcal{O}(\frac{\log^{\frac{s-1}{2}} N}{N})$; this value is attained for instance for the Hammersley point set. In the case of the sphere various authors such as Kuijlaars, Rakhmanov, Saff, Zhou and others considered minimal energy problems for different kinds of functionals (cf. [17, 18, 25, 26]). In the spherical case explicit constructions for the point sets are presently not known. In [13] spherical designs were shown to be minimal energy point sets for certain functionals. (Just recall that a spherical t–design is a set of integration points giving exact integration for polynomials up to order t; for more details, see [4]).

Thus the problem of estimating the average case complexity of numerical integration is reduced to finding the minimal energy point configuration for the corresponding diaphony. In particular, the average case ε–complexity is given by the smallest N such that $\Phi(N) \leq \varepsilon$, provided that Φ is decreasing.

Acknowledgement. The authors are indebted to Persi Diaconis for pointing out the references [20, 21, 22].

References

[1] W. R. Bloom, H. Heyer: *Harmonic Analysis of Probability Measures on Hypergroups*, de Gruyter, 1995.

[2] Z. Ciesielski: *On Levy's Brownian motion with several-dimensional time*, in: *Probability Winter School Karpacz, Poland*, 1975, Lect. Notes Math. **472**, 29–56, Springer, Berlin–New York, 1975.

[3] Z. Ciesielski, A. Kamont: *On the fractional Lévy's random field on the sphere*, East J. Approx. **1** (1995), 111–123.

[4] P. Delsarte, J. M. Goethals, J. J. Seidel: *Spherical codes and designs*, Geometriae Dedicata **6** (1977), 363–388.

[5] P. Diaconis: *Bayesian numerical analysis*, in: *Statistical Decision Theory and Related Topics IV, Papers from the 4th Purdue symposium on statistical decision theory, West Lafayette, IN*, S. S. Gupta and J. O. Berger (eds.), 163–175, Springer, Berlin–New York, 1988.

[6] J. Dieudonné: *Special Functions and Linear Representations of Lie Groups*, Vol. 42 of *CBMS, Regional conference series in mathematics*. Amer. Math. Soc., Providence, Rhode Island, 1980.

[7] M. D. Donsker: *Justification and extension of Doob's heuristic approach to the Kolmogorov–Smirnov theorems*, Ann. Math. Stat. **23** (1952), 277–281.

[8] J. L. Doob: *Heuristic approach to the Kolmogorov–Smirnov theorems*, Ann. Math. Stat. **20** (1949), 393–403.

[9] M. Drmota, R. F. Tichy: *Sequences, Discrepancies, and Applications*, Lect. Notes Math. **1651**, Springer, Berlin–New York, 1997.

[10] B. Golubov, A. Efimov, V. Skvantsov: *Walsh Series and Transforms*. Kluwer, Dordrecht, 1991.

[11] P. J. Grabner, B. Klinger, R. F. Tichy. *Discrepancies of point sequences on the sphere and numerical integration*, in: *Multivariate Approximation, Recent Trends and Results*, W. Haußmann et al. (eds.), 95–112, Akademie Verlag, Berlin, 1997.

[12] P. J. Grabner, O. Strauch, R. F. Tichy: *Maldistribution in higher dimensions*, Math. Pannonica **8** (1997), 215–223.

[13] P. J. Grabner, R. F. Tichy: *Spherical designs, discrepancy and numerical integration*, Math. Comp. **60** (1993), 327–336.

[14] P. Hellekalek, H. Leeb: *Dyadic diaphony*, Acta Arith. **80** (1997), 187–196 to appear.

[15] E. Hewitt, K. A. Ross: *Abtract Harmonic Analysis*, Springer, Berlin, 1970.

[16] A. Kolmogorov: *Sulla determinazione empirica di una legge di distribuzione*, Giorn. Ist. Ital. Attuari **4** (1933), 83–91.

[17] A. B. J. Kuijlaars, E. B. Saff: *Asymptotics for minimal discrete energy on the sphere*, Trans. Amer. Math. Soc. **350** (1996), 523–538.

[18] A. B. J. Kuijlaars, E. B. Saff: *Distributing many points on a sphere*, Math. Intell. **19** (1997), 5–11.

[19] L. Kuipers, H. Niederreiter: *Uniform Distribution of Sequences*, Wiley, New York, 1974.

[20] S. L. Lauritzen: *An application of continuous functionals and unitary representations of groups to physical geodesy*, in: *Papers "Open House in Probability"*, 1971, Vol. **21**, 167–178. Mat. Inst. Aarhus Univ., 1972.

[21] S. L. Lauritzen: *The Probabilistic Background of Some Statistical Methods in Physical Geodesy*, Meddelelse No. **48**, Geodætisk Institut, København, Danmark, 1973.

[22] S. L. Lauritzen: *Random orthogonal set functions and stochastic models for the gravity potential of the earth*, Stochastic Processes Appl. **3** (1975), 65–72.

[23] W. Magnus, F. Oberhettinger, R.P. Soni: *Formulas and Theorems for the Special Functions of Mathematical Physics*, Springer, Berlin–New York, 1966.

[24] C. Müller: *Spherical Harmonics, Lect. Notes Math.* Vol. **17**, Springer, Berlin–New York, 1966.

[25] E. A. Rakhmanov, E. B. Saff, Y. M. Zhou: *Minimal discrete energy on the sphere*, Math. Res. Lett. **1** (1994), 647–662.

[26] E. A. Rakhmanov, E. B. Saff, Y. M. Zhou: *Electrons on the sphere*, in: *Computational Methods in Function Theory*, 293–309. World Scientific, River Edge, New Jersey, 1995.

[27] W. Schempp, B. Dreseler: *Einführung in die harmonische Analyse*, Teubner, Stuttgart, 1980.

[28] F. Schipp, W. R. Wade, P. Simon: *Walsh Series. An Introduction to Dyadic Harmonic Analysis*, Adam Hilger, Bristol–New York, 1990.

[29] I. J. Schoenberg: *Metric spaces and positive definite functions*, Trans. Amer. Math. Soc. **44** (1938), 522–536.

[30] I. J. Schoenberg: *Positive definite functions on spheres*, Duke Math. J. **9** (1942), 96–108.

[31] O. Strauch: L^2-*Discrepancy*, Math. Slovaca **44** (1994), 601–632.

[32] J. F. Traub, G. W. Wasilkowski, H. Woźniakowski: *Information Based Complexity*, Academic Press, New York, 1980.

[33] J. F. Traub, H. Woźniakowski: *A General Theory of Optimal Algorithms*, Academic Press, New York, 1980.

[34] H. Woźniakowski: *Average case complexity of multivariate integration*, Bull. Amer. Math. Soc. **24** (1991), 185–194.

[35] P. Zinterhof: *Über einige Abschätzungen bei der Approximation von Funktionen mit Gleichverteilungsmethoden*, Österr. Akad. Wiss. SB II, **185** (1976), 121–132.

Addresses:

PETER J. GRABNER, ROBERT F. TICHY
Institut für Mathematik A
Technische Universität Graz
Steyrergasse 30
A–8010 Graz
Austria

PIERRE LIARDET
DSA
Université de Provence
CMI, Château Gombert
39, rue Joliot–Curie
F–13453 Marseille, Cedex 13
France

Cubature Formulas on Spheres

Hermann König

Dedicated to Manfred Reimer on the occasion of his 65th birthday

Abstract

We study explicit cubature formulas on spheres S^{n-1} in $\mathbb{R}^n$ or $\mathbb{C}^n$ of degrees 4 and 5 with only $O(n^2)$ nodes and positive weights. Some general facts are presented for formulas of higher degrees.

1   General Facts on Cubature Formulas

Let $n, k \in \mathbb{N}$ and σ_n denote the rotation invariant, normalized Haar measure on $S^{n-1} \subseteq \mathbb{K}^n$, $\mathbb{K} \in \{\mathbb{R}, \mathbb{C}\}$. By $\mathcal{P}^{\mathrm{hom}}_{2k,n}$ we denote the space of homogeneous polynomials of degree $2k$ in n variables if $\mathbb{K} = \mathbb{R}$ and the space of all polynomials $q(z_1, \ldots, z_n, \overline{z_1}, \ldots, \overline{z_n})$ which are homogeneous of degree k in each set of variables $(z_1, \ldots, z_n)$ and $(\overline{z_1}, \ldots, \overline{z_n})$ if $\mathbb{K} = \mathbb{C}$.

A *cubature formula* of degree $2k$ on S^{n-1} consists of N nodes $x_1, \ldots, x_N \in S^{n-1}$ and N weights $\lambda_1, \ldots, \lambda_N \in \mathbb{R}$ with $\sum\limits_{s=1}^{N} \lambda_s = 1$ such that

$$\int\limits_{S^{n-1}} p(y) d\sigma_n(y) = \sum_{s=1}^{N} \lambda_s \, p(x_s), \ p \in \mathcal{P}^{\mathrm{hom}}_{2k,n}. \tag{1.1}$$

Only those cubature formulas with positive weights $\lambda_s > 0$ are of numerical importance. For $0 \le l < k$ and $q \in \mathcal{P}^{\mathrm{hom}}_{2l,n}$, $p(x) = q(x) < x, x >^{k-l}$, $x \in \mathbb{K}^n$ defines $p \in \mathcal{P}^{\mathrm{hom}}_{2k,n}$ with $p|_{S^{n-1}} = q|_{S^{n-1}}$. Thus (1.1) holds for all *even* polynomials of degree $\le 2k$ in n variables. Adding the points $(-x_s)$ and taking weights $\lambda_s/2$ on each of x_s and $(-x_s)$ yields a cubature formula with $2N$ nodes such that (1.1) holds for *all* polynomials of degree $\le (2k+1)$ in n variables.

We start with a characterization of cubature formulas which essentially goes back to Goethals, Seidel [3] and Reznick [12].

Advances in Multivariate Approximation; W. Haußmann, K. Jetter and M. Reimer (eds.)
Mathematical Research, Vol. 107, pp. 201–211, ISBN 3-527-40236-5
© WILEY-VCH, Berlin 1999

Proposition 1. *Let $n, k, N \in \mathbb{N}$, $x_1, \ldots, x_N \in S^{n-1}$ and $\lambda_1, \ldots, \lambda_N \in \mathbb{R}$ with $\sum_{s=1}^{N} \lambda_s = 1$. Let*

$$c_{nk} := \int_{S^{n-1}} |x_1|^{2k} d\sigma_n(x).$$

Then the following are equivalent :

$$(1) \quad \int_{S^{n-1}} p(y) d\sigma_n(y) = \sum_{s=1}^{N} \lambda_s \, p(x_s), \quad p \in \mathcal{P}_{2k,n}^{\mathrm{hom}}.$$

$$(2) \quad c_{nk} = \sum_{s,t=1}^{N} \lambda_s \lambda_t \, |<x_s, x_t>|^{2k}.$$

$$(3) \quad c_{nk} \|x\|_2^{2k} = \sum_{s=1}^{N} \lambda_s \, |<x, x_s>|^{2k}, \quad x \in \mathbb{K}^n.$$

By $\| \cdot \|_p$, we denote the l_p–norm, $\|y\|_p = \left(\sum_i |y_i|^p \right)^{1/p}$. The space $\mathbb{K}^m$, equipped with the l_p–norm, is written as l_p^m. If the cubature formula (1) has positive weights, (3) defines an isometric (linear) imbedding,

$$l_2^n \hookrightarrow l_{2k}^N, \quad x \mapsto \left((\lambda_s / c_{nk})^{1/2k} <x, x_s> \right)_{s=1}^{N}.$$

In fact, all such isometric imbeddings can be written in this form. For general systems of nodes and weights, c_{nk} is always smaller than the double sum in (2), see the tensor product "trick" below. This reformulates the cubature problem into an optimization problem. Further (2) can be used to verify that (x_s, λ_s) defines a cubature formula of degree $2k$ on S^{n-1} by *checking only one identity.*

Sketch of the proof.

$(1) \Rightarrow (3)$. Fix $x \in \mathbb{K}^n$ and apply (1) to $p(y) = |<x, y>|^{2k}$. Using the rotation invariance of σ_n, we find

$$\sum_{s=1}^{N} \lambda_s \, |<x, x_s>|^{2k} = \int_{S^{n-1}} |<x, y>|^{2k} d\sigma_n(y)$$

$$= \|x\|_2^{2k} \int_{S^{n-1}} |y_1|^{2k} \, d\sigma_n(y) = c_{nk} \|x\|_2^{2k}.$$

$(3) \Rightarrow (2)$. Take $x = x_t$ and sum over t.

$(2) \Rightarrow (1)$. For $x \in \mathbb{K}^n$, let $x^{\otimes k} = x \otimes \cdots \otimes x \in \mathbb{K}^{n^k}$ denote the k–fold tensor product. Then $< x^{\otimes k}, y^{\otimes k} > = < x, y >^k$ for $x, y \in \mathbb{K}^n$. As in [3], consider

$$\zeta := \sum_{s=1}^{N} \lambda_s \, x_s^{\otimes k} \otimes \overline{x_s}^{\otimes k} - \int_{S^{n-1}} (x^{\otimes k} \otimes \overline{x}^{\otimes k}) \, d\sigma_n(x) \in \mathbb{K}^{(n^{2k})}.$$

The rotation invariance of σ_n yields after an elementary calculation

$$0 \leq <\zeta, \zeta> = \sum_{s,t=1}^{N} \lambda_s \lambda_t \, | < x_s, x_t > |^{2k} - c_{nk}.$$

This shows that in (2), in general, one has "$\leq$". By the way, if $2k$ is replaced by $p \notin 2\mathbb{N}$, a corresponding statement is false. Assuming (2), we get $\zeta = 0$. Expressing ζ in coordinates, all monomials of degree $2k$ in $\mathcal{P}_{2k,n}^{\mathrm{hom}}$ are integrated exactly by the cubature formula represented by (x_s, λ_s). Since the monomials form a basis of $\mathcal{P}_{2k,n}^{\mathrm{hom}}$, (1) follows. $\square$

By Proposition 1 and the remarks after (1.1), if $l_2^n \hookrightarrow l_{2k}^N$ imbeds isometrically, also $l_2^n \hookrightarrow l_{2l}^N$ imbeds isometrically for $1 \leq l < k$.

Example: $\mathbb{K} = \mathbb{R}$, $n = 2$, $N = k + 1$. Take $x_s := \exp(\pi i s / N) \in \mathbb{C} = \mathbb{R}^2$, $\lambda_s := 1/N$ for $s = 0, \ldots, k$. Then (2) of Proposition 1 is satisfied:

$$\sum_{s,t=0}^{k} \lambda_s \lambda_t \, | < x_s, x_t > |^{2k} = \frac{1}{N} \sum_{s=0}^{k} \cos(\pi s/N)^{2k}$$

$$= \frac{1}{\pi} \int_{0}^{\pi} \cos(\pi t)^{2k} \, dt = c_{2k}.$$

The second equality holds by applying trigonometric interpolation. Thus $l_2^2 \hookrightarrow l_{2k}^{k+1}(\mathbb{R})$. For $k = 2$, the explicit formula for finding a circle in the unit ball of l_4^3 is

$$|x_1|^4 + |-\frac{x_1}{2} + \frac{\sqrt{3}}{2} x_2|^4 + |-\frac{x_1}{2} - \frac{\sqrt{3}}{2} x_2|^4 = \frac{9}{8} (|x_1|^2 + |x_2|^2)^2, \quad (x_1, x_2) \in \mathbb{R}^2.$$

Corollary 1. *A complex cubature formula of degree $2k$ in n variables with N nodes yields a real cubature formula of degree $2k$ in $2n$ variables with $N(k+1)$ nodes.*

Proof. This follows in view of the isometric imbeddings

$$l_2^{2n}(\mathbb{R}) = l_2^n(\mathbb{C}) \hookrightarrow l_{2k}^N(\mathbb{C}) = l_{2k}^N\left(l_2^2(\mathbb{R})\right) \hookrightarrow l_{2k}^N\left(l_{2k}^{k+1}(\mathbb{R})\right) = l_{2k}^{N(k+1)}(\mathbb{R}).$$

$\square$

Example: Let $k = 2$. If the nodes on $S^{n-1}(\mathbb{C})$ of a cubature formula of degree 4 have the form

$$x_s = \frac{1}{\sqrt{n}}\left(\exp\left(2\pi\, i\, f(s,j)/p\right)\right)_{j=1}^n \in S^{n-1}(\mathbb{C}); \quad s = 1,\ldots,N,$$

we get $3N$ nodes on $S^{2n-1}(\mathbb{R})$ of a cubature formula of degree 4:

$$\frac{1}{\sqrt{n}}\begin{pmatrix} \cos\left(2\pi\, f(s,j)/p + 2\pi\, l/3\right) \\ -\sin\left(2\pi\, f(s,j)/p + 2\pi\, l/3\right) \end{pmatrix} \in S^{n-1}(\mathbb{R}); \quad s = 1,\ldots,N; \; l = 0,1,2.$$

Examples of nodes x_s of the above type are given below.

2 Cubature Formulas with Few Nodes

We denote by $N_{\mathbb{K}}(n,k)$ the minimal number of nodes of a cubature formula of degree $2k$ in n variables on $S^{n-1}(\mathbb{K})$. General upper and lower bounds are given below; in the real case they can be found in [8] and [12]. The upper bound is classical, see [9]. For a proof, see [7].

Proposition 2. *For any $2 \leq n$, $k \in \mathbb{N}$, $L_{\mathbb{K}}(n,k) \leq N_{\mathbb{K}}(n,k) \leq U_{\mathbb{K}}(n,k)$, where*

$$L_{\mathbb{K}}(n,k) := \begin{cases} \dbinom{n+k-1}{k} & \mathbb{K} = \mathbb{R}, \\[2em] \dbinom{n+[\frac{k+1}{2}]-1}{[\frac{k+1}{2}]}\dbinom{n+[\frac{k}{2}]-1}{[\frac{k}{2}]} & \mathbb{K} = \mathbb{C}, \end{cases}$$

$$U_{\mathbb{K}}(n,k) := \begin{cases} \binom{n+2k-1}{2k} & \mathbb{K} = \mathbb{R}, \\ \\ \binom{n+k-1}{k}^2 & \mathbb{K} = \mathbb{C}. \end{cases}$$

Clearly, $L_{\mathbb{K}}(n,k) \sim n^k$, $U_{\mathbb{K}}(n,k) \sim n^{2k}$, up to constants depending on k as $n \to \infty$. We are interested in situations when the lower bound is (almost) attained. Define the k–th order polynomial $C_{n,k}$ by

$$C_{n,k}(x) := \begin{cases} P_k^{((n-1)/2,(n-1)/2)}(x) & \mathbb{K} = \mathbb{R}, \\ \\ P_{k/2}^{(n-1,0)}(2x^2 - 1) & \mathbb{K} = \mathbb{C}, \ k \text{ even}, \\ \\ x P_{(k-1)/2}^{(n-1,1)}(2x^2 - 1) & \mathbb{K} = \mathbb{C}, \ k \text{ odd}, \end{cases}$$

in terms of the Jacobi polynomials $P_l^{(\alpha,\beta)}$. The equality case $L_{\mathbb{K}}(n,k) = N_{\mathbb{K}}(n,k)$ was analyzed by Bannai [1], Hogger [5], see also Delsarte, Goethals, Seidel [2]. The result was

Proposition 3. *Let $2 \leq n, k \in \mathbb{N}$ and assume that $L_{\mathbb{K}}(n,k) = N_{\mathbb{K}}(n,k) =: N$. Then any cubature formula $(x_s, \lambda_s)_{s=1}^N$ with N nodes and weights of degree $2k$ satisfies $\lambda_s = 1/N$ for all s and for any $1 \leq s \neq t \leq N$, the number $|<x_s, x_t>|$ is a zero of the polynomial $C_{n,k}$. For $k = 2$, the coincidence of*

$$N = L_{\mathbb{K}}(n,2) = N_{\mathbb{K}}(n,2) = \begin{cases} n(n+1)/2 & \mathbb{K} = \mathbb{R}, \\ \\ n^2 & \mathbb{K} = \mathbb{C}, \end{cases}$$

is equivalent to the existence of N "equiangular" points $(x_s)_{s=1}^N$ with

$$|<x_s, x_t>| = \begin{cases} 1/\sqrt{n+2} & \mathbb{K} = \mathbb{R}, \\ \\ 1/\sqrt{n+1} & \mathbb{K} = \mathbb{C}, \end{cases} \tag{2.1}$$

which is the maximal possible number of points with (2.1).

The configuration of points (x_s) is thus very rigid in the case of Proposition 3. Various special examples are discussed e.g. in [2], [8]. The most spectacular is

$N_{\mathbb{R}}(24,5) = 98280$. For $k > 5$ or ($k > 2$ and $n > 24$) such configurations do not exist.

We now study a sequence of examples of cubature formulas of degree 4 and 5 on S^{n-1} where the number of nodes comes close to the lower bound and where they form "almost" equiangular point systems.

Proposition 4.　*Let*

　i) $n = p^l$, *p odd prime number, $l \in \mathbb{N}$ or*

　ii) $n = p^l + 1$, *p prime number, $l \in \mathbb{N}$.*

Then there is a complex cubature formula of degree 4 in n variables with N nodes where $N = n(n+1)$ in case i) and $N = n^2 + 1$ in case ii). The vectors x_s are of exponentional type as discussed at the end of Section 1. They yield real cubature formulas of degree 4 in $(2n)$ variables with $(3N)$ nodes, $N \sim 3/4\,(2n)^2$.

Proof.　　i) Let $s = (s_1, s_2) \in \{1, \dots, n\}^2 \equiv \{1, \dots, n^2\}$,

$$x_s := \frac{1}{\sqrt{n}} \left(\exp\left(\frac{2\pi i}{n}(s_1 j + s_2 j^2) \right) \right)_{j=1}^n \in S^{n-1}(\mathbb{C}).$$

Gaussian sum calculations yield "almost equiangularity"

$$| < x_s, x_t > | = \left\{ \begin{array}{cc} 1/\sqrt{n} & s_2 \neq t_2, \ \ s_1 \neq t_1 \\ 0 & s_2 = t_2, \ \ s_1 \neq t_1 \\ 1 & s = t \end{array} \right\}.$$

For $s = n^2 + j$, $1 \leq j \leq n$, take $x_s = e_j$. Hence $s = 1, \dots, n(n+1) = N$.

For all s, take $\lambda_s = 1/N$. One finds that (2) of Proposition 1 is satisfied;

$$\frac{1}{N} \sum_{t=1}^N | < x_s, x_t > |^4 = \frac{2}{N} = c_{n2} = \int_{S^{n-1}(\mathbb{C})} |y_1|^4 \, d\sigma_n(y)$$

holds for any $s \in \{1, \ldots, N\}$.

ii) For $n = p^l + 1$ and $M = n^2 - n + 1$, there exist integers $0 \leq d_1 < \ldots < d_n < M$ such that all numbers $1, \ldots, M - 1$ occur as residues mod M of the $M - 1 = n(n-1)$ differences $(d_i - d_j)$,

$$\{(d_i - d_j)(M) \mid i \neq j\} = \{1, \ldots, M - 1\}, \tag{2.2}$$

see [6]. For $s = 1, \ldots, M$ define

$$x_s := \frac{1}{\sqrt{n}} \left(\exp\left(\frac{2\pi i}{M} d_j s \right) \right)_{j=1}^{n} \in S^{n-1}(\mathbb{C}).$$

These vectors are equiangular (and *almost* the maximal number n^2 possible): an easy calculation using (2.2) shows that

$$| < x_s, x_t > | = \sqrt{n-1}/n, \ 1 \leq s \neq t \leq M.$$

For $s = M + j$, $1 \leq j \leq n$, let $x_s := e_j$. Then $s = 1, \ldots, M + n = n^2 + 1 = N$. For $1 \leq s \leq M$, let $\lambda_s = (n/(n+1))1/M$, for $M < s \leq N$, let $\lambda_s = (n/(n+1))1/n^2$. Then $\sum_{s=1}^{N} \lambda_s = 1$ and (2) of Proposition 1 holds,

$$\sum_{t=1}^{N} \lambda_t | < x_s, x_t > |^4 = \frac{2}{n(n+1)} = c_{n2}$$

for all $s = 1, \ldots, N$. $\qquad\qquad\square$

In the real case, the following examples exist, cf. [7], [12].

Proposition 5. *Let $n, m \in \mathbb{N}$ with i) $n = 4^m$, ii) $n = 2^m$. Then there exists a* real *cubature formula on S^{n-1} of degree 4 in n variables with N nodes where*

$$\text{i) } N = n(n+2)/2, \quad \text{ii) } N = n^2 + n.$$

Proof. Case i) is discussed in [7]. We consider ii) here. It was noticed by Schechtman using results of Hajela [4] and written down in [11]. Let $m, l \in \mathbb{N}$ and

$$r_j : \{-1, 1\}^{ml} \to \{-1, 1\}, \ (t_1, \ldots, t_{ml}) \mapsto t_j$$

denote the *Rademacher* functions and for $x \subseteq \{1, \ldots, ml\}$,

$$W_x := \prod_{j \in x} r_j$$

denote the 2^{ml} *Walsh* functions. If $m \geq \log_2 l + 2$, by Hajela [4] there exists a set $A \subseteq 2^{\{1,\ldots,ml\}}$ of 2^m subsets of $\{1, \ldots, ml\}$, $|A| = 2^m$, such that for any distinct $x_1, \ldots, x_{2j} \in A$ and $1 \leq j \leq l$

$$W_{x_1} \ldots W_{x_j} \neq W_{x_{j+1}} \ldots W_{x_{2j}}.$$

Hence for all such $x_1, \ldots, x_{2j}$, $1 \leq j \leq l$,

$$\sum_{t \in \{-1,1\}^{ml}} W_{x_1}(t) \ldots W_{x_{2j}}(t) = 0. \tag{2.3}$$

The proof relies on a finite field construction in the vector space $\mathbb{F}_{2^m}$ over $\mathbb{F}_2$. Take $l = 2$, $A \subseteq 2^{\{1,\ldots,2m\}}$, $|A| = 2^m$ to get for scalars $(a_x)_{x \in A} \subseteq \mathbb{R}$, using (2.3),

$$\Big\| \sum_{x \in A} a_x W_x \Big\|_4^4 = \sum_{t \in \{-1,1\}^{2m}} \Big| \sum_{x \in A} a_x W_x(t) \Big|^4$$

$$= \sum_{x_1,\ldots,x_4 \in A} a_{x_1} a_{x_2} a_{x_3} a_{x_4} \sum_{t \in \{-1,1\}^{2m}} W_{x_1}(t) W_{x_2}(t) W_{x_3}(t) W_{x_4}(t)$$

$$= n^2 \left(3 \sum_{x \neq y \in A} a_x^2 a_y^2 + \sum_{x \in A} a_x^4 \right)$$

$$= n^2 \left(3 \left(\sum_{x \in A} a_x^2 \right)^2 - 2 \left(\sum_{x \in A} a_x^4 \right) \right).$$

Thus

$$\|(a_x)_{x \in A}\|_2^4 = \frac{1}{3n^2} \Big\| \sum_{x \in A} a_x W_x \Big\|_4^4 + \frac{2}{3} \|(a_x)_{x \in A}\|_4^4.$$

Hence an isometric imbedding of l_2^n into $l_4^{n^2+n}$ can be given by

$$(a_x)_{x \in A} \mapsto \left(\frac{1}{\sqrt[4]{3}\,\sqrt{n}} \sum_{x \in A} a_x W_x, (a_x)_{x \in A} \right). \qquad \qquad \square$$

The cubature formulas of degree 4 studied in Proposition 4 and 5 have *positive* (and mostly almost equal) weights with the asymptotically minimal order of nodes $N = O(n^2)$. For $k = 3$, the minimal possible order of N is $O(n^3)$. We only know of such formulas when *some* of the weights are negative.

Proposition 6. *Let $n \in \mathbb{N}$ be a prime power, $N = n^3 + 2n^2 - n - 1$. Then there is a complex cubature formula of degree 6 with N nodes, $(N - n)$ positive and n negative weights on $S^{n-1}(\mathbb{C})$.*

Proof. Similarly as in the proof of Proposition 4, we use the B_3–sequences: by [6] there exist $1 \leq d_1 < d_2 \ldots < d_n \leq M := n^3 - 1$ such that $(d_{j_1} + d_{j_2} + d_{j_3})(M)$ are distinct for all $1 \leq j_1 \leq j_2 \leq j_3 \leq n$. This is also proved by a finite field construction. Take

$$x_s = \frac{1}{\sqrt{n}} \left(\exp\left(\frac{2\pi i}{M} d_j s \right) \right)_{j=1}^n \in S^{n-1}(\mathbb{C}); \quad s = 1, \ldots, M.$$

Then for $x = (x_j)_{j=1}^n \in \mathbb{C}^n$

$$\sum_{s=1}^M |<x, x_s>|^6$$

$$= \frac{1}{n^3} \sum_{\substack{j,k,l \\ o,p,q}} x_j \overline{x}_k x_l \overline{x}_o x_p \overline{x}_q \sum_{s=1}^M \exp\left(\frac{2\pi i}{M}[(d_j + d_l + d_p) - (d_k + d_o + d_q)]t \right)$$

$$= \frac{M}{n^3} \left(6 \sum_{\substack{(j,k,l)\,all \\ disjoint}} |x_j|^2 |x_k|^2 |x_l|^2 + 9 \sum_{j \neq k} |x_j|^4 |x_k|^2 + \sum_j |x_j|^6 \right)$$

$$= \frac{M}{n^3} \left(6 \left(\sum_j |x_j|^2 \right)^3 - 9 \sum_{j \neq k} |x_j|^4 |x_k|^2 - 5 \sum_j |x_j|^6 \right).$$

For the second equality one analyzes the equality case $d_j + d_l + d_p = d_k + d_o + d_q$. To the M vectors, we add the $2n(n-1)$ vectors $x_s = \frac{1}{\sqrt{2}}(e_j + i^\alpha e_k)$ for $1 \leq j < k \leq n$, $0 \leq \alpha \leq 3$, which gives a total amount of $L = M + 2n(n-1)$

vectors x_s. Since

$$\sum_{j<k}\sum_{\alpha=0}^{3}|x_j+i^\alpha x_k|^6 = 4(n-1)\sum_j |x_j|^6 + 36\sum_{j\neq k}|x_j|^4\,|x_k|^2,$$

$$\frac{n^3}{M}\sum_{s=1}^{M}|<x,x_s>|^6 + \sum_{s=M+1}^{L}|<x,x_s>|^6 - (n-6)\sum_{j=1}^{n}|<x,e_j>|^6$$

$$= 6\left(\sum_{j=1}^{n}|x_j|^2\right)^3. \tag{2.4}$$

The Euclidean norm is thus represented by $N = L + n$ vectors with L positive and n negative (for $n > 6$) sixth powers. This is (3) of Proposition 1 for $k = 3$. $\qquad\square$

In the **real** case, for $n = 4^m$, probably an analogous fact as in Proposition 6 holds, by using the codes of Delsarte, Goethals $DG(m,d)$, $d = (m-2)/2$ exhibited in Mac Williams, Sloane [10, p. 461, Theorem 19]. These generalize the Kerdock code.

Let $k \geq 2$. If (3) of Proposition 1 holds with $\lambda_s > 0$,

$$x \mapsto c_{nk}^{-1/2k}(<x,x_s>)_{s=1}^{N}$$

defines an isometric imbedding $i : l_2^n \hookrightarrow L_p^N(\lambda)$ for $p = 2k$. If the mapping $P : L_2^N(\lambda) \to L_2^N(\lambda)$ is the orthogonal projection onto $i(l_2^n)$, we find for any sequence $\zeta \in L_p^N(\lambda)$, $p = 2k \geq 2$,

$$\|P\zeta\|_{L_p^n(\lambda)} = \frac{c_{nk}^{1/2k}}{c_{n1}^{1/2}}\|P\zeta\|_{L_2^n(\lambda)}$$

$$\leq \frac{c_{nk}^{1/2k}}{c_{n1}^{1/2}}\|\zeta\|_{L_2^n(\lambda)} \leq \frac{c_{nk}^{1/2k}}{c_{n1}^{1/2}}\|\zeta\|_{L_p^n(\lambda)}.$$

For fixed k, $\displaystyle\sup_{n\in\mathbb{N}}\frac{c_{nk}^{1/2k}}{c_{n1}^{1/2}} < \infty$. Hence the image space $i(l_2^n)$ in $L_p^N(\lambda)$ is well-complemented.

References

[1] E. Bannai: *On extremal finite sets in the sphere and other metric spaces*, London Math. Soc. Lect. Notes Ser. **131** (1986), 13–38.

[2] P. Delsarte, J. M. Goethals, J. J. Seidel: *Spherical codes and designs*, Geom. Dedicata **6** (1977), 363–388.

[3] J. M. Goethals, J. J. Seidel: *Cubature formulas, polytopes and spherical designs*, in: *The Geometric Vein*, Coxeter Festschrift, Springer 1981, 203–218.

[4] D. Hajela: *Construction techniques for some thin sets in duals of compact Abelian groups*, Ann. Inst. Fourier **36** (1986), 137–166.

[5] S. G. Hogger: *t–designs in projective spaces*, Europ. J. Combin. **3** (1982), 233–254.

[6] H. Halberstam, K. Roth: *Sequences*, Springer, 1982.

[7] H. König: *Isometric imbeddings of Euclidean spaces into finite-dimensional l_p–spaces*, Banach Center Publ. **34** (1995), 79–87.

[8] Yu. Lubich, L. Vaserstein: *Isometric imbeddings between classical Banach spaces, cubature formulas, and spherical designs*, Geom. Dedicata **47** (1993), 327–362.

[9] V. Milman: *A few observations on the connections between local theory and some other fields*, Lect. Notes Math. **1317** (1988), 283–289.

[10] F. Mac Williams, N. Sloane: *The theory of error–correcting codes II*, North Holland, 1977.

[11] L. Rabinovitch: *Explicit isometric and almost isometric imbeddings*, M. Sc. Thesis, Weizmann Institute, Rehovot, Israel, 1977.

[12] B. Reznick: *Sums of even powers of real linear forms*, Memoirs Amer. Math. Soc. **96** (1992), no. 463.

Address:
HERMANN KÖNIG
Mathematisches Seminar
Universität Kiel
D–24098 Kiel
Germany

Numerical Calculation of Spherical Designs

Ulrike Maier

Dedicated to Professor Dr. M. Reimer on the occasion of his 65th birthday

Abstract

Spherical designs play an important role in several geometric problems on the sphere and for multivariate quadrature formulae with equal weights. Although these designs are known to exist for arbitrary dimension and some attempts for their construction have been undertaken, no general analytic method for their calculation is known yet. Even the exact number of points for these designs is still unknown. In this paper an approach for the numerical calculation of spherical designs using methods of multiobjective optimization is described.

1 Introduction

In applications, for instance in optics or astrophysics, there is a need for quadrature formulae with a high degree of exactness that are easily evaluated. This leads to the concept of Gaussian quadratures which in the multivariate case are closely related to quadrature formulae with equal weights ([5], [22]). The points for equally weighted quadrature formulae are known as *spherical t–designs* ([8], [13]).

Since the number of points in Gaussian formulae has to coincide with the dimension of a corresponding space of polynomials, it can be shown that these formulae – and in the notation of spherical designs so–called *tight* spherical designs – rarely exist in the multivariate case ([3]–[5], [22], [25]).

Although at first sight this result might be discouraging, one can ask if it is nevertheless possible to find point systems that enable quadratures with equal weights where the number of points is not restricted to the dimension of the polynomial space any more.

This leads to the problem of finding spherical t–designs. The existence of such point systems was proved by Seymour, Zaslavsky in 1984 ([24]). Unfortunately, the proof of this result is not constructive. For the number of points only a lower bound and conjectures for upper bounds are known ([8], [19]).

Advances in Multivariate Approximation; W. Haußmann, K. Jetter and M. Reimer (eds.)
Mathematical Research, Vol. 107, pp. 213–226, ISBN 3-527-40236-5
© WILEY-VCH, Berlin 1999

There are many attempts for the construction of spherical designs. Hong [18] in 1982 characterized spherical designs for the unit circle S^1; Mimura [20] in 1990 found a construction for spherical 2–designs using special monomial basis functions of degree 2. Bajnok [1,2] constructed spherical designs by chosing designs on the interval [−1,1] and on circles and from that built product spherical designs. His point systems have numbers that are far from being minimal. Finally, Reznick in 1995 [23] attacked the question of existence and nonexistence of spherical 5–designs in a constructive way.
All these constructions have in common that they give results for very special situations but cannot be used as a general method to obtain spherical t–designs.

So it is natural to try a numerical calculation of these point systems in order to get a better idea of the distribution of these points on the sphere.
Grabner and Tichy [15] have presented some calculations in 1993. They exploited a special property of spherical designs (see property (v) in Theorem 2.2) and used a Monte Carlo method and a gradient method to calculate point systems on the sphere.
The most convincing numerical approach so far is an approach of Hardin and Sloan [16, 17] who have not only presented a program known as **gosset** but have also given a table with group classifications for their resulting point systems for point numbers $N \leq 100$. They used the same property as Grabner and Tichy and remarked in [17] that the computation was terminated when "... no further improvements were found after several months of computing." Hardin and Sloan in their calculation used a Hookes and Reeves *pattern search* optimizer. For $N \leq 25$ and $N \in \{27, 30, 32, 48\}$ the existence of spherical designs can also be found algebraically.

In this article a different numerical approach for the computation of spherical designs using methods of optimization (and especially multiobjective optimization) is described. Section 2 contains a definition of spherical designs and a summary of some of their properties. In Section 3 a short introduction to multiobjective optimization is given which in Section 4 is applied to the problem of the computation of spherical designs. Finally, in Section 5 some numerical results are presented.
In this article the following notation is used. For $r \in \mathbb{N}$ the unit sphere in $\mathbb{R}^r$ is denoted by S^{r-1}. $\mathbb{P}_k^r(S^{r-1})$ denotes the linear space of restrictions of polynomials of degree $\leq k$ in r variables to the sphere. $\mathbb{H}_k^{*r}(S^{r-1})$ is the linear space of spherical harmonics, i.e. homogeneous, harmonic polynomials of

degree k restricted to the sphere whereas $\mathbb{H}_k^r(S^{r-1})$ is the space of harmonic polynomials on the sphere. The reproducing kernel of $\overset{*}{\mathbb{H}}_k^r(S^{r-1})$ is given by

$$G_k(x,y) = \mu(S^{r-1})^{-1} \cdot C_k^{(r-2)/2}(x \cdot y), \qquad (1.1)$$

where $\mu(S^{r-1})$ is the surface measure of the sphere S^{r-1}, $C_k^{(r-2)/2}$ denotes the *Gegenbauer polynomial* of degree k with index $(r-2)/2$ and $x \cdot y$ denotes the usual scalar product of the vectors x and y.

2 Spherical Designs

The definition of spherical t–designs was introduced by Delsarte, Goethals and Seidel in 1977 ([8]).

Definition 2.1. ([8])
Let $X \subset S^{r-1}$. X is a spherical t–design *if*

$$\frac{1}{|X|} \sum_{x \in X} f(x) = \frac{1}{\mu(S^{r-1})} \int_{S^{r-1}} f(\xi) d\mu(\xi)$$

for every $f \in \mathbb{P}_t^r(S^{r-1})$. Here, $|X|$ is the number of points of the set X.

Spherical designs can be characterized by several equivalent properties, some of which are summarized in the next theorem.

Theorem 2.2. ([13])
For a finite nonempty subset X of S^{r-1}, $|X| = N$, the following conditions are equivalent.

(i) *X is a spherical t-design.*

(ii) *$|X|^{-1} \sum_{x \in X} h(x) = \mu(S^{r-1})^{-1} \int_{S^{r-1}} h(\xi) d\mu(\xi)$ for all $h \in \mathbb{P}_k^r(S^{r-1})$, for $k = t$ and $k = t - 1$.*

(iii) *$\sum_{x \in X} f(x) = 0$, for all $f \in \overset{*}{\mathbb{H}}_k^r(S^{r-1})$, $k = 1, \ldots, t$.*

(iv) *$\sum_{x,y \in X} G_k(x,y) = 0$, for $k = 1, \ldots, t$, G_k as in (1.1).*

(v) *$N^{-2} \sum_{x,y \in X} (x \cdot y)^k = \begin{cases} 0 & k \text{ odd} \\ \frac{1 \cdot 3 \cdots (k-1)}{r(r+2)\ldots(r+k-2)} & k \text{ even} \end{cases}$
holds for $k = t$ and $k = t - 1$.*

Theorem 2.2 suggests the idea of exploiting some of the properties of spherical designs to get efficient algorithms. Grabner and Tichy as well as Hardy and Sloan have used property (v) for their numerical approaches. In the following, property (iii) of Theorem 2.2 will be used to obtain spherical designs numerically.

Since the basis functions for the space of spherical harmonics can be given explicitly, property (iii) can be used directly for the numerical calculation of spherical designs. Unfortunately, it is not possible to solve the resulting systems of equations exactly because algebraic packages using Groebner techniques are only able to handle very low numbers of variables.

So the aim is to get approximations of the solutions by using optimization techniques. The nonlinear optimization problem has as constraints the nonlinear relations $x_i \in S^{r-1}$ $(1 \le i \le N)$ which can be written as $\|x_i\|_2 = 1$. A formulation of the objective function involving spherical coordinates leads to an optimization problem with simple box constraints and $(r-1)N$ instead of rN unknowns for cartesian coordinates. For $r = 3$ this means that one has to deal with $2N$ unknowns. In this case with

$$x(\phi, \psi) = (\cos(\phi)\sin(\psi), \sin(\phi)\sin(\psi), \cos(\psi))$$

an orthonormal basis of $\overset{*}{\mathrm{I\!H}}{}^{3}_{k}(S^2)$ is given by

$$\tilde{C}_{k0}(x(\phi,\psi)) \;=\; \sqrt{\frac{2k+1}{4\pi}} \cdot P_k(\cos\psi), \tag{2.1}$$

$$\tilde{C}_{k\nu}(x(\phi,\psi)) \;=\; \sqrt{\frac{(k-\nu)!}{(k+\nu)!} \cdot \frac{2k+1}{2\pi}} \cdot \cos\nu\phi \cdot \sin^\nu\psi \cdot P^{(\nu)}_k(\cos\psi), \tag{2.2}$$

$$\tilde{S}_{k\nu}(x(\phi,\psi)) \;=\; \sqrt{\frac{(k-\nu)!}{(k+\nu)!} \cdot \frac{2k+1}{2\pi}} \cdot \sin\nu\phi \cdot \sin^\nu\psi \cdot P^{(\nu)}_k(\cos\psi). \tag{2.3}$$

Here, P_k are the Legendre–polynomials and $P^{(\nu)}_k$ denote their ν–th derivative, $\nu = 1,\ldots,k$. For arbitrary dimension r, an analogous basis for $\overset{*}{\mathrm{I\!H}}{}^{r}_{k}(S^{r-1})$ can e.g. be found with the aid of Theorem 9.10 in [21]. The polynomials P_k then have to be substituted by the reproducing kernel G_k of $\overset{*}{\mathrm{I\!H}}{}^{r}_{k}(S^{r-1})$ (see (1.1)).

Inserting (2.1)–(2.3) in property (iii) of Theorem 2.2 now leads to the following system of $2k+1$ nonlinear equations in the variables $x_1 = x(\phi_1, \psi_1), \ldots, x_N = x(\phi_N, \psi_N)$.

$$\sum_{j=1}^{N} \tilde{C}_{k0}(x_j) = 0, \tag{2.4}$$

$$\sum_{j=1}^{N} \tilde{C}_{k\nu}(x_j) = 0, \quad \nu = 1, \ldots, k, \tag{2.5}$$

$$\sum_{j=1}^{N} \tilde{S}_{k\nu}(x_j) = 0, \quad \nu = 1, \ldots, k. \tag{2.6}$$

Also with optimization methods some problems arise since standard optimization techniques only deal with <u>one</u> objective function in contrast to $2k+1$ in the equations (2.4)–(2.6). This leads to the concept of *multiobjective* or *multicriteria* optimization, still a field of intensive research nowadays.

3 Multiobjective Optimization

As mentioned above, in multiobjective optimization problems with several objective functions are handled. In more detail one has the following:
Given a vector of objectives $F(x) = \{F_1(x), F_2(x), \ldots, F_m(x)\}$, $x \in X \subset \mathbb{R}^r$, the task is to

$$\text{minimize} \quad F(x) \quad \text{subject to} \quad \left\{ \begin{array}{ll} G_i(x) = 0, & i = 1, \ldots, m_e, \\ G_i(x) \le 0, & i = m_e + 1, \ldots, m, \\ x_l \le x \le x_u \end{array} \right.$$

where m_e denotes the number of equality constraints.

One of the difficulties arising is that there is no unique solution to the problem if any of the components of $F(x)$ are competing. In multiobjective optimization, this effect is known as *noninferiority* or *Pareto optimality* ([6], [7]). More exactly, let

$$X := \{x \in \mathbb{R}^r \,|\, G_i(x) = 0, i = 1, \ldots, m_e,$$

$$G_i(x) \le 0, i = m_e + 1, \ldots, m, \ x_l \le x \le x_u\}$$

and
$$Y := \{y \in \mathbb{R}^m \,|\, y = F(x) \text{ subject to } x \in X\}.$$

Definition 3.1. *A point x^* is a <u>noninferior solution</u> if for some neighborhood of x^* there does not exist a Δx such that $x^* + \Delta x \in X$ and*

$$
\begin{aligned}
F_i(x^* + \Delta x) &\leq F_i(x^*), \quad i = 1, \dots, m, \\
F_j(x^* + \Delta x) &< F_j(x^*), \quad \text{for some} \quad j \in \{1, \dots, m\}.
\end{aligned}
$$

The aim of multiobjective optimization is to find such noninferior solutions. Many strategies are known for attacking this problem (see e.g. the references in [6]). One of the obvious methods is to create a single objective function from the vector of objective functions in order to make standard optimization methods applicable to the problem.

For the socalled *weighted sum method* the problem is changed to

$$
\operatorname*{minimize}_{x \in X} F(x) = \sum_{i=1}^{m} w_i\, F_i^2(x)\,.
$$

This method has the disadvantage that certain solutions may not be accessible if the set Y is not convex ([7]).

Another typical approach is the socalled *ε–constrained method*. Here, one of the components of the vector of objective functions is chosen as primary objective F_p, whereas the other components are regarded as constraints. In that way the problem becomes

$$
\operatorname*{minimize}_{x \in X} \; F_p(x) \quad \text{subject to} \quad F_i \leq \varepsilon_i, \quad i = 1, \dots, m, \quad i \neq p,
$$

with some constants ε_i. This can give solutions that have not been detected by the weighted sum method but it also has a serious disadvantage. It is a difficult task to prioritize the objectives and to select suitable bounds ε_i for the constraints. (For a geometric explanation of the mode of operation of this kind of algorithms, see [6], [7].)

In trying to avoid these disadvantages and to exploit the multivariate structure of the underlying optimization problem, one of the most promising approaches

so far is the so–called *goal attainment method* ([6], [12]).
In addition to the objective vector $F(x) = \{F_1(x), F_2(x), \ldots, F_m(x)\}$ a vector
of goals $F^* = \{F_1^*, F_2^*, \ldots, F_m^*\}$ and a vector of weight coefficients
$w = \{w_1, w_2, \ldots, w_m\}$ are chosen. Then the problem is reformulated to

$$\text{minimize} \quad \gamma \quad \text{such that} \quad F_i(x) - w_i\gamma \leq F_i^*, \quad i = 1, \ldots, m.$$
$$\gamma \in \mathrm{I\!R}, x \in X$$

Here the weight coefficients control the relative degree of under– or over–
achievement of the goals. The objectives have to be continuous and it may
happen that one only gets local solutions, a disadvantage which is shared with
most of the commonly used optimization methods.

4 Application to Spherical Designs

The problem of numerically calculating spherical designs has been attacked
with the above mentioned goal attainment method applied to the nonlinear
system of equations (2.4)–(2.6). The objective vector $F(x)$ consists of the
$2k+1$ sums given by (2.4)–(2.6) and the goals F_i^*, $i = 1, \ldots, 2k+1$, are chosen
to be zero.

4.1 Practical Aspects

In each step of the algorithm, Gegenbauer polynomials have to be evaluated.
Since the recurrence relation of the Gegenbauer polynomials is known to be numerically very stable, these evaluations are done with the aid of the recurrence
relation

$$(\mu + 1)C_{\mu+1}^\lambda(\xi) - 2(\mu + \lambda)\xi C_\mu^\lambda(\xi) + (\mu + 2\lambda - 1)C_{\mu-1}^\lambda(\xi) = 0, \qquad (4.1)$$

$C_{-1}^\lambda := 0$, $C_0^\lambda := 1$ (see e.g. [21]). Furthermore, for the derivatives in (2.5) and
(2.6) the following formulae are used:

$$\frac{d}{d\xi}C_\mu^0(\xi) = 2C_{\mu-1}^1, \qquad \frac{d}{d\xi}C_\mu^\lambda(\xi) = 2\lambda C_{\mu-1}^{\lambda+1} \quad \text{for} \quad \mu \in \mathrm{I\!N}, \lambda \neq 0.$$

For equation (2.4) the parameters in (4.1) have to be $\lambda = (r-2)/2$ and $\mu = k$
whereas in the equations (2.5) and (2.6) the parameters are $\lambda = (r-2)/2 + \nu$
and $\mu = k - \nu$ for $\nu = 1, \ldots, k$.

The evaluation of the Gegenbauer polynomials via (4.1) and the evaluation of the equations (2.4)–(2.6) has been written as Matlab code and has been combined with the goal attainment method (see Section 3) which is available in Matlab as program `attgoal`. It has turned out that the weights w_i, $i = 1, \ldots, 2k + 1$, and the number of point evaluations have to be chosen very differently depending on the starting points. In principle, the weights w_i can be chosen arbitrarily, but after some experiments the choices mentioned in Section 4.2 turned out to be reasonable. There might be other good choices for the weights (and the number of function evaluations). The accuracy that can be obtained for (2.4)–(2.6) depends very much on the choice of the weights.

An introductory comparison of `attgoal` with other Matlab procedures for standard optimization (`fmins`, `fminu`, `constr`) and the program `fsolve` for the solution of nonlinear equations showed that `attgoal` for the given problem of computing spherical designs is nevertheless far superior to the other procedures implemented in Matlab.

4.2   Starting Points

First, as starting systems, points are chosen which minimize the potential function

$$
E(x_1, \ldots, x_N) = \begin{cases}
\displaystyle\sum_{i,j=1}^{N} \log(\|x_i - x_j\|_2), & r = 2, \\
\displaystyle\sum_{i,j=1}^{N} \frac{1}{\|x_i - x_j\|_2^{r-2}}, & r \geq 3.
\end{cases}
\tag{4.2}
$$

For $r = 3$ such point systems are available from an earlier work of the author with J. Fliege ([10], [11]). To these starting systems `attgoal` is applied. The parameters for `attgoal` are chosen to be $w_i = 1/500$ for the weights and $options(14) = 50$ for the number of function evaluations. The number of function evaluations is increased by fifty every additional 10 points whereas the weights w_i then are doubled each time. The zeros in (2.4)–(2.6) are approximated within an accuracy of 10^{-10}.

Starting systems of the second chosen type are uniform random points which are produced by the random generator `rand` of Matlab. In contrast to the approach of Hardin and Sloan, no symmetry is required for the points. In this case the computation of designs is undertaken in two phases:

In Phase I, only a rough approximation of a global optimum is computed. This is done by a stochastic optimization method (a standard *simulated annealing* approach for continuous global optimization problems) in order to diminish the problem of sticking in a local optimum of the objective vector. In Phase II, the previously computed approximation is refined with a highly accurate non-linear local optimization method (the goal attainment method implemented in Matlab). As has been observed by several authors, stochastic optimization methods approach a global optimum very rapidly, but converge very slowly once they are in a certain neighbourhood of an optimum (see e.g. [9]). The change to another, locally faster method should therefore result in an overall better performance.

For the first phase the public domain code of Goffe, Ferrier and Rogers [14] for simulated annealing has been translated into Matlab code. Here, as starting temperature $T = 5.0$ is chosen and the temperature is reduced after $5N$ steps. To get the function value in each step, the `attgoal` algorithm has been incorporated in this program because the simulated annealing algorithm works with <u>one</u> objective function only. In this phase the weights are chosen to be $w_i = 1/100$, $i = 1, \ldots, 2k + 1$.

In the second phase in the beginning the weights are chosen to be $w_i = 1/100$, $i = 1, \ldots, 2k+1$, and the number of function evaluations $options(14)$ are set to $options(14) = 12 \cdot N$. The weights are doubled each time the number of points increases by five. The accuracy is first set to 10^{-8}. In an additional run the weights are changed to $w_i = 100 \cdot N$ whereas the number of function evaluations is also changed to $options(14) = 100 \cdot N$ and the accuracy is changed to 10^{-10}. The interplay between the choice of the weights and the number of function evaluations has turned out to be the most difficult part in the calculations.

5   Numerical Results

The optimization codes for the optimal potential points were run with MAT-LAB V on a Sun SPARCstation 20 with 64MB main memory and SunOS 5.5.1 operating system. For uniform random starting points the computations were done on a Sun UltraSparc2 with 256 MB main memory and 200MHz under the same operating system. Computation times were obtained with the Matlab `tic/toc` command.

In the case where minimal points of the potential function (4.2) were taken as starting points, spherical designs were calculated for $N = 1, \ldots, 100$ and $N = (m + 1)^2$, $m = 10, \ldots, 26$, so the maximum number of points calculated so far is 729. The computation times in this case are still moderate with about 6 hours for the case of 729 points.

In contrast to this, the Simulated Annealing code is very time consuming. For $N = 50$ the calculation of spherical designs in the first phase already needed more than 18 hours. Therefore, only point numbers up to $N = 50$ have been inspected so far. The second phase for these points only needs a few minutes (about 1 minute for $N = 50$).

5.1   Results for Minimal Potential Points

Figures 1 and 2 show results for the minimal points of the potential function as starting systems. Figure 1 shows the deviation from zero in (2.4)–(2.6) for the starting systems (upper two figures) and the end systems after application of `attgoal` (lower two figures). Note the different scaling for the starting and end points. For the minima of the potential function in the worst case the deviation from zero is about $7 \cdot 10^{-3}$, whereas most errors are within an order of 10^{-3} to 10^{-7}. For the minima of the design equations (2.4)–(2.6) in the worst case the deviation from zero is less than 10^{-7} whereas in the other cases the errors are of order 10^{-9} to 10^{-13}.

For $k = 1$, i.e. 1–designs, the points gained by `attgoal` approximate the zeros in (2.4)–(2.6) up to six orders better than the point systems which minimize the potential function.

For $k > 1$ the results for the potential points are sometimes better.

t	optimal potential points	attgoal points
1	$N \in \{1, \ldots, 100\}$	$N \in \{1, \ldots, 100\}$
2	$N \in \{4, 6, 12, 16, 22, 24, 28, 32, 40, 44, 48, 72\}$	$N \in \{4, 6, 16, 22, 28, 32\}$
3	$N \in \{6, 12, 24, 32, 44, 48, 72\}$	$N \in \{6, 32\}$
4	$N \in \{12, 32, 72\}$	$N \in \{32\}$
5	$N \in \{12, 32, 72\}$	$N \in \{32\}$
6	$- - -$	$- - -$

Table 1: point numbers $N \leq 100$ where t–designs occur

Table 1 lists a comparison of the point numbers where higher designs occur for starting and end points. The main reason for the differences of the starting and end points seems to be based on the fact that for the time being the systems of equations (2.4)–(2.6) are attacked for $k = 1$, only. A combination of the equations for several k, especially for neighbouring k, can probably avoid this. This will be investigated in some further research.

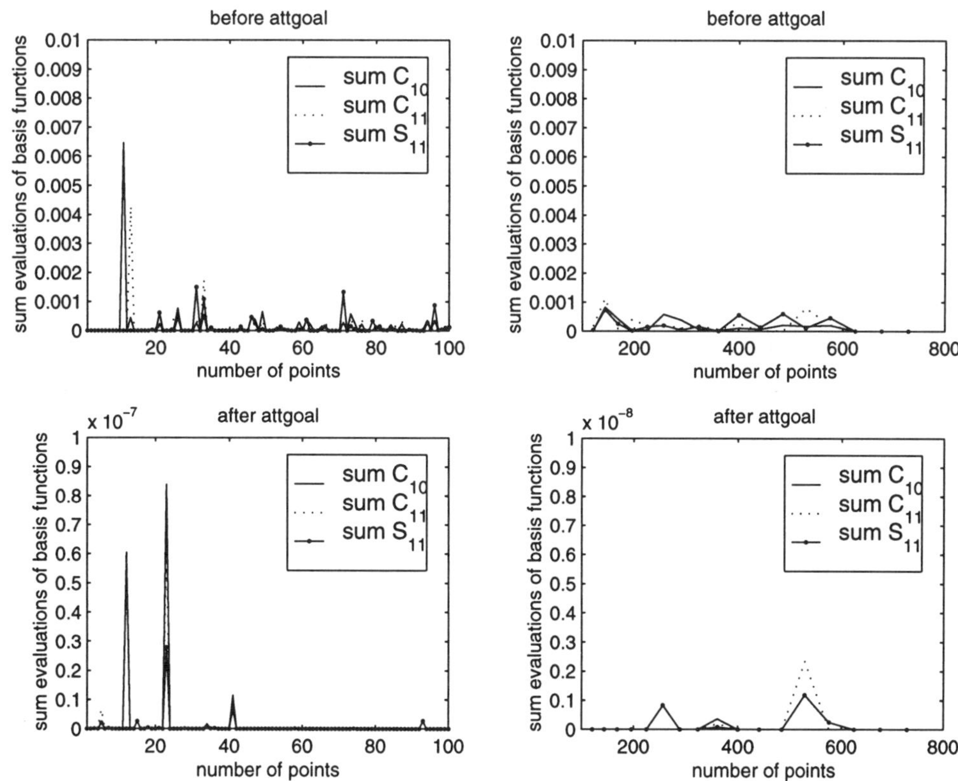

Figure 1: deviation from zero for the points minimizing the potential function and for the points after `attgoal`

Figure 2 shows typical distributions of spherical designs on the sphere.
The numerical results show that the points minimizing the potential function are good starting points. They are already nearly spherical designs. This result confirms a theoretical result of Korevaar and Meyers [19] who proved that optimal point systems for the potential function asymptotically coincide with spherical designs. (For details concerning the calculation of points that minimize the potential function see [10], [11], [26]).

5.2 Results for Uniform Random Points

For uniform random starting points, t–designs for $t = 1$ were calculated for $N = 1, \ldots, 50$ with the two–phase approach described in Section 4.2. After the first phase, simulated annealing, the deviation from zero in (2.4) − (2.6) lies

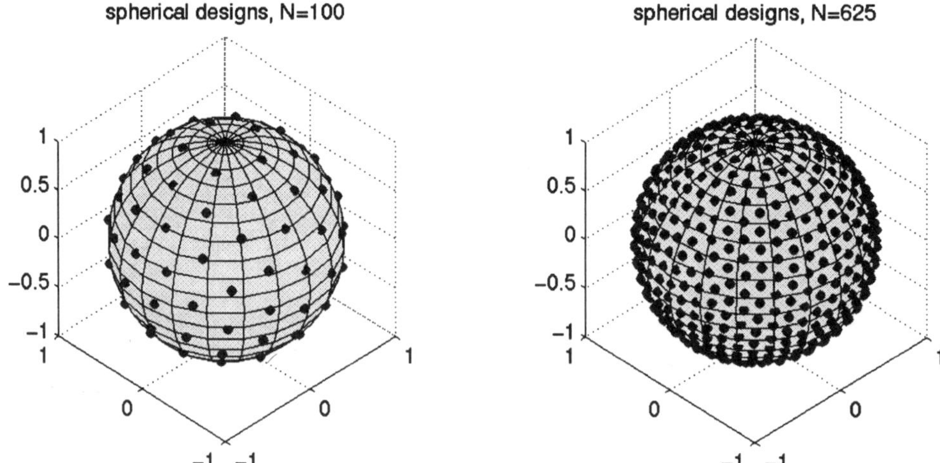

Figure 2: spherical designs gained by `attgoal` from minimal potential points

between 0.1 and 5.0. In contrast to this, for the resulting points the deviation from zero is of order 10^{-8} up to 10^{-13}. Nevertheless, the resulting points are distributed very irregularly on the sphere .

For randomly distributed starting points no higher designs, i.e. designs for $t > 1$, can be found. Another interesting phenomenon sometimes occurring is that after the application of simulated annealing some points coincide. The algorithm `attgoal` in general is able to separate these points a little bit.
The results indicate that for $t = 1$ many (also irregularly distributed) point systems are 1–designs. This becomes clearer from the fact that property (2.4) just means that the points sum up to zero. Certainly, a more regular distribution of the points can be obtained if equations for $t = 2$ (or higher) are considered additionally. Probably, it is also recommendable to require some symmetry for the points. These aspects will be inspected in the future.

References

[1]   B. Bajnok: *Construction of spherical 4– and 5–designs*, Graphs and Combinatorics **7** (1991), 219–233.

[2]   B. Bajnok: *Construction of spherical t–designs*, Geom. Dedicata **43** (1992), 167–179.

[3]   E. Bannai, R. M. Damerell: *Tight spherical designs I*, J. Math. Soc. Japan **31** (1979), 199–207.

[4]   E. Bannai, R. M. Damerell: *Tight spherical designs II*, J. London Math. Soc. (2) **21** (1980), 13–30.

[5]   L. Bos: *Some remarks on the Fejér–problem for Lagrange interpolation in several variables*, J. Approx. Theory **60** (1990), 133–140.

[6]   M. A. Branch, A. Grace: *Optimization Toolbox, User's Guide*, The Mathworks Inc. 1996.

[7]   I. Das, J. Dennis: *A closer look at drawbacks of minimizing weighted sums of objectives for Pareto set generation in multicriteria optimization problems*, Structural Optimization **14**(1) (1997), 63–69.

[8]   P. Delsarte, J. M. Goethals, J. J. Seidel: *Spherical codes and designs*, Geom. Dedicata **6** (1977), 363–388.

[9]   V. Fabian: *Simulated annealing simulated. Approximation theory and applications*, Comput. Math. Appl. **33** (1997), 81–94.

[10]   J. Fliege, U. Maier: *Charge distribution of points on the sphere and corresponding cubature formulae*, in: *Multivariate Approximation: Recent Trends and Results*, W. Haußmann et al. (eds.), 147–159, Akademie–Verlag, Berlin 1997.

[11]   J. Fliege, U. Maier: *The distribution of points on the sphere and corresponding cubature formulae*, IMA J. Numer. Anal., to appear.

[12]   F. W. Gembicki: *Vector Optimization for Control with Performance and Parameter Sensitivity Indices*, Ph. D. Thesis, Case Western Reserve Univ., Cleveland, Ohio, 1974.

[13]   J. M. Goethals, J. J. Seidel: *Spherical designs*, Proc. Symp. Pure Mathematics **34** (1979), 255–272.

[14]   W. L. Goffe, G. D. Ferrier, J. Rogers: SIMANN code, Electronically distributed as `ftp://netlib2.cs.utk.edu/opt/simann.f`

[15]   P. J. Grabner, R. F. Tichy: *Spherical designs, discrepancy and numerical integration*, Math. Comp. **60** (201), (1993), 327–336.

[16]   R. H. Hardin, N. J. A. Sloane: *McLaren's improved snub cube and other new spherical designs in three dimensions*, Discr. Comput. Geom. **15** (1996), 429–441.

[17]   R. H. Hardin, N. J. A. Sloane: *A new approach to the construction of optimal designs*, J. Statistical Planning and Inference **37** (1993), 339–369.

[18]   Y. Hong: *On spherical t–designs in* $\mathbb{R}^2$, Europ. J. Combinatorics **3** (1983), 255–258.

[19]   J. Korevaar, J. L. H. Meyers: *Spherical Faraday cage for the case of equal point charges and Chebyshev-type quadrature on the sphere*, Integral Transforms and Special Functions **1**(2) (1993), 105–117.

[20]   Y. Mimura: *A construction of spherical 2–designs*, Graphs and Combinatorics **6** (1990), 369–372.

[21]   M. Reimer: *Constructive Theory of Multivariate Functions*, BI Wissenschaftsverlag, Mannheim–Wien–Zürich 1990.

[22]   M. Reimer: *On the existence-problem of Gauss-quadrature on the sphere*, in: *Approximation by Solutions of Partial Differential Equations*, B. Fuglede et al. (eds.), 169–184, Kluwer, Dordrecht 1992.

[23]   B. Reznick: *Some constructions of spherical 5–designs*, Linear Alg. Appl. **226–228** (1995), 163–196.

[24]   P. D. Seymour, T. Zaslavsky: *Averaging sets: a generalization of mean values and spherical designs*, Adv. Math. **52** (1984), 213–240.

[25]   I. Sloan: *Interpolation and Hyperinterpolation on the Sphere*, in: *Multivariate Approximation: Recent Trends and Results*, W. Haußmann et al. (eds.), 255–268, Akademie-Verlag, Berlin 1997.

[26]   Y. Zhou: *Arrangement of Points on the Sphere*, Ph. D. Thesis, Tampa, Florida, 1995.

Address:

ULRIKE MAIER
Fachbereich Mathematik
Universität Dortmund
D–44221 Dortmund
Germany

On Bivariate Spline Spaces

Günther Nürnberger and Frank Zeilfelder

*Dedicated to Professor Dr. Manfred Reimer on the
occasion of his 65th birthday*

Abstract

We briefly describe some results on the dimension, the approximation
order and on interpolation of spaces of splines defined on triangulations,
and state some open problems.

1   Main Results and Problems

We consider spline spaces of the following type. Let Δ be a regular triangulation of a simply connected polygonal domain Ω. The *space of splines of degree q and smoothness r* (with respect to Δ) is defined by $S_q^r(\Delta) = \{s \in C^r(\Omega) : s|_T \in \Pi_q, T \in \Delta\}$. Here, $\Pi_q = \text{span}\ \{x^i y^j : i + j \leq q\}$ denotes the *space of bivariate polynomials of total degree q*.

In the literature, results are known on the dimension, the approximation order and on interpolation of $S_q^r(\Delta)$ for the case when Δ is a rectangular partition with one or two diagonals (see the survey [18] and the references therein). On the other hand, the spaces $S_q^r(\Delta)$ have a much more complex structure if Δ is an arbitrary triangulation. In this note, we briefly describe some recent developments concerning properties of such spaces, in particular for $r = 1$, and state some open problems.

We first consider the case $r = 1$. For $q \geq 5$ the dimension of $S_q^1(\Delta)$ was determined by Morgan–Scott [15] and for $q = 4$ by Alfeld–Piper–Schumaker [2] (using graph theory). On the other hand, the dimension is not known for quadratic and cubic splines. It was shown by Morgan–Scott that, surprisingly, the dimension of $S_2^1(\Delta)$ depends on the exact geometry of the triangulation Δ. G. Strang (1973) conjectured that $\dim S_3^1(\Delta) = 2V_I + 3V_B + \sigma + 1$. Here, V_I (respectively V_B) is the total number of interior (respectively boundary) vertices of Δ, and σ is the number of singular vertices of Δ.

Advances in Multivariate Approximation; W. Haußmann, K. Jetter and M. Reimer (eds.)
Mathematical Research, Vol. 107, pp. 227–230, ISBN 3-527-40236-5
© WILEY-VCH, Berlin 1999

Problem 1. Determine the dimension of $S_q^1(\Delta)$ for $q = 2, 3$.

It is known that for $q \geq 5$ the space $S_q^1(\Delta)$ has *optimal approximation order*, i.e., $\mathrm{dist}(f, S_q^1(\Delta)) \leq K h^{q+1}$, $f \in C^{q+1}(\Omega)$, where h is the maximal diameter of the triangles in Δ and K is a constant independent of h. On the other hand, for $q \in \{2, 3, 4\}$, the approximation order of $S_q^1(\Delta)$ is only known for special triangulations Δ (Sablonnière [19], Lai [12], Chui–Hong [5]).

Problem 2. Determine the approximation order of $S_q^1(\Delta)$ for $q \in \{2, 3, 4\}$.

We now consider interpolation. A set $\{z_1, \ldots, z_N\} \subset \Omega$, where $N = \dim S_q^1(\Delta)$, is called *Lagrange interpolation set* for $S_q^1(\Delta)$, if for each $f \in C(\Omega)$, a unique spline $s_f \in S_q^1(\Delta)$ exists such that $s_f(z_i) = f(z_i), i = 1, \ldots, N$. If also derivatives are involved, we speak of a *Hermite interpolation set*.
For arbitrary triangulations Δ, Morgan–Scott [15] described a Hermite–Birkhoff interpolation scheme for $S_q^1(\Delta), q \geq 5$. Later, for arbitrary Δ, Davydov–Nürnberger [7] developed an algorithm for constructing Lagrange and Hermite interpolation sets for $S_q^1(\Delta), q \geq 4$. In the case $q = 4$, it may be necessary to slightly modify Δ in exceptional cases. The question arises if the modifications on exceptional cases can be avoided.

Problem 3. Construct interpolation sets for $S_4^1(\Delta)$, where Δ is an arbitrary triangulation.

The construction of interpolation sets for quadratic and cubic spline spaces is even more complex since not even the dimensions of these spaces are known. Methods of constructing interpolation sets for $S_q^1(\Delta), q \geq 2$, were developed in [16], [17] for the following classes of triangulations Δ: First, for convex quadrangulations Δ with two diagonals ([16]) and second, triangulations Δ constructed for given points in the plane in a suitable way ([17]). Moreover, interpolation sets for $S_3^1(\Delta)$ were constructed in [8] (see also [9]) for triangulations Δ which consist of nested polygons whose vertices are connected by line segments. The problem arises to construct interpolation sets for more general classes of triangulations.

Problem 4. Construct interpolation sets for $S_q^1(\Delta), q = 2, 3$, for general classes of triangulations Δ.

We finally discuss the case when $r \geq 2$. The dimension of $S_q^r(\Delta)$ was determined by Alfeld–Schumaker [1] for $q \geq 4r + 1$, and for $4r \geq q \geq 3r + 2$ by Hong [11]. The problem arises to determine the dimension of $S_q^r(\Delta)$ in the case when $q < 3r + 2$.

Results on the approximation order were given by the following authors. For $q \geq 3r + 2$, it was proved by de Boor–Höllig [3] (by using abstract methods), Chui–Hong–Jia [6], and Lai–Schumaker [13] (by using quasi interpolation methods) that $S_q^r(\Delta)$ has optimal approximation order. A Hermite–Birkhoff interpolation scheme which yields optimal approximation order was constructed in [10]. On the other hand, for $q < 3r + 2$ de Boor–Jia [4] proved that $S_q^r(\Delta)$ does not have optimal approximation order, in general. Moreover, it was shown by Lai–Schumaker [14] that $S_{3r}^r(\Delta)$ has optimal approximation order, where Δ is a quadrangulation with diagonals. In general, the problems arise to determine the approximation order of $S_q^r(\Delta)$ and to construct interpolation sets for $q < 3r + 2$.

References

[1] P. Alfeld, L. L. Schumaker: *On the dimension of bivariate spline spaces of smoothness r for degree $d \geq 4r + 1$*, Constr. Approx. **3** (1987), 189–197.

[2] P. Alfeld, B. Piper, L. L. Schumaker: *An explicit basis for C^1 quartic bivariate splines*, SIAM J. Numer. Anal. **24** (1987), 891–911.

[3] C. de Boor, K. Höllig: *Approximation order of smooth bivariate pp functions*, Math. Z. **197** (1988), 343–363.

[4] C. de Boor, Q. Jia: *A sharp upper bound on the approximation order of smooth bivariate pp functions*, J. Approx. Theory **72** (1993), 24–33.

[5] C. K. Chui, D. Hong: *Swapping edges of arbitrary triangulations to achieve the optimal order of approximation*, SIAM J. Numer. Anal. **34** (1997), 1572–1582.

[6] C. K. Chui, D. Hong, Q. Jia: *Stability of optimal-order approximation by bivariate splines over arbitrary triangulations*, Trans. Amer. Math. Soc. **347** (1995), 3301–3318.

[7] O. Davydov, G. Nürnberger: *Interpolation by C^1 splines of degree $q \geq 4$ on triangulations*, submitted.

[8] O. Davydov, G. Nürnberger, F. Zeilfelder: *Interpolation by cubic splines on triangulations*, in: *Approximation Theory IX*, C. K. Chui and L. L. Schumaker (eds.), Vanderbilt University Press, Nashville 1998.

[9] O. Davydov, G. Nürnberger, F. Zeilfelder: *Interpolation by splines on triangulations,* in: *Approximation Theory,* M. Buhmann and M.W. Müller (eds.), 1998, to appear.

[10] O. Davydov, G. Nürnberger, F. Zeilfelder: *Bivariate spline interpolation with optimal approximation order,* submitted.

[11] D. Hong: *Spaces of bivariate spline functions over triangulations,* Approx. Theory Appl. **7** (1991), 56–75.

[12] M. J. Lai: *Scattered data interpolation and approximation using bivariate C^1 piecewise cubic polynomials,* Computer Aided Geometric Design **13** (1996), 81–88.

[13] M. J. Lai, L. L. Schumaker: *On the approximation power of bivariate splines,* Adv. Comp. Math. **9** (1998), 251–279.

[14] M. J. Lai, L. L. Schumaker: *On the approximation power of splines on triangulated quadrangulations,* SIAM J. Numer. Anal. **36** (1999), 143–159.

[15] J. Morgan, R. Scott: *A nodal basis for C^1 piecewise polynomials of degree $n \geq 5$,* Math. Comp. **29** (1975), 736–740.

[16] G. Nürnberger, F. Zeilfelder: *Spline interpolation on convex quadrangulations,* in: *Approximation Theory IX,* C. K. Chui and L. L. Schumaker (eds.), Vanderbilt University Press, 259–266, Nashville 1998.

[17] G. Nürnberger, F. Zeilfelder: *Interpolation by spline spaces on classes of triangualtions,* preprint.

[18] G. Nürnberger, O. V. Davydov, G. Walz, F. Zeilfelder: *Interpolation by bivariate splines on crosscut partitions,* in: *Multivariate Approximation and Splines,* G. Nürnberger et al. (eds.) Internat. Ser. Numer. Math. **125**, 189–204, Birkhäuser, Basel–Boston–Berlin 1997.

[19] P. Sablonnière: *Error bounds for Hermite interpolation by quadratic splines on an α triangulation,* IMA J. Numer. Anal. **7** (1987), 495–508.

Address:

GÜNTHER NÜRNBERGER, FRANK ZEILFELDER
Fakultät für Mathematik und Informatik
University of Mannheim
D–68131 Mannheim
Germany

Spherical Polynomial Approximations: A Survey

Manfred Reimer

Abstract

This is a survey and synopsis on selected topics in spherical polynomial approximation. So different problems, such as approximation of the Radon inverse, best approximation to monomials and magnitude of the corresponding coefficient functionals, interpolation, hyperinterpolation and quadrature with positive weights, can be treated and sometimes solved by reproducing kernel techniques or by polynomials furnished by them. The methods work also in the construction of point distributions with a large minimal distance. These results are compared with theoretical results of geometry.

1 Introduction

Univariate polynomial approximations have been investigated in the past intrinsically and with a long chain of important and beautiful results. Some of them can be lifted to the multivariate case by product arguments. Our interest, however, is directed to more genuine multivariate problems, such as approximations on the unit ball B^r or on the unit sphere S^{r-1}. Throughout this work we assume $r \in \mathbb{N} \setminus \{1\}$. Both problems are closely related by the maps

$$S^{r-1} \ni x = (x_1, ..., x_r) \mapsto \overline{x} = (x_1, ..., x_{r-1-s}) \in B^{r-1-s}, \qquad (1.1)$$

$$\overline{C}(S^{r-1}) \ni F \mapsto \overline{F} \in C(B^{r-1-s}), \qquad (1.2)$$

where $\overline{C}(S^{r-1})$ consists of all $F \in C(S^{r-1})$ not depending on the last $s + 1$ arguments, $s \in \{0, ..., r-1\}$, and where $\overline{F}(\overline{x}) := F(x)$. For these functions

$$\int_{S^{r-1}} F(x)dx = \omega_s \int_{B^{r-1-s}} \overline{F}(\overline{x})(1 - |\overline{x}|)^{\frac{s-1}{2}} d\overline{x} \qquad (1.3)$$

holds with $\omega_{\kappa-1} := measure(S^{\kappa-1})$. Hence, if we define the inner product

$$< F, G > = < F, G >_r := \int_{S^{r-1}} F(x)G(x)dx, \qquad (1.4)$$

Advances in Multivariate Approximation; W. Haußmann, K. Jetter and M. Reimer (eds.)
Mathematical Research, Vol. 107, pp. 231–252, ISBN 3-527-40236-5
© WILEY-VCH, Berlin 1999

on $C(S^{r-1})$ then this induces an inner product on B^{r-1-s} in a natural way.

For topological reasons, it is often more convenient to treat a problem on the sphere than on the ball. For instance, it is favorable to calculate quadrature rules for B^3 numerically as a problem on S^3, i.e. in $\mathbb{R}^4$, [14], or to identify orthogonal projections onto a space of polynomial restrictions to B^r with partial sums of a Laplace series on a higher dimensional sphere, hence bringing down to the ball, all the theory known there, as L. Koschmieder did first with respect to *Cesàro's* summation method, [13]. This idea can be used, for instance, in the solution of the inversion problem of *Radon's* transform, the basic problem of mathematical tomography, [19].

So there is no need of a further eulogy for the sphere as a subject of mathematical research. In what follows, we give a synopsis of several selected topics.

2   Special Polynomials

The classical univariate orthogonal polynomials, for instance the Gegenbauer polynomials C_μ^λ, can be defined by the use of well known generating functions. This method has been generalized in the definition of multivariate polynomials $V_m^{(s)}$ by Appell and Kampé de Fériet [1]. We shall discuss the use of the families $V_m^{(+1)}$ and $V_m^{(-1)}$ in approximation theory, though in a different normalisation. But first let us define the following spaces of real $r-$variate polynomials with degree $\mu \in \mathbb{N}_0$ (where $|m| := m_1 + ... + m_r$ for $m \in \mathbb{N}_0^r$):

$$\begin{aligned}
\mathbb{P}_\mu^r &:= \{P \text{ polynomial with } (total) \text{ degree } \mu \}, \\
\mathbb{P}_m^r &:= \{P \in \mathbb{P}_\mu^r\colon (single) \text{ degree in } x_j \text{ is } m_j\}, \quad m \in \mathbb{N}_0^r, |m| = \mu, \\
\overset{*}{\mathbb{P}}{}_\mu^r &:= \{P \in \mathbb{P}_\mu^r\colon P \text{ homogeneous with degree } \mu\}, \\
\mathbb{H}_\mu^r &:= \{P \in \overset{*}{\mathbb{P}}{}_\mu^r\colon P \text{ harmonic}\}.
\end{aligned}$$

In addition, if V is any space of functions and if D is contained in their set of definition, then $V(D)$ denotes the space of their restrictions to D.

Now assume $Q_\mu \in \mathbb{P}_\mu^1 \setminus \mathbb{P}_{\mu-1}^1$ satisfies $Q_\mu(-\xi) = (-1)^\mu Q_\mu(\xi)$. Then

$$|t|^\mu Q_\mu \left(\frac{tx}{|t|}\right) = \sum_{|m|=\mu} A(x) t^m$$

defines a multivariate polynomial family $A_m \in \mathbb{P}^r_m$, $m \in \mathbb{N}_0$. These polynomials generate an additional family of polynomials $\overset{*}{A}_m \in \overset{*}{\mathbb{P}}{}^r_\mu$, given by

$$\overset{*}{A}_m (x) := |x|^\mu A_m \left(\frac{x}{|x|} \right).$$

In particular, the families P_m and R_m may be defined by

$$|t|^\mu \overset{*}{\Gamma}_\mu \left(\frac{tx}{|t|} \right) = \sum_{|m|=\mu} P_m(x)t^m, \tag{2.1}$$

$$|t|^\mu G_\mu \left(\frac{tx}{|t|} \right) = \sum_{|m|=\mu} R_m(x)t^m, \tag{2.2}$$

where

$$\overset{*}{\Gamma}_\mu (xy) = \frac{1}{\omega_{r-1}} C_\mu^{\frac{r}{2}}(xy) \text{ and } G_\mu(xy) = \frac{2\mu + r}{r\omega_{r-1}} C_\mu^{\frac{r-2}{2}} (xy) \tag{2.3}$$

are the reproducing kernel functions with respect to the inner product (1.4) of the spaces $\overset{*}{\mathbb{P}}{}^r_\mu (S^{r-1})$ and $\overset{*}{\mathbb{H}}{}^r_\mu (S^{r-1})$, respectively. As indicated, the P_m and R_m correspond with the polynomials $V_m^{(+1)}$ and $V_m^{(-1)}$, respectively, and they have remarkable properties. For instance, the corresponding homogeneous polynomials $\overset{*}{R}_m$ are harmonic, i.e. $\overset{*}{R}_m \in \mathbb{H}^r_\mu$ for $|m| = \mu$, [17], while together with the monomials $M_n(x) = x^n$ the $\overset{*}{P}_m$ form a biorthogonal system, i.e.

$$<\overset{*}{P}_m, M_n> = \delta_{m,n} \tag{2.4}$$

holds for $|m| = |n|$, with several important implications.

3 Orthogonal Projections and Summation

From (2.1), (2.2) and (2.3) it follows that $P_m(x) = const \cdot R_{(m,0,0)}(x, *, *)$ is valid. Because $\overset{*}{R}_{(m,0,0)} \in \overset{*}{\mathbb{H}}{}^{r+2}_\mu$, $\mu = |m|$, this implies, together with (1.3), with $s = 1$ and $r + 2$ instead of r, that the P_m are orthogonal with $\mathbb{P}^r_{\mu-1}$ w.r.t. $(F, G) := \int_{B^r} F(x)G(x)dx$, so that

$$\mathbb{P}^r_\mu = \bigoplus_{\nu=0}^{\mu} V_\nu, \tag{3.1}$$

$$V_\nu := span\{P_n : |n| = \nu\} \text{ for } \nu \in \mathbb{N}_0,$$

is an orthogonal decomposition. For $F \in C(B^r)$ the orthogonal projections

$$\Omega_\mu : \; C(B^r) \to V_\mu(B^r), \tag{3.2}$$

$\mu \in \mathbb{N}_0$, give rise to the series

$$\sum_{\mu=0}^{\infty} F_\mu, \quad F_\mu := \Omega_\mu F, \tag{3.3}$$

which can be identified with a *Laplace* series in $r + 2$ dimensions. Therefore, if the matrix method $A = (a_{jk})_{j,k=0,1,\dots}$ sums Laplace series of continuous functions on S^{r+1} in the uniform norm, then

$$F = A - \sum_{\mu=0}^{\infty} F_\mu \tag{3.4}$$

also holds, now in the uniform norm on B^r. Koschmieder [13] was the first to use this idea in connection with *Cesàro's* method (C, k). Actually, by a result of Kogbetliantz [11], (C, k) satisfies our assumptions for $k \geq \frac{r+1}{2}$, the partial sum operators even becoming positive for $k \geq r + 1$.

Application to Tomography. Let R_0 denote *Radon's transform* as a mapping

$$R_0 : C(B^r) \to C(Z^r), \;\; Z^r := [-1, 1] \times S^{r-1}.$$

It can be normalized by defining $R := R_0/R_0(1)$. Then R maps polynomials onto polynomials, [17]. Now let $\tilde{C}_\mu^{\frac{r}{2}} := C_\mu^{\frac{r}{2}}/C_\mu^{\frac{r}{2}}(1)$. Then, for fixed $a \in S^{r-1}$,

$$[R \overset{*}{\Gamma}_\mu (a\cdot)](s,t) = \tilde{C}_\mu^{\frac{r}{2}}(s) \overset{*}{\Gamma}_\mu (at) \tag{3.5}$$

holds for $(s,t) \in Z^r$ by an important result of Davison and Grünbaum, [6]. It follows from (2.1) that

$$(RP_m)(s,t) = \tilde{C}_\mu^{\frac{r}{2}}(s)P_m(t) \tag{3.6}$$

is valid for $|m| = \mu$, so that (3.4) implies

$$(R\,F)(s,t) = A - \sum_{\mu=0}^{\infty} \tilde{C}_\mu^{\frac{r}{2}}(s)F_\mu(t), \tag{3.7}$$

uniformly for $(s, t) \in Z^r$. This equation contains the whole information about *Radon's transform*. For instance, orthogonality of the Gegenbauer polynomials and reproduction arguments can be used to reconstruct F from (3.7) in the form

$$F(x) = \lim_{\mu \to \infty} \int_{Z^r} (R_0 F)(s, t) K_\mu^A(s, tx) d(s, t),$$

uniformly for $x \in B^r$, where K_μ^A is a bivariate kernel defined by A, which is a polynomial if A is row–finite, and especially a polynomial with degree μ in each of the arguments if A is *subdiagonal*, such as (C, k). For details see [17].

4   Best Approximations

Every polynomial can be written in the form

$$P(x) = \sum c_m(P) x^m.$$

We are interested in the magnitude of the coefficient functionals $c_m(\cdot)$ in different norms.

2–Norm: As $\overset{*}{\mathbb{P}}_\mu^r$ is isometrically isomorphic with $\overset{*}{\mathbb{P}}_\mu^r (S^{r-1})$, (1.4) defines an inner product on $\overset{*}{\mathbb{P}}_\mu^r$ itself. Let $\| \cdot \|_2$ be the norm induced by $< \cdot, \cdot >$. From (2.1), (2.4) it follows that

$$c_m(P) = < P, \overset{*}{P}_m >$$

holds for $P \in \overset{*}{\mathbb{P}}_\mu^r$ and $|m| = \mu$. This means that $\overset{*}{P}_m$ is the representer of the coefficient functional

$$c_m : \overset{*}{\mathbb{P}}_\mu^r \to \mathbb{R}.$$

Theorem 4.1 *Let $P \in \overset{*}{\mathbb{P}}_\mu^r$, $\|P\|_2 \leq 1$. Then $|c_m(P)| \leq \| \overset{*}{P}_m \|_2$ for $|m| = \mu$. Equality occurs if and only if $P = \pm \overset{*}{P}_m / \| \overset{*}{P}_m \|_2$. I.e. $\|c_m\|_2 = \| \overset{*}{P}_m \|_2$.*

The proof is standard, not so the value of $\|c_m\|_2$. Using the coefficients of

$$\overset{*}{\Gamma}_\mu(\xi) = \sum_{\nu=0}^{\lfloor\frac{\mu}{2}\rfloor} \alpha_{\mu-2\nu}\xi^{\mu-2\nu}$$

we can express it in the form

$$\|c_m\|_2^2 = \sum_{\substack{n\leq m \\ n\equiv m(2)}} \binom{|n|}{n}\left(\frac{|\frac{m-n}{2}|}{\frac{m-n}{2}}\right)^2 \alpha_{|n|},$$

see [17]. A similar result holds for the restriction of c_m to $\overset{*}{\mathrm{IH}}{}^r_\mu$, where $\overset{*}{R}_m$ occurs instead of $\overset{*}{P}_m$ as the representer.

Uniform Norm: It is well known that in the univariate case the Chebyshev polynomials T_μ are distinguished by the fact that they let the leading coefficient become maximum among all polynomials with uniform norm on B^1 not exceeding unity. An equivalent statement is that T_μ is best approximated in $\mathrm{IP}^1_{\mu-1}(B^1)$ by zero. A generalisation of this result to the space IP^2_m, $|m| = \mu$, was given first by Gearhart, [9], who proved that the polynomials

$$U_{m_1}(x_1)U_{m_2}(x_2) + U_{m_1-2}(x_1)U_{m_2-2}(x_2) = 2^{m_1+m_2}x_1^{m_1}x_2^{m_2} + TLD$$

(TLD = $terms\ of\ lower\ degree$) are best approximated by zero in the uniform norm in $\mathrm{IP}^2_{\mu-1}(B^2)$. Gearhart proved in addition that the best approximant is not uniquely determined.

The solution of Gearhart is somewhat of a product solution. So it is worth while to know that the polynomials

$$R_m(x) = c_m x^m + TLD, \quad m \in \mathrm{IN}^2_0,$$

are also best approximated by zero in $\mathrm{IP}^2_{\mu-1}(B^2)$, [17]. The proof uses the fact that $R_m(x)$, $x \in B^r$, occurs in (2.2) as the coefficient of a homogeneous polynomial with uniform norm $G_\mu(1)$ on B^2. By a result of Kellogg, [10], this implies

$$|R_m(x)| \leq \binom{\mu}{m}G_\mu(1)$$

with equality occurring on S^1, only. As $R_m(\cos\phi, \sin\phi)$ has the form $const \cdot \cos(\mu\phi + \alpha)$ the extreme points are known. They support an alternating $extremal\ signature$ w.r.t. $\mathrm{IP}^2_{\mu-1}(B^2)$ in the sense of $Rivlin$ and $Shapiro$, [23], with length 2μ, which finishes the proof. The result yields

Theorem 4.2 *Let* $P \in \mathbb{P}^2_m(B^2)$, $\|\mathrm{P}\|_\infty \leq 1$. *Then* $|c_m(P)| \leq 2^\mu$, $\mu = |m|$. *Equality holds for* $P = R_m/R_m(1)$, *implying* $\|c_m\|_\infty = 2^\mu$ *for* $|m| = \mu$.

Apart from constant factors, the bivariate polynomials R_m occur also as the remainder in some *Kergin interpolation* to the monomials with nodes in S^1, Bos [3]. This is an example where Kergin interpolants are best–approximating in the uniform norm.

Unfortunately, the result of *Theorem* 4.2 cannot be generalized to the case $r \geq 3$. For instance, if the monomial $x_1 x_2 x_3$ is best–approximated in $\mathbb{P}^3_2(B^3)$, then irrational coefficients occur, indicating that the remainder cannot be produced by a rational generating function, such as the R_m's are, see (2.2) again. So far, for $r \geq 3$ best approximation to monomials M_m in $\mathbb{P}_{|m|-1}(B^r)$ in the uniform norm is an *open problem*.

The reason for this could be sought in a more *complicated* structure of extremal signatures occurring in higher dimensions. But the contrary can also be true, as the following shows:

Example: If in $\overset{*}{\mathbb{P}}{}^r_\mu (S^{r-1})$ the (restriction of the) monomial $M_{\mu e_1}$ is to be best approximated in $span\{M_m : |m| = \mu, \ m \neq \mu e_1\}$, zero is a best approximant, and this can be decided by an extremal signature based on a *single* point, namely on e_1. Obviously, the most *simple* extremal signature occurs.

5 Interpolation and Quadrature

In what follows, let $\mathbb{P}$ denote one of the spaces $\mathbb{P}^r_\mu, \overset{*}{\mathbb{P}}{}^r_\mu$. Let $N := dim \, \mathbb{P} \left(S^{r-1}\right)$ and let $P(xy)$ be its reproducing kernel w.r.t. $< \cdot, \cdot >$. A family $T = \{t_1, ..., t_N\}$ of nodes $t_j \in S^{r-1}$ is a *fundamental system* if the corresponding evaluation functionals are linearly independent. In this case, the *Lagrange polynomials* $L_1, ..., L_N \in \mathbb{P}(S^{r-1})$ are defined by $L_j(t_k) = \delta_{j,k}$ for $j, k \in \{1, ..., N\}$. Together with the polynomials $P_k \in \mathbb{P}(S^{r-1})$, $P_k(\cdot) := P(t_k \cdot)$, $k = 1, ..., N$, they form a biorthonormal system,

$$< L_j, P_k > = \delta_{j,k},$$

$j, k \in \{1, ..., N\}$. It follows that

$$P(xy) = \sum_{j=1}^{N} P_j(x)L_j(y), \quad x, y \in S^{r-1}, \tag{5.1}$$

$$L_j = \sum_{k=1}^{N} < L_j, L_k > P_k, \tag{5.2}$$

$$P_k = \sum_{j=1}^{N} < P_k, P_j > L_j \tag{5.3}$$

hold, where the values of $< P_j, P_k >= P(t_j t_k) =: P_{j,k}$ can actually be calculated numerically. The same holds for the matrices

$$\mathbf{P} := (P_{j,k}), \qquad \mathbf{L} := (< L_j, L_k >) = \mathbf{P}^{-1}, \tag{5.4}$$

which are positive definite, with strong implications for the numerical evaluation of the interpolatory projection

$$C(S^{r-1}) \ni F \mapsto \Lambda F = \sum_{j=1}^{N} F(t_j)L_j \in \mathbb{P}(S^{r-1}), \tag{5.5}$$

which is connected with the interpolatory quadrature

$$\sum_{j=1}^{N} A_j F(t_j) = \int_{S^{r-1}} F(x)dx \quad \text{for} \quad F \in \mathbb{P}, \tag{5.6}$$

$A_j = \int_{S^{r-1}} L_j(x)dx$. With $A := (A_1, ..., A_N)'$, $E := (1, ..., 1)'$, the weights can be calculated safely from

$$\mathbf{P}A = E. \tag{5.7}$$

A measure for the size of Λ is the *Lebesgue constant* $\|\Lambda\|_\infty$.

Lebesgue–Constants. In finite dimensional rotation invariant spaces of spherical polynomials,

$$\|\Lambda\|_\infty \leq \frac{N}{\sqrt{\omega_{r-1}\lambda_1}} \tag{5.8}$$

holds if λ_1 is the smallest eigenvalue of $\mathbf{P}$, [17]. Besides, the orthogonal projection Π is a *minimal projection* in the uniform norm (generalisation of *Berman's* result, see Daugavet [5]). In case of the spaces $\mathbb{P}_\mu^r(S^{r-1})$, $r \geq 3$, and with $\Lambda = \Lambda_\mu^r$, $\Pi = \Pi_\mu^r$,

$$\|\Lambda_\mu^r\|_\infty \geq \|\Pi_\mu^r\|_\infty \sim \frac{2}{\pi^{\frac{3}{2}}} \cdot \frac{\Gamma(\frac{r-1}{4})\Gamma(\frac{r}{4})}{\Gamma(\frac{r-1}{2})^2} \cdot \mu^{\frac{r-2}{2}} \tag{5.9}$$

holds as $\mu \to \infty$, [5] again, so that pointwise convergence cannot take place in the uniform norm. The divergence of $\|\Pi_\mu^r\|_\infty$ was known before to Kogbetliantz [11, p. 141].

Despite of this more negative result, we may ask for favorable interpolation nodes at fixed degree. T is called *extremal* if $det(\mathbf{P})$ attains its maximum value. In this case

$$\|\Lambda\|_\infty \leq N, \tag{5.10}$$

$$\|L_1\|_\infty = \ldots = \|L_N\|_\infty = 1 \tag{5.11}$$

are valid. Extremal fundamental systems can be calculated by a *Remez − type* algorithm, [22]. Many numerical evaluations of the bound (5.8) have shown that $\|\Lambda\|_\infty$ is far overestimated by (5.10), [14], [22], [28]. To the extent that the numerical results were safe, all weights A_j have been found to be positive. In case of $\mu = 2$, positivity is proved for arbitrary r, [18].

In case of the space $\mathrm{I\!P}_\mu^3(S^2)$, where $N = (\mu + 1)^2$ holds, it is reasonable to display, alternatingly, $1, 3, \ldots, 2\mu + 1$ of the nodes (dimensions of the constituent spaces of spherical harmonics) equidistantly on certain $\mu + 1$ parallel circles on S^2, including one pole. Sündermann [28] proved, actually, that these nodes form a fundamental system, with favorable Lebesgue constants.

Such more or less arbitrary constructions are necessary, as the best which may be wanted, namely *Gauß nodes*, do not exist for $(r, \mu) \geq (3, 3)$, see Bannai and Damerell [2] and Bos [4].

Gauß Nodes with respect to $\mathrm{I\!P}_\mu^r(S^{r-1})$ are nodes whose interpolatory quadrature (5.6) is exact on $\mathrm{I\!P}_{2\mu}^r(S^{r-1})$. So a quadrature with $(r, \mu) \geq (3, 3)$ and this degree of exactness must have the form

$$QF = \sum_{j=1}^M a_j F(t_j), \tag{5.12}$$

where M is greater than the dimension $N = N_\mu^r$ of $\mathrm{I\!P}_\mu^r(S^{r-1})$, which is

$$N_\mu^r = \binom{\mu + r - 1}{r - 1} + \binom{\mu + r - 2}{r - 1}. \tag{5.13}$$

In special situations, the defect $M - N$ has been estimated from below by Yudin [31]. Möller's ideal theoretical method, [15], can be expected to yield further results. Even if the weights a_j are required to be positive, such quadratures do exist and can, for instance, be constructed recursively by means of univariate Gauß quadratures, Stroud [27]. The resulting formulae have positive weights and are supported by $M = M_\mu^r$ nodes, where constants $k_r > 0$ exist such that $M_\mu^r \leq k_r N_\mu^r$ holds for arbitrary $\mu \in \mathrm{I\!N}_0$. Improvements concern the value of

k_r. Formulae (5.12) with positive coefficients have received much attention in *hyperinterpolation*, see below. They also play an important role in the problem of *covering the sphere* by caps.

Stability. We may ask how dependent interpolation is with respect to the choice of the nodes. Actually, $det \, \mathbf{P}$ should not vanish close to an extremal fundamental system, for example. An answer can be given as follows: We can *average* $det \, \mathbf{P} = det \, \mathbf{P}(T)$ as a function of T. The average has the remarkably large value $\frac{N!}{\omega_{r-1}^N}$. If T_N is an extremal fundamental system, this yields

$$\frac{1}{\omega_{r-1}} \frac{N}{e} \sim \frac{1}{\omega_{r-1}} (N!)^{\frac{1}{N}} \leq (det \, \mathbf{P}(T_N))^{\frac{1}{N}} \leq \frac{1}{\omega_{r-1}} N$$

as $N = N_\mu^r$ tends to infinity. So the average is near to the maximum. For details see [20].

Covering the Sphere by Caps. A cap with center $t \in S^{r-1}$ and radius ϕ is a set

$$C(t,\phi) = \{x \in S^{r-1} |\, tx \geq \cos(\phi)\}, \quad 0 < \phi \leq \pi.$$

A spherical $\mu - design$ is a family $T = \{t_1, ..., t_M\}$ of nodes $t_j \in S^{r-1}$ which supports an equally weighted quadrature (5.12) which is exact on $\mathbb{P}_\mu^r(S^{r-1})$. Yudin [30] has shown that a spherical 2μ or $2\mu - 1$-design always gives rise to a covering of the sphere by the caps $C(t_j, \phi)$, $j = 1, ..., M$, where $\eta = \cos(\phi)$, $0 < \phi < \pi$, is the greatest root of the Gegenbauer polynomial $C_\mu^{\frac{r-2}{2}}$ or, what is the same, of G_μ. The result is not dependent on the weights being equal.

Theorem 5.1 *Let T support a quadrature with positive weights which is exact on* $\mathbb{P}_{2\mu-1}$. *Then*

$$S^{r-1} \subset \bigcup_{j=1}^{M} C(t_j, \phi).$$

Proof. We give a modified version of Yudin's proof under the weakened assumptions. Actually, for $x \in S^{r-1}$ we get

$$0 = \int_{S^{r-1}} G_\mu(tx) \frac{G_\mu(tx)}{tx - \eta} dt$$

(orthogonality), and applying the quadrature (5.12) we obtain

$$0 = \sum_{j=1}^{M} a_j \frac{G_\mu^2(t_j x)}{t_j x - \eta}.$$

Assume $t_j x < \eta$ holds for $j = 1, ..., M$. Then this is also true in a neighbourhood of x, implying $G_\mu(t_j y) = 0$ for $j = 1, ..., M$ and y in this neighbourhood. Finally connect x with t_1 by a great circle and move x on it towards this node. Then $t_1 x$ does not continue to be a root of G_μ, implying a contradiction. So, for arbitrary x there exists t_j with $t_j x \geq \eta$, which finishes the proof.

Hyperinterpolation. Let $\mathbb{P} := \mathbb{P}_\mu^r$, $P := \overset{*}{\Gamma}_\mu + \overset{*}{\Gamma}_{\mu-1}$ be the reproducing kernel of $\mathbb{P}(S^{r-1})$, $\Pi : C(S^{r-1}) \to \mathbb{P}(S^{r-1})$ the orthogonal projection, i.e.

$$(\Pi F)(x) = \int_{S^{r-1}} F(t)P(xt)dt, \quad x \in S^{r-1}.$$

Hyperinterpolation, introduced by Sloan [25] by means of a little–known orthonormal system, also arises if the integral in this formula is evaluated by a quadrature (5.12) which is exact on $\mathbb{P}_{2\mu}$. The *hyperinterpolation* operator $H : C(S^{r-1}) \to \mathbb{P}(S^{r-1})$ is defined in this case by

$$(HF)(x) := \sum_{j=1}^{M} a_j F(t_j) P_j(x) \quad , x \in S^{r-1}.$$

Obviously, H is exact on $\mathbb{P}_\mu^r(S^{r-1})$. Moreover, Sloan proved that if the weights of Q are positive, then $\|H\|_{(\infty,2)} \leq \sqrt{\omega_{r-1}}$. In this way, pointwise convergent sequences of hyperinterpolation operators can be constructed.

6   Nodes with a Great Minimal Distance

Hitherto the node systems investigated have been intended to form a fundamental system for some polynomial space, where applications to geometry, such as to a covering problem for the sphere, were not excluded. Node distributions which occur in the sequel need not have such a property, but are expected to be *uniformly displaced* on the sphere in a sense to be made more precise. A general principle of distributing M arbitrary nodes $t_1, ..., t_M$ on S^{r-1} is the following: Assume that the continuous function $\Phi : (0, 2] \to \mathbb{R}$ is monotonically decreasing. Then define the t_j's to be a solution of the optimisation problem

$$\sum_{j=0}^{M} \sum_{k=0}^{M} \Phi(|t_j - t_k|) \to \min !.$$

Several authors have used potentials in the definition of Φ, see e.g. [7], [12]. Their results are rather useful in the treatment of special problems, for instance in astrophysics, [26]. Now let us recall that $|x - y|^2 = 2(1 - xy)$ holds for $x, y \in S^{r-1}$. Therefore, a similar approach is to solve the problem

$$\sum_{j=0}^{M} \sum_{k=0}^{M} \Phi(t_j t_k) \to \ \min !, \tag{6.1}$$

but where Φ is now assumed to be continuous on the compact interval $[-1, +1]$ and to attain its maximum value at $\xi = 1$, only. A necessary condition that the minimum in (6.1) is attained is given by

$$\sum_{\substack{k=1 \\ k \neq j}}^{M} \Phi(t_j t_k) \leq \sum_{\substack{k=1 \\ k \neq j}}^{M} \Phi(x t_k), \tag{6.2}$$

holding for arbitrary $x \in S^{r-1}$ and $j = 1, ..., M$. Now, averaging these inequalities over S^{r-1} we obtain, with an arbitrary $t \in S^{r-1}$,

$$\sum_{\substack{k=1 \\ k \neq j}}^{M} \Phi(t_j t_k) \leq \frac{M - 1}{\omega_{r-1}} \cdot \int_{S^{r-1}} \Phi(xt) dx. \tag{6.3}$$

In what follows we need the reproducing kernel $P_\mu = \frac{1}{\omega_{r-1}} \{ C_\mu^{\frac{r}{2}} + C_{\mu-1}^{\frac{r}{2}} \}$ of $\mathbb{P}_\mu^r(S^{r-1})$, see e.g. [17]. Also let $M = N = N_\mu^r$ and $\tilde{P}_\mu := P_\mu / P_\mu(1)$, so that $\tilde{P}_\mu(1) = 1$. We discuss three choices of Φ:

Choice 1: $\Phi := P_\mu$. For arbitrary $t_1, \ldots, t_N \in S^{r-1}$ let $V \in \mathbb{P}_\mu^r(S^{r-1})$ be defined by $V := \sum_{k=1}^{N} \Phi(\cdot t_k)$. By the reproducing property of Φ we obtain now $0 \leq \|V - P(1)\|^2 = \|V\|^2 - N P(1)$ and

$$\|V\|^2 = \sum_j \sum_k \Phi(t_j t_k).$$

If the nodes form a μ–design, $V(x) = P(1)$ is valid and $\|V\|^2$ attains its minimum value w.r.t. the nodes. Conversely:

Theorem 6.1 *Assume $\|V\|$ attains its minimum value w.r.t. the nodes. Then either the nodes $t_1, \ldots, t_N$ form a spherical μ–design, or the corresponding evaluation functionals are linearly dependent.*

Proof. For $\mu = 0$ the statement is evident. So let $\mu \in \mathbb{N}$. By the assumptions, the nodes solve problem (6.1), so that (6.2) is valid. For $x \in S^{r-1}$, $V(x)$ may be represented in the form $V(x) = W(x)$, where

$$W(x) = \sum_{|m| \in \{\mu, \mu-1\}} c_m x^m,$$

see [17]. Because of (6.2), $V(x) - \Phi(xt_j)$ and hence $W(x) - \Phi(xt_j)$ is attaining its minimum value at $x = t_j$. Simultaneously, $\Phi(xt_j)$ is maximum, such that both functions satisfy *Lagrange's condition*. I.e. $(\operatorname{grad} W)(t_j) = \lambda_j t_j$ for $j = 1, \ldots, N$ holds for some $\lambda_j \in \mathbb{R}$, implying that

$$\sum_{|m| \in \{\mu, \mu-1\}} c_m [m_\nu t_j^{m-e_\nu+e_\kappa} - m_\kappa t_j^{m+e_\nu-e_\kappa}] = 0$$

for $j = 1, \ldots, N$ and arbitrary $\nu, \kappa \in \{1, \ldots, r\}$. Now we assume that the nodes form a fundamental system. Then we get

$$\sum_{|m| \in \{\mu, \mu-1\}} c_m [m_\nu x^{m-e_\nu+e_\kappa} - m_\kappa x^{m+e_\nu-e_\kappa}] = 0$$

for arbitrary x. Define $c_n := 0$ for $n \not\geq 0$. With $g := (\operatorname{grad} W)(x)$ this can be written in the form $g x^T = x g^T$. Multiplication from the right side by x yields $g = (gx) \cdot x$, i.e.

$$(\operatorname{grad} W)(x) = \lambda(x) \cdot x$$

holds with some real $\lambda(x)$ for arbitrary $x \in S^{r-1}$. Now it is easy to see that W must be constant along every great circle contained in S^{r-1}. Hence it must be constant everywhere, so that $V(x) = const$ is valid on S^{r-1}. Integration over S^{r-1} finally yields

$$V(x) = \frac{N}{\omega_{r-1}} \quad \text{for } x \in S^{r-1}.$$

Now let $F \in \mathbb{P}_\mu^r(S^{r-1})$. Using the reproducing property of $\Phi(\cdot t_k)$, we get

$$\int_{S^{r-1}} F(x)dx = \frac{\omega_{r-1}}{N} \int_{S^{r-1}} F(x)V(x)dx = \frac{\omega_{r-1}}{N} \sum_{k=1}^{N} F(t_k).$$

I.e. $t_1, \ldots, t_N$ form a $\mu - design$.

Choice 2: $\Phi := \tilde{P}_\mu^2$. In this case, inequality (6.3) can be evaluated easily, with the result

$$\sum_{\substack{k=1 \\ k \neq j}}^{N_\mu^r} \tilde{P}_\mu^2(t_j t_k) \leq 1 - \frac{1}{N_\mu^r}. \tag{6.4}$$

Though the matrix occurring is already diagonal dominant, this is just guaranteeing that the nodes are pairwise distinct.

Choice 3: $\Phi := \tilde{P}^4_\mu$. Now Φ is steeper at 1 and things change:

Theorem 6.2. *Let* $\Phi = \tilde{P}^4_\mu$, $M = N^r_\mu$. *Then the solutions of* (6.1) *satisfy for* $j = 1, ..., N^r_\mu$ *and with an arbitrary* $t \in S^{r-1}$

$$\sum_{\substack{k=1 \\ k \neq j}}^{N^r_\mu} \tilde{P}^4_\mu(t_j t_k) \leq A^r_\mu < 1, \tag{6.5}$$

where

$$A^r_\mu := P^{-3}_\mu(1) \cdot \int_{S^{r-1}} P^4_\mu(xt)dx,$$

$$\lim_{\mu \to \infty} A^r_\mu = A^r_\infty := 2^{4-r} \frac{\Gamma(r)^2}{\Gamma(\frac{r}{2})^4} \int_0^1 \int_0^\xi [(1-\xi^2)(1-\eta^2)]^{\frac{r-2}{2}} \eta^{r-1} d\eta d\xi < 1. \tag{6.6}$$

For the proof see [21]. It uses the *linearisation formulae* of *Rogers* and *Ramanujan*, [8], and an asymptotic representation of their coefficients.

It follows from *Theorem* 6.2 that $\tilde{P}_\mu(t_j t_k) \leq \sqrt[4]{A^r_\mu} < 1$ holds for $j \neq k$, so that the nodes are pairwise distinct. But we can draw further information on the minimal distance between them, as follows. Define ϕ^r_μ to be the smallest positive root of

$$\tilde{P}_\mu\left(\cos(\phi^r_\mu)\right) = \sqrt[4]{A^r_\mu}. \tag{6.7}$$

Then

$$\arccos(t_j t_k) \geq \phi^r_\mu \tag{6.8}$$

is valid for arbitrary pairs of different nodes. In order to determine the asymptotics of ϕ^r_μ we need the following lemma.

Lemma 6.3. *Assume the functions* $f : [a,b] \to \mathbb{R}$ *and* $f_\kappa : [a,b] \to \mathbb{R}$, $\kappa \in \mathbb{N}$, *are continuous and strictly monotonically decreasing, and assume*

$$\lim_{\kappa \to \infty} f_\kappa(x) = f(x)$$

holds for $x \in [a,b]$. *Also let*

$$\lim_{\kappa \to \infty} f_\kappa(x_\kappa) = f(\xi)$$

hold for some $\xi \in (a,b)$ and $x_\kappa \in [a,b]$, $\kappa \in \mathbb{N}$. Then

$$\lim_{\kappa \to \infty} x_\kappa = \xi.$$

Proof. By the assumptions, the sequence x_κ has at least one limit point in $[a,b]$. Assume $\eta \neq \xi$ is such a limit point, say $\eta < \xi$. Choose $\zeta \in (\eta, \xi)$. Then a subsequence x_{κ_ι} exists with $a \leq x_{\kappa_\iota} < \zeta$ and

$$\lim_{\iota \to \infty} x_{\kappa_\iota} = \eta.$$

By the monotonicity of the functions involved we obtain

$$f(\xi) = \lim_{\iota \to \infty} f_{\kappa_\iota}(x_{\kappa_\iota}) \geq \lim_{\iota \to \infty} f_{\kappa_\iota}(\zeta) = f(\zeta) > f(\xi),$$

which is contradictory. Similarly no limit point $\eta > \xi$ exists, i.e. ξ is the unique limit point.

Now we return to our original problem. Let $\nu := \frac{r-1}{2}$ and define the *normalized Bessel function*

$$Z_\nu(x) := \Gamma(\nu + 1) \left(\frac{2}{x} \right)^\nu J_\nu(x) \tag{6.9}$$

of the first kind, so that, together with $Z_\nu(0) = 1$,

$$\begin{aligned}
Z_\nu(x) &= \Gamma(\nu + 1) \sum_{k=0}^{\infty} (-1)^k \frac{1}{k! \Gamma(k + \nu + 1)} \left(\frac{x}{2} \right)^{2k}, \\
Z_\nu'(x) &= -\frac{x}{\nu + 1} Z_{\nu+1}(x)
\end{aligned}$$

hold. Let j_ν be the smallest positive root of J_ν. Then Z_ν is a strictly decreasing mapping of $[0, j_\nu]$ onto $[0, 1]$ as, by the *interlacing property* of the roots of the Bessel functions, Z_ν' does not vanish in this interval. In addition, recall that

$$\lim_{\mu \to \infty} \tilde{P}_\mu \left(\cos \frac{x}{\mu} \right) = Z_\nu(x) \tag{6.10}$$

uniformly on compact sets, and because $0 < A_\infty^r < 1$ we may define $c_r \in (0, j_\nu)$ to be the unique solution of

$$Z_\nu(c_r) = \sqrt[4]{A_\infty^r}. \tag{6.11}$$

Also let $c_\mu^r := \mu\,\phi_\mu^r$, so that (6.7) implies together with (6.6) that

$$\lim_{\mu\to\infty} \tilde{P}_\mu\left(\cos\left(\frac{1}{\mu}c_\mu^r\right)\right) = Z_\nu(c_r).$$ (6.12)

We are going to apply *Lemma* 6.3 to the functions

$$f := Z_\nu, \quad f_\mu := \tilde{P}_\mu\left(\cos\frac{\cdot}{\mu}\right), \quad \mu \in \mathbb{N}_0.$$

Now let $a := 0$ and choose b such that $c_r < b < j_\nu$. Then all these functions are defined and strictly monotonically decreasing on $[a, b]$ for $\mu > \mu_1$, while by (6.10), (6.12)

$$\lim_{\mu\to\infty} f_\mu(b) = f(b) < f(c_r) = \lim_{\mu\to\infty} f_\mu(c_\mu^r)$$

holds. This implies $a < c_\mu^r < b$ for $\mu \geq \mu_2 \geq \mu_1$, so that *Lemma* 6.3 can now be applied. So we obtain $c_\mu^r \to c_r$, i.e.

$$\phi_\mu^r \sim \frac{c_r}{\mu}, \quad \mu \to \infty.$$ (6.13)

For the first values of A_∞^r and c_r see *Table* 1.

In what follows we determine the asymptotics of c_r. By replacing in the integrand of (6.6) one factor η by ξ we get the estimate $A_\infty^r < B_r$, where B_r can be evaluated by formal integration. Together with Legendre's reduction formula for the gamma function we obtain

$$B_r = \frac{2}{\sqrt{\pi}} \cdot \frac{r-1}{r} \cdot \frac{[\Gamma(r)]^2 \cdot [\Gamma(\frac{r-1}{2})]^2}{[\Gamma(\frac{r}{2})]^3 \cdot \Gamma(\frac{3r-1}{2})}.$$

By inserting in (6.6) an additional factor $\xi\eta$ we get likewise $A_\infty^r > \frac{r-1}{3r-1}B_r$. This yields, by the use of Stirling's formula,

$$\lim_{\nu\to\infty}(A_\infty^r)^{\frac{1}{\nu}} = \lim_{\nu\to\infty}(B_r)^{\frac{1}{\nu}} = \frac{16}{27}.$$ (6.14)

r	A_∞^r	c_r	$2\,c_r/\pi$	c_r/r
2	0.6666667	0.7719	0.4914	0.3860
3	0.4596204	1.2265	0.7808	0.4088
4	0.3238095	1.6447	1.0471	0.4112
5	0.2314601	2.0465	1.3028	0.4093
6	0.1671662	2.4386	1.5525	0.4064
7	0.1216686	2.8243	1.7980	0.4035
8	0.0890881	3.2054	2.0407	0.4007
9	0.0655467	3.5831	2.2811	0.3981
10	0.0484167	3.9579	2.5197	0.3958
11	0.0358815	4.3306	2.7569	0.3937
12	0.0266661	4.7014	2.9930	0.3918
13	0.0198652	5.0706	3.2280	0.3900
14	0.0148298	5.4385	3.4622	0.3885
15	0.0110910	5.8052	3.6957	0.3870
16	0.0083083	6.1709	3.9285	0.3857
17	0.0062328	6.5356	4.1607	0.3844
18	0.0046819	6.8996	4.3924	0.3833
19	0.0035210	7.2629	4.6237	0.3823
20	0.0026507	7.6255	4.8545	0.3813
21	0.0019975	7.9875	5.0850	0.3804
22	0.0015066	8.3489	5.3151	0.3795
23	0.0011373	8.7099	5.5449	0.3787
24	0.0008591	9.0704	5.7744	0.3779
25	0.0006495	9.4304	6.0036	0.3772
26	0.0004913	9.7901	6.2326	0.3765
27	0.0003718	10.1494	6.4613	0.3759
28	0.0002816	10.5084	6.6899	0.3753
29	0.0002134	10.8671	6.9182	0.3747
30	0.0001618	11.2254	7.1463	0.3742
31	0.0001227	11.5835	7.3743	0.3737
32	0.0000931	11.9413	7.6021	0.3732
33	0.0000707	12.2989	7.8297	0.3727
34	0.0000537	12.6562	8.0572	0.3722
35	0.0000408	13.0134	8.2846	0.3718
∞	–	–	–	0.3490

Table 1

Now recall

$$\sqrt{\nu(\nu+2)} < j_\nu < \sqrt{2(\nu+1)(\nu+3)}, \tag{6.15}$$

Watson [29, pp. 485, 486]. In particular, this implies that $Z_\nu(\nu\xi)$ is a strictly monotonically decreasing function for $0 \le \xi \le 1$. Further,

$$J_\nu(\nu\xi) \sim \frac{\xi^\nu \exp(\nu\sqrt{1-\xi^2})}{\sqrt{2\pi\nu}(1-\xi^2)^{\frac{1}{4}}(1+\sqrt{1-\xi^2})^\nu}$$

holds for $\xi \in (0,1)$ as $\nu \to \infty$, [29, p. 227]. Inserting this in (6.9) and again using Stirling's formula, we obtain

$$\lim_{\nu\to\infty}(Z_\nu(\nu\xi))^{\frac{1}{\nu}} = f(\xi), \tag{6.16}$$

where

$$f(\xi) = \frac{2}{e} \cdot \frac{e^{\sqrt{1-\xi^2}}}{1+\sqrt{1-\xi^2}} \,.$$

It it evident that (6.16) is also valid for $\xi = 0$. For $\xi = 1$ the same follows from Cauchy's marvellous formula

$$J_\nu(\nu) \sim \frac{\Gamma(\frac{1}{3})}{2^{\frac{2}{3}}3^{\frac{1}{6}}\pi\nu^{\frac{1}{3}}},$$

$\nu \to \infty$, see Watson [29, p. 231], which actually yields

$$\lim_{\nu\to\infty}(Z_\nu(\nu))^{\frac{1}{\nu}} = \frac{2}{e}$$

by the definition (6.9). So (6.16) is valid on the whole interval $0 \le \xi \le 1$. Note that f is strictly monotonically decreasing on $[0,1]$ and that

$$f(1) < \sqrt[4]{\frac{16}{27}} < f(0).$$

Now define $\xi_0 \in (0,1)$ to be the unique solution of the equation $f(\xi_0) = \sqrt[4]{\frac{16}{27}}$, and let

$$f_\nu := (Z_\nu(\nu\cdot))^{\frac{1}{\nu}} \,.$$

Then f and the functions f_ν are defined and strictly monotonically decreasing on $[0,1]$, while we get with (6.11) and (6.14)

$$\lim_{\nu\to\infty} f_\nu\left(\frac{c_r}{\nu}\right) = f(\xi_0). \tag{6.17}$$

Together with (6.16), this yields

$$\lim_{\nu \to \infty} f_\nu(1) = f(1) < f(\xi_0) = \lim_{\nu \to \infty} f_\nu(\frac{c_r}{\nu}),$$

implying that $0 < \frac{c_r}{\nu} < 1$ for $\nu \geq \nu_1$. Now we can apply *Lemma* 6.3 and obtain $\lim_{\nu \to \infty} \frac{1}{\nu} c_r = \xi_0$ and hence

$$c_r \sim \tfrac{1}{2}\xi_0 \cdot r \quad \text{as} \quad r \to \infty. \tag{6.18}$$

We gather our results of (6.8), (6.13) and (6.18) in

Corollary 6.4 *Under the assumptions of Theorem 6.2, every pair of different nodes* t_j, t_k *satisfies*

$$\arccos(t_j t_k) \geq \phi_\mu^r \sim \frac{c_r}{\mu}$$

as $\mu \to \infty$. *In addition* $c_r \sim \tfrac{1}{2}\xi_0 \cdot r$ *holds as* $r \to \infty$, *if* ξ_0 *is defined to be the unique solution of the equation*

$$\frac{e^{\sqrt{1-\xi^2}}}{1 + \sqrt{1-\xi^2}} = \frac{e}{\sqrt[4]{27}}.$$

Its value is $\xi_0 = 0.69800985\ldots$.

Packing the Sphere by Nonoverlapping Caps:
It is worth while to compare the nodes of *Theorem* 6.2 with theoretical results. For extremal fundamental systems the minimal distance between different nodes is bounded from below by $\arccos(t_j t_k) \geq \frac{\pi}{2\mu}$, [17]. So, as *Table* 1 shows, the result of *Corollary* 6.3 is comparatively weak for $r \leq 3$, but things change for increasing r. For $r = 35$ and large μ we already get $\arccos(t_j t_k) > 8 \cdot \frac{\pi}{2\mu}$.

Actually the nodes define a system of N_μ^r nonoverlapping caps on S^{r-1} with radius $\tfrac{1}{2}\phi_\mu^r$. By results of Shannon [24] and Rankin [16] the maximum number $N_r(\phi)$ of nonoverlapping caps with radius ϕ, $0 < \phi \leq \frac{\pi}{2}$, which can be placed on S^{r-1} satisfies the (asymptotic) inequalities

$$\frac{1}{\sin 2\phi} < (N_r(\phi))^{\frac{1}{r-1}} \lesssim \frac{1}{\sqrt{2}\sin \phi} \tag{6.19}$$

as $r \to \infty$. So, let us identify $\phi := \tfrac{1}{2}\phi_\mu^r$, so that $N_\mu^r \leq N_r(\phi_\mu^r)$ holds. Then, using (5.13) and *Corollary* 6.4, the upper bound of (6.19) yields

$$0.34900492\ldots = \frac{\xi_0}{2} \leftarrow \frac{c_r}{r} \lesssim \frac{\sqrt{2}}{e} = 0.5202600\ldots \tag{6.20}$$

as $r \to \infty$. Here the limiting value of $\frac{c_r}{r}$ exceeds 67% of the upper bound. The lower bound of (6.19) does not lead to a safe comparison. But if we were allowed to identify N_μ^r and $N_r(\phi_\mu^r)$, then this would yield

$$0.3678794\ldots = \frac{1}{e} \lesssim \frac{c_r}{r}, \tag{6.21}$$

i.e. $\frac{\xi_0}{2}$ would fall short of the lower bound of (6.21) by about 5% only, which indicates that $N_r(\phi_\mu^r)$ is overestimating N_μ^r only slightly. Note that the lower bound of (6.19) has been obtained nonconstructively by using binary codes, whilst our node system can really be calculated.

References

[1]   P. Appell, J. Kampé de Fériet: *Fonctions hypergéometriques et hypersphériques. Polynomes d'Hermite,* Gauthier–Villars: Paris 1926.

[2]   E. Bannai, E. M. Damerell: *Tight spherical designs I,* J. Math. Soc. Japan **31** (1979), 199–207.

[3]   L. Bos: *On Kergin interpolation on the disk,* J. Approx. Theory **37** (1983), 251–261.

[4]   L. Bos: *Some remarks on the Fejér problem for Lagrange interpolation in several variables,* J. Approx. Theory **60** (1990), 133–140.

[5]   I. K. Daugavet: *Some applications of the Marcinkiewicz–Berman identity,* Vestnik Leningrad Univ. Math. **1** (1974), 321–327.

[6]   M. E. Davison, F. A. Grünbaum: *Tomographic reconstruction with arbitrary directions,* Comm. Pure Appl. Math. **34** (1983), 428–448.

[7]   J. Fliege, U. Maier: *The distribution of points on the sphere and corresponding cubature formulae,* IMA J. Numer. Anal. **14** (1998), 1–24.

[8]   G. Gasper: *Rogers' linearization formula for the continuous q-ultraspherical polynomials and quadratic transformation formulas,* SIAM J. Math. Anal. **16** (1985) 1061–1071.

[9]   W. B. Gearhart: *Some Chebyshev approximations by polynomials in two variables,* J. Approx. Theory **8** (1973), 195–209.

[10]   O. D. Kellogg: *On bounded polynomials in several variables*, Math. Z. **27** (1927), 55–64.

[11]   E. Kogbetliantz: *Recherches sur la sommabilité de séries ultra-sphériques par le méthode des moyennes arithmétiques*, J. Math. Pure Appl. Ser. **9**, V. **3** (1924), 107–187.

[12]   A. V. Kolushov, V. A. Yudin: *Extremal dispositions of points on the sphere*, Analysis Mathematica **23** (1997), 25–34.

[13]   L. Koschmieder: *Über die C–Summierbarkeit gewisser Reihen von Didon und Appell*, Math. Ann. **104** (1931), 387–402.

[14]   U. Linde, B. Sündermann, M. Reimer: *Numerische Berechnung extremaler Fundamentalsysteme für Polynomräume über der Vollkugel*, Computing **43**, 37–45 (1989).

[15]   H. M. Möller: *Lower bounds for the number of nodes in cubature formulae*, in: Numerische Integration, ISNM **45**, 221–230, Birkhäuser, Basel–Boston–Stuttgart 1979.

[16]   R. A. Rankin: *The closest packing of spherical caps in n dimensions*, Glasgow Math. J. **2** (1955), 139–144.

[17]   M. Reimer: *Constructive theory of multivariate functions*, BI-Wiss.—Verl. Mannheim–Wien–Zürich 1990.

[18]   M. Reimer: *Quadrature rules for the surface integral of the unit sphere on extremal fundamental systems*, Math. Nachr. 169 (1994), 235–241.

[19]   M. Reimer: *Radon–transform, Laplace–series and matrix–transforms*, Comm. Appl. Anal. **1** (1997), 337–349.

[20]   M. Reimer: *The average size of certain Gram–determinants and interpolation on non–compact sets*, in: Multivariate Approximation and Splines, 235–243, Birkhäuser, Basel–Boston–Stuttgart 1997.

[21]   M. Reimer: *Asymptotic behaviour of the four–norms of the polynomial reproducing kernel functions on the sphere*, Analysis **18** (1998), 85–95.

[22]   M. Reimer, B. Sündermann: *A Remez–type algorithm for the calculation of extremal fundamental systems for polynomial spaces on the sphere*, Computing **37** (1986), 43–58.

[23]   Th. Rivlin: *The Chebyshev polynomials*, J. Wiley, New York etc. 1974.

[24]   C. E. Shannon: *Probabiliy error for optimal codes in a Gaussian channel*, BSTJ **38** (1959), 611–656.

[25]   I. H. Sloan: *Interpolation and hyperinterpolation on the sphere*, in: *Multivariate Approximation*, W. Haußmann et al. (eds.) Math. Research **101**, 255–268, Akademie Verlag, Berlin 1997.

[26]   J. Steinacker, E. Thamm, U. Maier: *Efficient integration of intensity functions on the unit sphere*, J. Quantitative Spectroscopy and Radiative Transfer **56** (1996) 97–107.

[27]   A. H. Stroud: *Approximate calculation of multiple integrals*, Prentice Hall, Englewood Cliffs 1971.

[28]   B. Sündermann: *Projektionen auf Polynomräume in mehreren Veränderlichen*, Diss. Dortmund 1983.

[29]   G. N. Watson: *A treatise on the theory of Bessel functions*, Cambridge Univ. Press, Cambridge 1966.

[30]   V. A. Yudin: *Covering a sphere and extremal properties of orthogonal polynomials*, Discrete Math. Appl. **5** (1995), 371–379.

[31]   V. A. Yudin: *Lower bounds for spherical designs*, Izvestiya Mathematics **61** (1997), 673–683.

Address:

MANFRED REIMER
Fachbereich Mathematik
Universität Dortmund
D–44221 Dortmund
Germany

Range Restricted Interpolation by Cubic C^1–Splines on Clough–Tocher Splits

Jochen W. Schmidt

Dedicated to Manfred Reimer on the occasion of his 65th birthday

Abstract

The well–known Clough–Tocher splines of quadratic precision are shown
to be also useful in range restricted interpolation of scattered data. For
interpolation subject to piecewise constant lower and upper bounds on
the function values, an algorithm for constructing desired interpolants is
offered, which is always successful if the bounds satisfy natural compatibility conditions.

1 Introduction

Piecewise cubic polynomials on Clough–Tocher triangulations Δ_{CT} were originally introduced for finite element methods [3]. However, these spline functions
soon turned out to be a useful tool also for the interpolation and approximation
of scattered data; see the survey paper [6] as well as [1], [2], [9], [11], and the
references therein. Extensions to cubic splines on splitted quadrangulations are
treated in [10]. Further we refer to the review paper [4]. This survey deals with
higher–degree polynomial splines on extended Clough–Tocher splits suitable in
treating special kinds of monotonicity and convexity preserving interpolation;
see in addition [5].

The present paper is also concerned with an important type of constrained interpolation, namely with the range restricted interpolation having the
nonnegative interpolation as a special case. Given is a set of data sites
$P_i = (x_i, y_i) \in \mathbb{R}^2$ and corresponding data values $z_i \in \mathbb{R}^1$, $i = 0, 1, \ldots, N$. We
denote the convex hull of the data sites by $\Omega = \mathrm{conv}\{P_0, P_1, \ldots, P_N\}$. If the
data values are nonnegative, the problem of nonnegative interpolation consists
in the construction of a $C^1(\Omega)$–function s which interpolates,

$$s(P_i) = z_i, \quad i = 0, 1, \ldots, N, \tag{1.1}$$

Advances in Multivariate Approximation; W. Haußmann, K. Jetter and M. Reimer (eds.)
Mathematical Research, Vol. 107, pp. 253–267, ISBN 3–527–40236–5
© WILEY–VCH, Berlin 1999

and which satisfies the nonnegativity requirement

$$s(P) \geq 0 \quad \text{for} \quad P \in \Omega. \tag{1.2}$$

More generally, we prescribe lower and upper bounds on the values of the interpolant, for example in the following form: Let Δ be an admissible triangulation of Ω, i.e., Δ is a partition of Ω into nondegenerate triangles. Exactly the data sites occur as vertices, and two different triangles are either disjoint or share a common vertex or a common edge. One choice is the well–known Delaunay triangulation. The vertices of a triangle $\Delta_n \in \Delta$ are denoted by P_i, P_j, and P_k. Suppose we are given lower and upper obstacles $-\infty \leq L_n \leq U_n \leq +\infty$ for all triangles $\Delta_n \in \Delta$, which satisfy the natural compatibility conditions

$$L_n \leq z_i, z_j, z_k \leq U_n \quad \text{for} \quad \Delta_n \in \Delta. \tag{1.3}$$

Then the present problem of range restricted interpolation is to find a $C^1(\Omega)$ function s which satisfies the interpolation requirement (1.1), and the range restrictions

$$L_n \leq s(P) \leq U_n \quad \text{for} \quad P \in \Delta_n, \ \Delta_n \in \Delta. \tag{1.4}$$

It is the main objective of this paper to show that the described problem (1.1), (1.4) can always be solved when using Clough–Tocher splits Δ_{CT} of Δ in their original form. It is remarkable that to this end the split points in Δ_{CT} subdividing each (macro–) triangle $\Delta_n \in \Delta$ into three microtriangles can be fixed in advance, independently of the data values.

While for splines on triangulations the Bernstein–Bézier representation is widely in use (see [6] and most of the papers cited above), we now prefer another description in terms of function values and first order derivatives at the vertices of the triangles; compare with [7], [8], [14] for computing related splines. Following this line we are able to give a simple existence and uniqueness proof for the underlying Hermite problem (see [13] for a corresponding suggestion) and, moreover, the range restrictions (1.4) are easy to meet.

A first algorithm for solving the problem (1.1), (1.4) of range restricted interpolation is proposed in [12]. Here, and in [17] for treating nonnegative least squares smoothing, quadratic C^1–splines on Powell–Sabin refinements Δ_{PS} of the triangulation Δ are seen to be suitable. A Powell–Sabin refinement Δ_{PS} requires subdividing each macrotriangle into six microtriangles and, in addition, the split points have to satisfy geometric restrictions not occurring in Clough–Tocher splits Δ_{CT}. Moreover, it seems to be more promising to attack other more complicated interpolation problems with Clough–Tocher splines than with Powell–Sabin splines. We explicitly mention the scattered data interpolation subject not to (1.4), but to piecewise linear continuous obstacles for the function values.

Finally we refer to the review paper [15] which is concerned with several types of restricted interpolation mainly for univariate data sets and also for bivariate ones.

2 C^1–Conditions for Cubic Splines on Triangulations

In this and the next section, the aim is to give a representation of cubic C^1–splines on Clough–Tocher splits Δ_{CT} which follows Heindl's description [8] for quadratic C^1–splines on Powell–Sabin refinements Δ_{PS} of Δ. In this way, we can propose a straightforward algorithm for computing a cubic C^1–interpolant on Δ_{CT} assuring, in addition, existence and uniqueness. It should be mentioned that we are also stimulated by Powell's recent remark [13] to be unable to find a simple proof for this claim in the literature.

Let Δ_n be an arbitrary triangle belonging to an admissible triangulation Δ and having the vertices P_i, P_j, and P_k. We denote the barycentric coordinates of a point $P = (x,y) \in \mathbb{R}^2$ with respect to Δ_n by $u = u(x,y)$, $v = v(x,y)$, and $w = w(x,y)$. They are defined by

$$P = P_i u + P_j v + P_k w, \quad u + v + w = 1. \tag{2.1}$$

Thus, Cramer's rule yields

$$u = |PP_jP_k|/D_n, \; v = |P_iPP_k|/D_n, \; w = |P_iP_jP|/D_n \tag{2.2}$$

where D_n is the determinant

$$D_n = |P_iP_jP_k| = \begin{vmatrix} 1 & 1 & 1 \\ x_i & x_j & x_k \\ y_i & y_j & y_k \end{vmatrix} \neq 0; \tag{2.3}$$

the others in (2.2) are defined analogously. For the partial derivatives with respect to x and y we obtain

$$\begin{aligned} u_x = (y_j - y_k)/D_n, \; v_x = (y_k - y_i)/D_n, \; w_x = (y_i - y_j)/D_n, \\ u_y = (x_k - x_j)/D_n, \; v_y = (x_i - x_k)/D_n, \; w_y = (x_j - x_i)/D_n, \end{aligned} \tag{2.4}$$

and

$$u_x v_y - u_y v_x = v_x w_y - v_y w_x = w_x u_y - w_y u_x = 1/D_n. \tag{2.5}$$

We describe a cubic polynomial on Δ_n by

$$s(P) = s(x,y) = z_i u^3 + z_j v^3 + z_k w^3 + 3a_{ij} u^2 v + 3a_{ji} uv^2 +$$
$$+ 3a_{ik} u^2 w + 3a_{ki} uw^2 + 3a_{jk} v^2 w + 3a_{kj} vw^2 + 6b_n uvw, \ P \in \Delta_n. \tag{2.6}$$

Obviously, the interpolation requirement (1.1) is incorporated into this representation. For determining the coefficients $a_{\mu\nu}$ we introduce the notation

$$g_i = \text{grad } s(P_i) \tag{2.7}$$

for the gradients at the data sites. In view of

$$\text{grad } s(P) = 3(z_i u^2 + 2a_{ij} uv + a_{ji} v^2 + 2a_{ik} uw + a_{ki} w^2 + 2b_n vw) \begin{bmatrix} u_x \\ u_y \end{bmatrix} +$$
$$+ 3(z_j v^2 + a_{ij} u^2 + 2a_{ji} uv + 2a_{jk} vw + a_{kj} w^2 + 2b_n uw) \begin{bmatrix} v_x \\ v_y \end{bmatrix} +$$
$$+ 3(z_k w^2 + a_{ik} u^2 + 2a_{ki} uw + a_{jk} v^2 + 2a_{kj} vw + 2b_n uv) \begin{bmatrix} w_x \\ w_y \end{bmatrix}, \tag{2.8}$$

we obtain in P_i

$$g_i = 3 \left(z_i \begin{bmatrix} u_x \\ u_y \end{bmatrix} + a_{ij} \begin{bmatrix} v_x \\ v_y \end{bmatrix} + a_{ik} \begin{bmatrix} w_x \\ w_y \end{bmatrix} \right), \tag{2.9a}$$

in P_j

$$g_j = 3 \left(a_{ji} \begin{bmatrix} u_x \\ u_y \end{bmatrix} + z_j \begin{bmatrix} v_x \\ v_y \end{bmatrix} + a_{jk} \begin{bmatrix} w_x \\ w_y \end{bmatrix} \right), \tag{2.9b}$$

and in P_k

$$g_k = 3 \left(a_{ki} \begin{bmatrix} u_x \\ u_y \end{bmatrix} + a_{kj} \begin{bmatrix} v_x \\ v_y \end{bmatrix} + z_k \begin{bmatrix} w_x \\ w_y \end{bmatrix} \right). \tag{2.9c}$$

If (2.5) is taken into account, by solving systems of two linear equations it follows from (2.9a)–(2.9c) that the coefficients $a_{\mu\nu}$ in (2.6) are

$$a_{\mu\nu} = z_\mu + \frac{1}{3} g_\mu(P_\nu - P_\mu) \quad \text{for} \ \ \mu,\nu \in \{i,j,k\}, \ \mu \neq \nu. \tag{2.10}$$

Here the inner product of the gradient g_μ and the vector $P_\nu - P_\mu$ is written as $g_\mu(P_\nu - P_\mu)$. Analogously, the inner product of vectors $P_i - P_j$ and $P_j - P_k$ is denoted by $(P_i - P_j)(P_j - P_k)$.

Next, for computing the remaining coefficient b_n it is usual to prescribe cross–boundary derivatives at the midpoint of each edge. Though this leads to an overdetermination, in preparing the subsequent considerations we describe the procedure, say for the edge P_iP_j. From (2.8) and (2.9a), (2.9b) it follows that for $P \in P_iP_j$

$$\begin{aligned}
\operatorname{grad} s(P) = {} & g_i u^2 + g_j v^2 + (g_i + g_j)uv \\
& + 6 \left(a_{ij} \begin{bmatrix} u_x \\ u_y \end{bmatrix} + a_{ji} \begin{bmatrix} v_x \\ v_y \end{bmatrix} + b_n \begin{bmatrix} w_x \\ w_y \end{bmatrix} - \frac{1}{6}(g_i + g_j) \right) uv.
\end{aligned} \tag{2.11}$$

With the vector

$$n_{ij} = \begin{bmatrix} y_i - y_j \\ x_j - x_i \end{bmatrix} \tag{2.12}$$

normal to the edge P_iP_j we require

$$\left(a_{ij} \begin{bmatrix} u_x \\ u_y \end{bmatrix} + a_{ji} \begin{bmatrix} v_x \\ v_y \end{bmatrix} + b_n \begin{bmatrix} w_x \\ w_y \end{bmatrix} - \frac{1}{6}(g_i + g_j) \right) n_{ij} = 0,$$

i.e., the cross boundary derivative shall be

$$\operatorname{grad} s(P) n_{ij} = g_i n_{ij}(u^2 + uv) + g_j n_{ij}(v^2 + uv) \quad \text{for} \ \ P \in P_iP_j. \tag{2.13}$$

Hence, on the edge P_iP_j we are led to the following relation for b_n,

$$\begin{aligned}
a_{ij}(P_j - P_k)(P_i - P_j) + a_{ji}(P_k - P_i)(P_i - P_j) \\
+ b_n \|P_i - P_j\|^2 = D_n(g_i + g_j)n_{ij}/6.
\end{aligned} \tag{2.14}$$

Further, since

$$\left(a_{ij} \begin{bmatrix} u_x \\ u_y \end{bmatrix} + a_{ji} \begin{bmatrix} v_x \\ v_y \end{bmatrix} + b_n \begin{bmatrix} w_x \\ w_y \end{bmatrix} \right) (P_i - P_j) = a_{ij} - a_{ji} \tag{2.15}$$

and

$$
\begin{aligned}
\frac{(y_i - y_j)n_{ij} + (x_i - x_j)(P_i - P_j)}{\|P_i - P_j\|^2} &= \begin{bmatrix} 1 \\ 0 \end{bmatrix}, \\
\frac{(x_j - x_i)n_{ij} + (y_i - y_j)(P_i - P_j)}{\|P_i - P_j\|^2} &= \begin{bmatrix} 0 \\ 1 \end{bmatrix},
\end{aligned}
\tag{2.16}
$$

with (2.10) we find immediately that the gradient (2.11) of s on P_iP_j depends on P_i, P_j, z_i, z_j, and on the gradients g_i, g_j, but not on P_k, z_k, and g_k.

Of course, a relation of the type (2.14) also holds for the edges P_iP_k and P_jP_k, leading to an overdetermination of the coefficient b_n. This drawback is known to be avoidable by changing to a Clough–Tocher split of the triangle Δ_n.

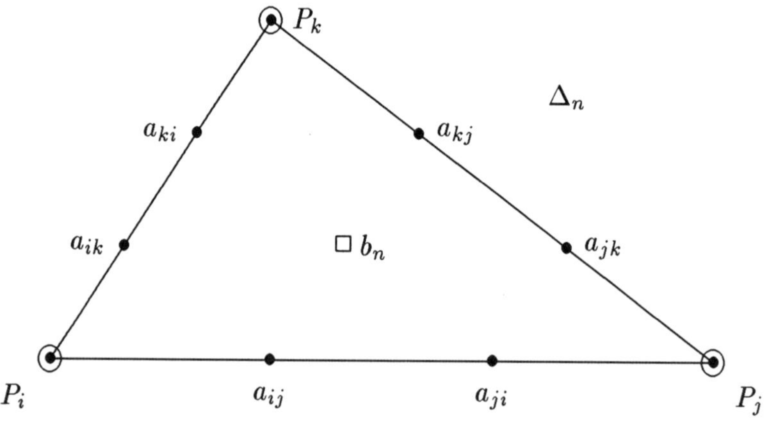

Figure 1: Cubic polynomial on a triangle Δ_n with the vertices P_i, P_j, P_k, $\odot$ position and gradient given, $\bullet$ to compute by (2.10), $\square$ to compute, e.g., by (2.14)

In order to formulate conditions for the C^1–continuity of cubic splines across an edge we consider two adjacent triangles Δ_n and Δ_m with the vertices P_i, P_j, P_k and P_i, P_j, P_ℓ, respectively. The barycentric coordinates with respect to the second triangle Δ_m are again denoted by u, v, and w. However, in the preceding formulas we then have to substitute $P_\ell = (x_\ell, y_\ell)$ for $P_k = (x_k, y_k)$. For example the determinant D_m is

$$D_m = |\,P_i P_j P_\ell\,| = \begin{vmatrix} 1 & 1 & 1 \\ x_i & x_j & x_\ell \\ y_i & y_j & y_\ell \end{vmatrix}.$$

In view of (2.6) and (2.10), we find for a point P from the common edge $P_i P_j$ of Δ_n and Δ_m

$$\lim_{Q \to P, Q \in \Delta_n} s(Q) = z_i u^3 + z_j v^3 + (3z_i + g_i(P_j - P_i))u^2 v +$$
$$+ (3z_j + g_j(P_i - P_j))uv^2 = \lim_{Q \to P, Q \in \Delta_m} s(Q) \tag{2.17}$$

implying the continuity of s on $\Delta_n \cup \Delta_m$. Further, the preceding considerations yield

$$\lim_{Q \to P, Q \in \Delta_n} \operatorname{grad} s(Q) = \lim_{Q \to P, Q \in \Delta_m} \operatorname{grad} s(Q) \tag{2.18}$$

for $P \in P_i P_j$ if with the coefficients (2.10)

$$a_{ij}(P_j - P_k)(P_i - P_j) + a_{ji}(P_k - P_i)(P_i - P_j)$$
$$+ b_n \| P_i - P_j \|^2 = D_n(g_i + g_j)n_{ij}/6, \tag{2.19a}$$

$$a_{ij}(P_j - P_\ell)(P_i - P_j) + a_{ji}(P_\ell - P_i)(P_i - P_j)$$
$$+ b_m \| P_i - P_j \|^2 = D_m(g_i + g_j)n_{ji}/6 \tag{2.19b}$$

hold. In other words, the C^1–smoothness across the common edge $P_i P_j$ of the triangles Δ_n and Δ_m is assured by (2.19a) and (2.19b).

Of course, if $P_i P_j$ is a boundary edge of the whole triangulation, then only one of the preceding conditions is required, for example (2.19a), so $P_i P_j$ belongs to Δ_n.

In our context, we need a second formulation for the C^1–property across $P_i P_j$; for this form see also [7]. In view of (2.11) and (2.15) we get for $P \in P_i P_j$

$$\lim_{Q \to P, Q \in \Delta_n} \operatorname{grad} s(Q)(P_i - P_j) = \lim_{Q \to P, Q \in \Delta_m} \operatorname{grad} s(Q)(P_i - P_j). \tag{2.20}$$

Further, if we require

$$\lim_{Q \to P, Q \in \Delta_n} \operatorname{grad} s(Q)(P_k - P_\ell) = \lim_{Q \to P, Q \in \Delta_m} \operatorname{grad} s(Q)(P_k - P_\ell), \tag{2.21}$$

we obtain

$$\lim_{Q \to P, Q \in \Delta_n} \operatorname{grad} s(Q) = \lim_{Q \to P, Q \in \Delta_m} \operatorname{grad} s(Q).$$

This together with (2.17) assures the C^1-continuity of s across the common edge $P_i P_j$ of the triangles Δ_n and Δ_m. Now, we rewrite straightforwardly the equation (2.21) as

$$\frac{1}{D_n}\left(a_{ij}|P_j P_k P_\ell| - a_{ji}|P_i P_k P_\ell| + b_n(D_m - D_n)\right) =$$
$$= \frac{1}{D_m}\left(a_{ij}|P_j P_k P_\ell| - a_{ji}|P_i P_k P_\ell| + b_m(D_m - D_n)\right).$$

Hence, from the fact that $D_n = |P_i P_j P_k|$ and $D_m = |P_i P_j P_\ell|$ are both nonzero and of opposite sign it follows

$$a_{ij}|P_j P_k P_\ell| - a_{ji}|P_i P_k P_\ell| + b_n|P_i P_j P_\ell| - b_m|P_i P_j P_k| = 0. \qquad (2.22)$$

Thus we have the

Proposition 2.1. *Let $P_i P_j$ be the common edge of two neighbouring triangles Δ_n and Δ_m with the vertices P_i, P_j, P_k and P_i, P_j, P_ℓ, respectively. If the coefficients defining a cubic spline s on $\Delta_n \cup \Delta_m$ satisfy in addition to (2.10) the equalities (2.19a) − (2.19b) or (2.22), then s is C^1-continuous across $P_i P_j$.*

3   Cubic C^1-Splines on Clough–Tocher Splits

In general, it is not possible to construct cubic splines of C^1-continuity on an original triangulation Δ of the data sites. However, if we refine Δ by subdividing each triangle into three subtriangles at an interior split point, then we can always find a cubic C^1-spline to interpolate given data values and gradients at vertices of Δ. This construction is called Clough–Tocher element and is usually described in terms of Bernstein–Bézier patches; see [6], [13], and other expositions. Our representation is based on Proposition 2.1. We use both C^1-characterizations, namely (2.19a) for boundary edges and (2.22) for inner edges. In this specific way we can derive a scheme which allows easily to meet the range restrictions (1.4).

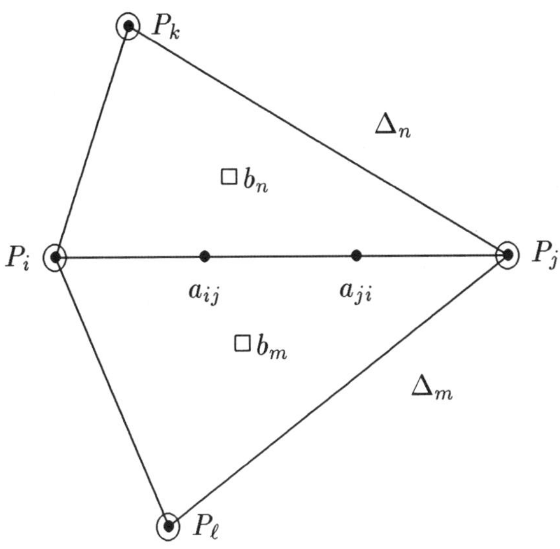

Figure 2: Illustration of the C^1–conditions across the edge $P_i P_j$ described in Proposition 2.1.

Let $\Delta_n \in \Delta$ be a triangle with the vertices P_i, P_j and P_k. In Δ_n an interior point

$$Q_n = \alpha_n P_i + \beta_n P_j + \gamma_n P_k, \ \alpha_n > 0, \ \beta_n > 0, \ \gamma_n > 0, \ \alpha_n + \beta_n + \gamma_n = 1 \quad (3.1)$$

is selected and is connected by straight lines to the vertices of Δ_n subdividing Δ_n into three microtriangles. The arising refined triangulation of Δ is denoted by Δ_{CF}.

Given are the data values z_i, z_j, z_k and the gradients g_i, g_j, g_k at the vertices of Δ_n. The coefficients of the cubic polynomials at the three micro–triangles are abbreviated as shown in Figure 3. Because of (2.10), in step 1 we obtain

$$\begin{aligned}
a_{\mu\nu} &= z_\mu + \frac{1}{3} g_\mu (P_\nu - P_\mu), \ \mu, \nu \in \{i,j,k\}, \ \mu \neq \nu, \\
a_{\mu n} &= z_\mu + \frac{1}{3} g_\mu (Q_n - P_\mu), \ \mu \in \{i,j,k\}.
\end{aligned} \quad (3.2)$$

In step 2, we compute b_{ni}, b_{nj}, and b_{nk} applying the C^1–condition (2.19a). The result is

$$b_{ni} = \frac{a_{jk}(Q_n - P_k)(P_j - P_k) + a_{kj}(P_j - Q_n)(P_j - P_k) + \alpha_n D_n(g_j + g_k)n_{jk}/6}{\|P_j - P_k\|^2},$$

$$b_{nj} = \frac{a_{ki}(Q_n - P_i)(P_k - P_i) + a_{ik}(P_k - Q_n)(P_k - P_i) + \beta_n D_n(g_k + g_i)n_{ki}/6}{\|P_k - P_i\|^2},$$

$$b_{nk} = \frac{a_{ij}(Q_n - P_j)(P_i - P_j) + a_{ji}(P_i - Q_n)(P_i - P_j) + \gamma_n D_n(g_i + g_j)n_{ij}/6}{\|P_i - P_j\|^2}.$$

$$(3.3)$$

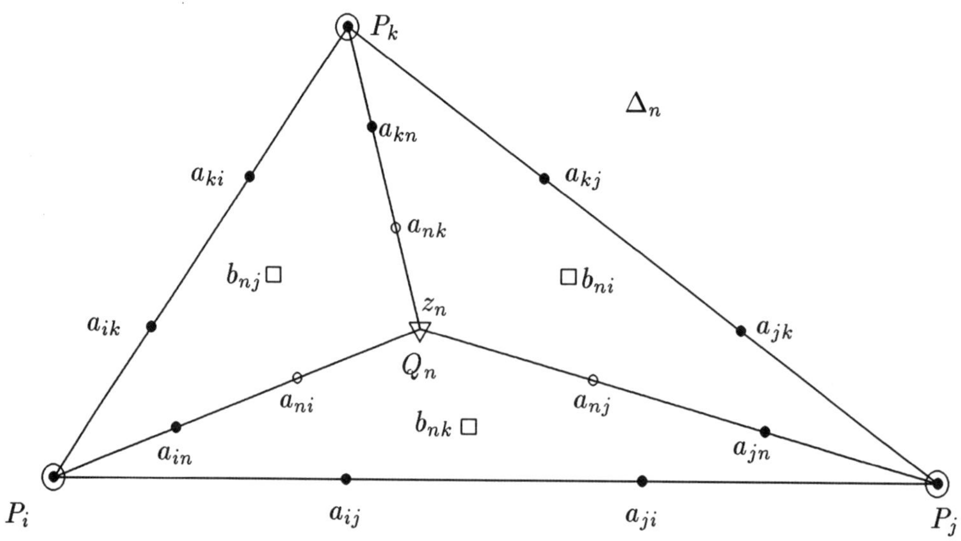

Figure 3: Clough–Tocher split of a triangle Δ_n, $\odot$ position and gradient given, $\bullet$ to compute by (3.2), $\square$ to compute by (3.3), $\circ$ to compute by (3.4), $\triangledown$ to compute by (3.5)

Next, in step 3 the coefficients a_{ni}, a_{nj} and a_{nk} are determined by means of the C^1–characterization (2.22). In this manner we get

$$a_{ni} = \frac{a_{in}|Q_n P_j P_k| + b_{nj}|P_i P_j Q_n| + b_{nk}|P_i Q_n P_k|}{|P_i P_j P_k|}$$

$$=\alpha_n a_{in} + \gamma_n b_{nj} + \beta_n b_{nk}, \tag{3.4}$$

$$a_{nj} = \beta_n a_{jn} + \alpha_n b_{nk} + \gamma_n b_{ni},$$

$$a_{nk} = \gamma_n a_{kn} + \beta_n b_{ni} + \alpha_n b_{nj}.$$

Finally, in step 4 we compute the function value z_n and the gradient g_n at the split point Q_n. Starting from the three equations

$$a_{n\nu} = z_n + \frac{1}{3} g_n (P_\nu - Q_n),\ \nu \in \{i, j, k\},$$

being special cases of (2.10) we find immediately

$$z_n = \alpha_n a_{ni} + \beta_n a_{nj} + \gamma_n a_{nk}, \tag{3.5}$$

and g_n is obtained by solving a system of two linear equations.

Thus, the cubic spline interpolant s is completely determined on any triangle Δ_n. It is C^1 on Δ_n and, in view of (3.3), it is also C^1 across the common edge $P_i P_j$ of any two adjacent triangles. Further, the cross boundary derivatives satisfy the relation (2.13) with the normal vectors n_{ij}. Hence we have proved

Proposition 3.1. *Let Δ be an admissible triangulation of the convex hull of the data sites $P_0, P_1, \ldots, P_N$, let Δ_{CT} be a Clough–Tocher refinement of Δ, and let data $z_i \in \mathbb{R}^1$, $g_i \in \mathbb{R}^2$ be arbitrarily given. Then, the interpolation problem*

$$s(P_i) = z_i,\ \operatorname{grad} s(P_i) = g_i,\ i = 0, 1, \ldots, N, \tag{3.6}$$

$$\operatorname{grad} s(\tfrac{1}{2}(P_i + P_j))n_{ij} = \tfrac{1}{2}(g_i + g_j)n_{ij}\ \textit{for the edges } P_i P_j\ \textit{of}\ \Delta \tag{3.7}$$

is uniquely solvable by a cubic C^1–spline s on Δ_{CT}. The interpolant s can be computed by the steps $(3.2) - (3.5)$ in this succession.

Moreover, like in the standard case the described interpolation process is straightforwardly seen to be of quadratic precision, i.e. quadratic polynomials are reproduced.

4   Range Restricted Interpolation with Clough–Tocher Splines

It is obvious that the range restrictions (1.4) are satisfied on Δ_n, i.e. the Clough–Tocher C^1–interpolant s is on Δ_n, lying between the lower bound L_n and the upper bound U_n if this holds true for the coefficients (3.2)–(3.5) of s. Especially, the compatibility requirements (1.3) should be valid. It turns out that some of these conditions are redundant, and therefore they can be omitted. First, from

$$L_n \le a_{\mu\nu} = z_\mu + \frac{1}{3}g_\mu(P_\nu - P_\mu) \le U_n, \ \mu,\nu \in \{i,j,k\}, \ \mu \ne \nu \qquad (4.1)$$

and (1.3) it follows directly that $L_n \le a_{\mu n} \le U_n, \mu \in \{i,j,k\}$. Further, with the coefficients (3.3) the inequalities

$$L_n \le b_{n\nu} \le U_n, \ \nu \in \{i,j,k\} \qquad (4.2)$$

are assumed. Then in view of (3.4) we obtain $L_n \le a_{n\nu} \le U_n, \nu \in \{i,j,k\}$, and these inequalities imply immediately $L_n \le z_n \le U_n$. Thus it follows

Theorem 4.1. *An interpolating cubic C^1-spline s on a Clough–Tocher split Δ_{CT} of an admissible triangulation Δ of the data sites $P_0, P_1, \ldots, P_N$ satisfies the range restrictions (1.4) if besides the compatibility (1.3) the inequalities (4.1) and (4.2) for all triangles $\Delta_n \in \Delta$ are valid.*

In order to guarantee the existence of range restricted interpolants it suffices to set the gradients $g_0, g_1, \ldots, g_N$ equal to the zero vector. Then the conditions (4.1) reduce to the compatibility (1.3). Further, we assume the angles in the microtriangles at the data sites to be acute. This requirement is met, for example, if the split points Q_n are equal to the incenters of the Δ_n, or if the Δ_n themselves are acute. In other words, we require

$$\begin{aligned}
(Q_n - P_k)(P_j - P_k) \ge 0, \ (P_j - Q_n)(P_j - P_k) \ge 0, \\
(Q_n - P_i)(P_k - P_i) \ge 0, \ (P_k - Q_n)(P_k - P_i) \ge 0, \\
(Q_n - P_j)(P_i - P_j) \ge 0, \ (P_i - Q_n)(P_i - P_j) \ge 0.
\end{aligned} \qquad (4.3)$$

Then, together with

$$(Q_n - P_k)(P_j - P_k) + (P_j - Q_n)(P_j - P_k) = \|P_j - P_k\|^2$$

and the two analogous equalities, also the conditions (4.2) are satisfied. Thus the following smooth existence result is obtained.

Theorem 4.2. *For bounds compatible in the sense* (1.3), *the problem of range restricted C^1–interpolation* (1.1), (1.4) *is always solvable by cubic splines on Clough–Tocher splits Δ_{CT} of Δ satisfying* (4.3).

The trivial choice $g_0 = g_1 = \cdots = g_N = 0$ always assuring the existence of desired interpolants, mostly results in visually unappealing splines. Therefore a second stage should follow the existence proof, namely the construction of more suitable gradients. This fairing step is usually done by minimizing a choice functional. One idea is the minimum norm modification of initial gradients $g_0^0, g_1^0, \ldots, g_N^0$ obtained, e.g., by numerical differentiation. This leads in our context to the program

$$\text{minimize} \sum_{i=0}^{N} \|g_i - g_i^0\|^2 \quad \text{s.t. } (4.1) \text{ and } (4.2) \text{ for all } \Delta_n \in \Delta. \qquad (4.4)$$

In view of Theorem 4.2 the feasible domain is nonempty. Another popular proposal is the minimization of the thin plate functional. For more details on fairing procedures we refer to [12] and to the volume Advanced Course on Fairshape cited by [4].

5   Boundary Constrained Clough–Tocher Splines

Often it is of interest to construct an interpolant which satisfies additional boundary constraints. For example, the function values and/or the normal derivatives may be required to be zero on the boundary of the convex hull Ω, or on a part of that. We refer to the paper [16] where this problem is attacked by means of Powell–Sabin splines.

When using Clough–Tocher splines, boundary constraints are easily incorporated. Let $P_i P_j$ be a boundary edge of Ω belonging to the triangle $\Delta_n \in \Delta$. Of course,

$$s(P) = 0 \quad \text{for} \quad P \in P_i P_j \qquad (5.1)$$

holds true if $z_i = z_j = a_{ij} = a_{ji} = 0$. Thus, in view of (3.2) this requirement is valid if and only if

$$z_i = z_j = 0, \; g_i(P_i - P_j) = 0, \; g_j(P_j - P_i) = 0. \qquad (5.2)$$

Next, the normal derivatives vanish on $P_i P_j$, i.e.,

$$\operatorname{grad} s(P) n_{ij} = 0 \quad \text{for} \quad P \in P_i P_j \tag{5.3}$$

if $g_i n_{ij} = g_j n_{ij} = 0$ and if b_{nk} is computed by (3.3); see (2.11), (2.13). Thus, both conditions (5.1) and (5.3) are satisfied if and only if

$$z_i = z_j = 0, \ g_i = g_j = 0, \tag{5.4}$$

implying $a_{ij} = a_{ji} = a_{ik} = a_{jk} = a_{in} = a_{jn} = b_{nk} = 0$.

References

[1] P. Alfeld: *A trivariate Clough–Tocher scheme for tetrahedral data*, Comput. Aided Geom. Design **1** (1984), 169–181.

[2] P. G. Ciarlet: *Sur l'élement de Clough et Tocher*, RAIRO Anal. Numér. **R2** (1974), 19–27.

[3] R. W. Clough, J. L. Tocher: *Finite element stiffness matrices for analysis of plates in bending*, in: *Proceedings of Conference on Matrix Methods in Structural Analysis*, 515–546, Wright–Patterson Air Force Base, Ohio, 1965.

[4] P. Costantini: *Shape-preserving interpolation with variable degree polynomial splines*, in: *Advanced Course on Fairshape*, J. Hoschek, P. Kaklis (eds.), 84–114, Teubner, Stuttgart 1996.

[5] P. Costantini, C. Manni: *On a class of polynomial triangular macro-elements*, J. Comp. Appl. Math. **73** (1996), 45–64.

[6] G. Farin: *Triangular Bernstein–Bézier patches*, Comput. Aided Geom. Design **3** (1986), 83–127.

[7] R. H. J. Gmelig Meyling: *Approximation by cubic C^1-splines on arbitrary triangulations*, Numer. Math. **51** (1987), 65–85.

[8] G. Heindl: *Interpolation and approximation by piecewise quadratic C^1-functions of two variables*, in: *Multivariate Approximation Theory*, W. Schempp, K. Zeller (eds.), 146–161, Birkhäuser, Basel 1979.

[9] P. Kashyap: *Improving Clough–Tocher interpolants*, Comput. Aided Geom. Design **13** (1996), 629–651.

[10] M.-J. Lai: *Scattered data interpolation and approximation using bivariate C^1 piecewise cubic polynomials*, Comput. Aided Geom. Design **13** (1996), 81–88.

[11] St. Mann: *Cubic precision Clough–Tocher interpolant*, Comput. Aided Geom. Design **16** (1999), 85–88.

[12] B. Mulansky, J. W. Schmidt: *Powell–Sabin splines in range restricted interpolation of scattered data*, Computing **53** (1994), 137–154.

[13] M. J. D. Powell: *A review of methods for multivariable interpolation at scattered data points*, in: *State of the Art in Numerical Analysis*, I. S. Duff, G. A. Watson (eds.), 283–309, Oxford Univ. Press, New York 1997.

[14] J. W. Schmidt: *Scattered data interpolation applying rational quadratic C^1 splines on refined triangulations*, Z. Angew. Math. Mech. **79** (1999), to appear.

[15] J. W. Schmidt, W. Heß: *Numerical methods in strip interpolations applying C^1, C^2, and C^3 splines on refined grids*, Mitt. Math. Gesellsch. Hamburg **16** (1997), 107–135.

[16] K. Willemans, P. Dierckx: *Constrained surface fitting using Powell–Sabin splines*, in: *Wavelets, Images and Surface Fitting*, P. J. Laurent, A. Le Méhauté, L. L. Schumaker (eds.), 511–520, A. K. Peters, Wellesley 1994.

[17] K. Willemans, P. Dierckx: *Nonnegative surface fitting with Powell–Sabin splines*, Numer. Algorithms **9** (1995), 263–276.

Address:

Jochen W. Schmidt
Institute of Numerical Mathematics
Technical University of Dresden
D–01062 Dresden
Germany

Some Error Estimates for Periodic Interpolation of Functions from Besov Spaces

Winfried Sickel and Frauke Sprengel

Dedicated to Professor M. Reimer on the occasion of his 65th birthday

Abstract

Using periodic Strang–Fix conditions, we can give an approach to error estimates for periodic interpolation on equidistant and sparse grids for functions from certain Besov spaces.

1   Introduction

We investigate the L_2–error of interpolation on equidistant and sparse grids for periodic functions from isotropic L_2–Besov spaces and L_2–Besov spaces of functions with dominating mixed smoothness properties.

The interpolation of periodic functions by translates of a given function and the corresponding error estimates have been analyzed by several authors (e.g. [3, 8, 14]) in the univariate as well as in the multivariate case. The periodic Strang–Fix conditions were introduced in [2, 14]. There, they were used to find L_2–error estimates for functions from isotropic L_2–Sobolev spaces.

The approximation of functions on sparse grids and the related field of hyperbolic approximation have a fairly long tradition (e.g. [4, 5, 28]) as well. For bivariate functions, the number of interpolation knots can be reduced to $\mathcal{O}(N \log_2 N)$ for the sparse grids where the equidistant grid has $\mathcal{O}(N^2)$ points. Nevertheless, the interpolation on sparse grids yields error estimates for functions with dominating mixed smoothness properties which are asymptotically only by a logarithmic term worse than the error estimates for the interpolation on the corresponding equidistant grids.

The aim of this paper is to give error estimates for periodic interpolation for functions from L_2–Besov spaces which extend the results for the L_2–Sobolev

Advances in Multivariate Approximation; W. Haußmann, K. Jetter and M. Reimer (eds.)
Mathematical Research, Vol. 107, pp. 269–287, ISBN 3-527-40236-5
© WILEY–VCH, Berlin 1999

spaces [2, 14, 16] on one hand. On the other hand, there already exist error estimates for interpolation on sparse grids for functions from Besov spaces [22]. But there, for the general L_p–case, we needed conditions on the cardinal fundamental interpolant from which the periodic fundamental interpolant was constructed via periodization. In the L_2–case, we do not need the long way around with cardinal interpolation but can use conditions on the periodic fundamental interpolant directly.

2   Besov Spaces

We start with recalling the definition and some basic properties of the function spaces to be dealt with. For this, we follow [19, Chap. 3]. By $\mathbb{T}^n$, we denote the n–dimensional torus represented by the cube

$$\mathbb{T}^n := \{x = (x_1, \ldots, x_n) \in \mathbb{R}^n \; ; \; |x_r| \leq \pi, \; r = 1, \ldots, n\}.$$

Let $D(\mathbb{T}^n)$ and $D'(\mathbb{T}^n)$ denote the set of all complex–valued, 2π–periodic (in each component), and infinitely differentiable functions and its dual space, respectively. The Fourier coefficients of a distribution $g \in D'(\mathbb{T}^n)$ are

$$c_k(g) := g(\mathrm{e}^{-ik\cdot})$$

for $k \in \mathbb{Z}^n$. With the help of the inner product in $L_2(\mathbb{T}^n)$,

$$\langle f, g \rangle_{\mathbb{T}^n} := \frac{1}{(2\pi)^n} \int\limits_{\mathbb{T}^n} f(x)\overline{g(x)}\,\mathrm{d}x,$$

the Fourier coefficients for functions $g \in L_1(\mathbb{T}^n)$ can be written as $c_k(g) = \langle g, \mathrm{e}^{ik\cdot} \rangle_{\mathbb{T}^n}$. Then any $f \in D'(\mathbb{T}^n)$ can be represented by its Fourier series

$$f = \sum_{k \in \mathbb{Z}^n} c_k(f)\,\mathrm{e}^{ik\cdot} \qquad (\text{convergence in } D'(\mathbb{T}^n))$$

and the Fourier coefficients satisfy an inequality of the type

$$|c_k(f)| \leq C_M (1 + |k|_2)^M, \qquad k \in \mathbb{Z}^n, \tag{2.1}$$

for some $M \in \mathbb{N}$. Here and in the sequel, $|k|_2 := (k_1^2 + k_2^2 + \cdots + k_n^2)^{1/2}$ is the Euclidian norm. Conversely, each formal Fourier series with polynomially bounded Fourier coefficients as in (2.1) can be interpreted as a periodic distribution in $D'(\mathbb{T}^n)$.

The Wiener algebra of functions with absolutely summable Fourier series we denote by $A(\mathbb{T}^n)$.

In the following, we restrict our definitions to the L_2–case because all the estimates in the forthcoming sections hold in this case only. We need the index sets

$$
\begin{aligned}
Q_0^n &= \{0\}, \\
Q_j^n &= \{k \in \mathbb{Z}^n \; ; \; |k_r| < 2^j, r = 1, \ldots, n\} \\
&\quad \setminus \{k \in \mathbb{Z}^n \; ; \; |k_r| < 2^{j-1}, r = 1, \ldots, n\}.
\end{aligned}
$$

Definition 2.1. *Let $1 \leq q \leq \infty$ and $s \in \mathbb{R}$. Then we define the isotropic periodic L_2–Besov space $B_{2,q}^s(\mathbb{T}^n)$ as*

$$
B_{2,q}^s(\mathbb{T}^n) := \Big\{ f \in D'(\mathbb{T}^n) \; ; \; \|f \mid B_{2,q}^s(\mathbb{T}^n)\| =
$$

$$
\Big(\sum_{j=0}^{\infty} 2^{jsq} \Big\| \sum_{k \in Q_j^n} c_k(f)\, e^{ik \cdot} \, \Big| L_2(\mathbb{T}^n) \Big\|^q \Big)^{1/q} < \infty \Big\}
$$

for $q < \infty$ and

$$
B_{2,\infty}^s(\mathbb{T}^n) := \Big\{ f \in D'(\mathbb{T}^n) \; ; \; \|f \mid B_{2,\infty}^s(\mathbb{T}^n)\| =
$$

$$
\sup_{j \in \mathbb{N}_0} 2^{js} \Big\| \sum_{k \in Q_j^n} c_k(f)\, e^{ik \cdot} \, \Big| L_2(\mathbb{T}^n) \Big\| < \infty \Big\},
$$

respectively.

For the definition of the spaces of functions with dominating mixed smoothness properties, we restrict ourselves to the two–dimensional situation. We put the index sets

$$
P_{j_1,j_2} = Q_{j_1}^1 \times Q_{j_2}^1, \quad j_1, j_2 \in \mathbb{N}_0.
$$

As a consequence, we have the splitting

$$
\mathbb{Z}^2 = \bigcup_{j_2=0}^{\infty} \bigcup_{j_1=0}^{\infty} P_{j_1,j_2} \quad \text{with} \quad P_{j_1,j_2} \cap P_{j_1',j_2'} = \emptyset \quad \text{if} \quad (j_1,j_2) \neq (j_1',j_2').
$$

Definition 2.2. *Let $1 \leq q \leq \infty$ and $r_1, r_2 \in \mathbb{R}$. Then the L_2–Besov space $S_{2,q}^{r_1,r_2} B(\mathbb{T}^2)$ of bivariate periodic functions with dominating mixed smoothness* *properties is defined as*

$$S_{2,q}^{r_1,r_2} B(\mathbb{T}^2) \;:=\; \Big\{ f \in D'(\mathbb{T}^2)\,;\; \|f \mid S_{2,q}^{r_1,r_2} B(\mathbb{T}^2)\| =$$

$$\Big(\sum_{j_1,j_2=0}^{\infty} 2^{(j_1 r_1 + j_2 r_2) q} \Big\| \sum_{k \in P_{j_1,j_2}} c_k(f)\, e^{ik\cdot} \mid L_2(\mathbb{T}^2) \Big\|^q \Big)^{1/q} < \infty \Big\}$$

for $q < \infty$ and

$$S_{2,\infty}^{r_1,r_2} B(\mathbb{T}^2) \;:=\; \Big\{ f \in D'(\mathbb{T}^2)\,;\; \|f \mid S_{2,\infty}^{r_1,r_2} B(\mathbb{T}^2)\| =$$

$$\sup_{j_1,j_2 \in \mathbb{N}_0} 2^{(j_1 r_1 + j_2 r_2)} \Big\| \sum_{k \in P_{j_1,j_2}} c_k(f)\, e^{ik\cdot} \mid L_2(\mathbb{T}^2) \Big\| < \infty \Big\},$$

respectively.

Equivalent definitions of the Besov spaces using the moduli of smoothness and further characterizations can be found in [18, 19, 21].

By construction, it holds that

$$B_{2,2}^0(\mathbb{T}^n) = L_2(\mathbb{T}^n) \qquad \text{and} \qquad S_{2,2}^{0,0} B(\mathbb{T}^2) = L_2(\mathbb{T}^2). \tag{2.2}$$

The Besov spaces of bivariate functions with dominating mixed smoothness properties can be characterized as tensor products

$$B_{2,q}^{s_1}(\mathbb{T}) \otimes_b B_{2,q}^{s_2}(\mathbb{T}) = S_{2,q}^{s_1,s_2} B(\mathbb{T}^2)$$

of the corresponding univariate Besov spaces for $q < \infty$ (equivalent norms). Here, the norm b which was used for the completion of the algebraic tensor product is the usual Besov norm $b := \| \cdot \mid S_{2,q}^{s_1,s_2} B(\mathbb{T}^2)\|$. In the sequel, we will use tensor product spaces where the the completion is taken due to the 2-nuclear norm α_2 (cf. [7])

$$\tilde{S}_{2,q}^{s_1,s_2} B(\mathbb{T}^2) := B_{2,q}^{s_1}(\mathbb{T}) \otimes_{\alpha_2} B_{2,q}^{s_2}(\mathbb{T}).$$

The 2-nuclear norms have the main advantage to be uniform crossnorms, cf. e.g. [7, 21]. This means (together with (2.2)) in particular that, for two operators $P \in \mathcal{L}(B_{2,q}^{s_1}(\mathbb{T}), L_2(\mathbb{T}))$ and $Q \in \mathcal{L}(B_{2,q}^{s_2}(\mathbb{T}), L_2(\mathbb{T}))$, the tensor product operator $P \otimes Q$ given by

$$(P \otimes Q)(f \otimes g) := P(f) \otimes Q(g)$$

is bounded, i.e. $P \otimes Q \in \mathcal{L}(\tilde{S}_{2,q}^{s_1,s_2} B(\mathbb{T}^2), L_2(\mathbb{T}^2))$, and its norm can be estimated as

$$\|P \otimes Q \mid \mathcal{L}(\tilde{S}_{2,q}^{s_1,s_2} B(\mathbb{T}^2), L_2(\mathbb{T}^2))\| \tag{2.3}$$
$$\leq \; C \, \|P \mid \mathcal{L}(B_{2,q}^{s_1}(\mathbb{T}), L_2(\mathbb{T}))\| \, \|Q \mid \mathcal{L}(B_{2,q}^{s_2}(\mathbb{T}), L_2(\mathbb{T}))\|$$

with some constant C independent of P and Q.

Remark. Now, the question arises how the spaces $\tilde{S}_{2,q}^{s_1,s_2} B(\mathbb{T}^2)$ are related to the usual Besov spaces $S_{2,q}^{s_1,s_2} B(\mathbb{T}^2)$. For $q = 2$, the L_2–Besov spaces equal the L_2–Sobolev spaces $B_{2,2}^s(\mathbb{T}) = H_2^s(\mathbb{T})$. In this case, we know from the results in [21, 25] that the spaces coincide

$$\tilde{S}_{2,2}^{s_1,s_2} B(\mathbb{T}^2) = S_{2,2}^{s_1,s_2} B(\mathbb{T}^2)$$

(equivalent norms). Because of the imbeddings $B_{2,q}^s(\mathbb{T}^n) \hookrightarrow B_{2,\infty}^s(\mathbb{T}^n)$ for $1 \leq q < \infty$ we may restrict our error estimates in the following sections to the most interesting case $q = \infty$. Here, the 2-nuclear norm of the spaces $\tilde{S}_{2,\infty}^{s_1,s_2} B(\mathbb{T}^2)$ turns out to be stronger than the original Besov norm

$$\tilde{S}_{2,\infty}^{s_1,s_2} B(\mathbb{T}^2) \hookrightarrow S_{2,\infty}^{s_1,s_2} B(\mathbb{T}^2)$$

what has been proved in [21].

3 Interpolation on Equidistant Grids

This section is devoted to error estimates for periodic interpolation on equidistant grids. We can apply the concept of periodic Strang–Fix conditions on the fundamental interpolant in order to find such error estimates.

Let N be a natural number and denote by

$$J_N = \left\{ k \in \mathbb{Z}^n \; ; \; -\frac{N}{2} \leq k_r < \frac{N}{2}, \; r = 1, \ldots, n \right\}$$

a related set of indices. Further

$$T_N = \left\{ \sum_{k \in J_N} \eta_k \, e^{ik\cdot} \; ; \; \eta_k \in \mathbb{C} \right\}$$

denotes a corresponding set of trigonometric polynomials. The discrete Fourier coefficients of a continuous function f are given by

$$c_k^N(f) = \frac{1}{N} \sum_{\ell \in J_N} f\left(\frac{2\pi\ell}{N}\right) e^{2\pi i k \ell / N}, \qquad k \in J_N.$$

Discrete Fourier coefficients and Fourier coefficients are connected by aliasing

$$c_k^N(f) = \sum_{\ell \in \mathbb{Z}^n} c_{k+\ell N}(f),$$

as long as $f \in A(\mathbb{T}^n)$. We consider interpolation on equidistant grids of type

$$\mathcal{T}_N = \left\{ \frac{2\pi k}{N} \; ; \; k \in J_N \right\}.$$

The continuous and 2π–periodic function Λ_N is called a fundamental interpolant for $\mathcal{T}_N$ if

$$\Lambda_N\left(\frac{2\pi k}{N}\right) = \delta_{0,k}, \qquad k \in J_N.$$

The associated Lagrange interpolation operator L_N is defined as

$$L_N f = \sum_{k \in J_N} f\left(\frac{2\pi k}{N}\right) \Lambda_N\left(\cdot - \frac{2\pi k}{N}\right).$$

The Fourier coefficients of $L_N f$ can be easily computed:

$$c_k(L_N f) = N^n\, c_k^N(f)\, c_k(\Lambda_N) = N^n\, c_k(\Lambda_N) \sum_{\ell \in \mathbb{Z}} c_{k+\ell N}(f)$$

for $f \in A(\mathbb{T}^n)$. Finally, we denote the N–th Fourier partial sum by

$$S_N f = \sum_{k \in J_N} c_k(f)\, e^{ik\cdot}.$$

For cardinal interpolation, one can use the Strang–Fix conditions [20, 26] on the fundamental interpolant in order to characterize the reproduction of polynomials and therefore the order of interpolation, too. Up to now there is no complete periodic counterpart.

But we can use the concept of periodic Strang–Fix conditions introduced by Pöplau [2, 14] for L_2–error estimates. Here, the behaviour of the fundamental interpolant is characterized by a certain decay of the Fourier coefficients of Λ_N.

Definition 3.1. *Let $\Lambda_N \in A(\mathbb{T}^n)$ be a fundamental interpolant with respect to $\mathcal{T}_N$. Then Λ_N satisfies the periodic Strang–Fix conditions of order $m > 0$ if for all $k \in J_N$ the inequalities*

$$|1 - N^n c_k(\Lambda_N)| \;\leq\; b_0 \, |k|_2^m \, N^{-m},$$

$$|N^n c_{k+\ell N}(\Lambda_N)| \;\leq\; b_\ell \, |k|_2^m \, N^{-m}, \qquad \ell \in \mathbb{Z}^n \setminus \{0\},$$

hold for some sequence $\{b_\ell\}_{\ell \in \mathbb{Z}^n} \in \ell_2(\mathbb{Z}^n)$ of non–negative numbers.

The periodic Strang–Fix conditions can be seen as the periodic counterpart of the strong Strang–Fix conditions for cardinal interpolation [6].

Theorem 3.2. *Let the fundamental interpolant $\Lambda_N \in A(\mathbb{T}^n)$ satisfy the periodic Strang–Fix conditions of order $m > 0$. Let $n/2 < s < m$. Then there exists a constant C (independent of N) such that*

$$\|f - L_N f \mid L_2(\mathbb{T}^n)\| \leq C \, N^{-s} \, \|f \mid B_{2,\infty}^s(\mathbb{T}^n)\|$$

holds for all $f \in B_{2,\infty}^s(\mathbb{T}^n)$.

Proof. STEP 1. We investigate the case $f \in \mathcal{T}_N$ first. Some computations and the periodic Strang–Fix conditions yield

$$\|f - L_N f \mid L_2(\mathbb{T}^n)\|^2$$

$$= \left\| \sum_{k \in \mathbb{Z}^n} \left(c_k(f) - N^n c_k^N(f) c_k(\Lambda_N)\right) e^{ik\cdot} \,\Big|\, L_2(\mathbb{T}^n) \right\|^2$$

$$= \left\| \sum_{k \in J_N} c_k(f) \, e^{ik\cdot} \Big((1 - N^n c_k(\Lambda_N)) \right.$$

$$\left. - \sum_{\ell \in \mathbb{Z}^n \setminus \{0\}} N^n c_{k+\ell N}(\Lambda_N) \, e^{i\ell N\cdot} \Big) \,\Big|\, L_2(\mathbb{T}^n) \right\|^2$$

$$= \sum_{k \in J_N} |c_k(f)|^2 \Big(|1 - N^n c_k(\Lambda_N)|^2 + \sum_{\ell \in \mathbb{Z}^n \setminus \{0\}} |N^n c_{k+\ell N}(\Lambda_N)|^2 \Big)$$

$$\leq \sum_{k \in J_N} |c_k(f)|^2 |k|_2^{2m} N^{-2m} \sum_{\ell \in \mathbb{Z}^n} b_\ell^2.$$

Let $2^{r-1} \leq N < 2^r$. Then

$$\sum_{k \in J_N} |k|_2^{2m} |c_k(f)|^2 \;=\; \sum_{\ell=0}^{r} \sum_{k \in Q_\ell^n} |k|_2^{2m} 2^{-\ell s} 2^{\ell s} |c_k(f)|^2$$

$$\leq\; 2^{2m} n^m \sum_{\ell=0}^{r} 2^{2(m-s)\ell} \, 2^{2ls} \sum_{k \in Q_\ell^n} |c_k(f)|^2.$$

We apply Hölder's inequality and obtain

$$\sum_{k \in J_N} |k|_2^{2m} |c_k(f)|^2 \;\leq\; 2^{2m} n^m \left(\sum_{\ell=0}^{r} 2^{2(m-s)\ell}\right) \sup_{\ell=0,\dots,r} 2^{2ls} \sum_{k \in Q_\ell^n} |c_k(f)|^2$$

$$\leq\; C_1 N^{2(m-s)} \|f \mid B_{2,\infty}^s(\mathbb{T}^n)\|^2.$$

This means that, for $f \in T_N$, we proved

$$\|f - L_N f \mid L_2(\mathbb{T}^n)\| \leq C_2\, N^{-s} \|f \mid B_{2,\infty}^s(\mathbb{T}^n)\|, \tag{3.1}$$

where C_2 does not depend on f.

STEP 2. We investigate the general case $f \in B_{2,\infty}^s(\mathbb{T}^n)$. Because of $s > n/2$ it holds that $B_{2,\infty}^s(\mathbb{T}^n) \hookrightarrow B_{2,1}^{n/2}(\mathbb{T}^n) \hookrightarrow A(\mathbb{T}^n)$. The interpolation is well-defined and aliasing is applicable. Using the periodic Strang–Fix conditions with $n/2 < s' < s$ and Cauchy–Schwarz inequality, it follows

$$\|L_N(f - S_N f) \mid L_2(\mathbb{T}^n)\|^2$$

$$= \sum_{k \in \mathbb{Z}^n} \left| N^n c_k(\Lambda_N) \sum_{\ell \in \mathbb{Z}^n} c_{k+\ell N}(f - S_N f) \right|^2$$

$$= \sum_{k \in J_N} \sum_{r \in \mathbb{Z}^n} \left| N^n c_{k+rN}(\Lambda_N) \sum_{\ell \in \mathbb{Z}^n} c_{k+rN+\ell N}(f - S_N f) \right|^2$$

$$\leq\; C_3 \sum_{k \in J_N} \sum_{r \in \mathbb{Z}^n} b_r^2 \,|k|_2^{2s'}\, N^{-2s'} \left| \sum_{\ell \in \mathbb{Z}^n} c_{k+\ell N}(f - S_N f) \right|^2$$

$$\leq\; C_3 N^{-2s'} \|\{b_r\} \mid \ell_2(\mathbb{Z}^n)\|^2$$
$$\sum_{k \in J_N} |k|_2^{2s'} \sum_{\ell \in \mathbb{Z}^n} |k + \ell N|_2^{2s'} |c_{k+\ell N}(f - S_N f)|^2 \sum_{r \in \mathbb{Z}^n} |k + rN|_2^{-2s'}.$$

Next we use that for $s' > n/2$

$$\sup_{k \in J_N} |k|_2^{2s'} \sum_{r \in \mathbb{Z}^n} |k + rN|_2^{-2s'} = \sup_{k \in J_N} \left|\frac{k}{N}\right|_2^{2s'} \sum_{r \in \mathbb{Z}^n} \left|\frac{k}{N} + r\right|_2^{-2s'} = C_4 < \infty.$$

This proves

$$\|L_N(f - S_N f) \mid L_2(\mathbb{T}^n)\|^2$$
$$\leq\; C_3 C_4\, N^{-2s'} \|\{b_r\} \mid \ell_2(\mathbb{Z}^n)\|^2 \sum_{k \in J_N} \sum_{\ell \in \mathbb{Z}^n} |k + \ell N|^{2s'} |c_{k+\ell N}(f - S_N f)|^2$$
$$\leq\; C_5\, N^{-2s'} \|f - S_N f \mid H_2^{s'}(\mathbb{T}^n)\|^2,$$

where $H_2^{s'}(\mathbb{T}^n)$ denotes the fractional order Sobolev space with the norm

$$\|f \mid H_2^{s'}(\mathbb{T}^n)\|^2 := \sum_{k \in \mathbb{Z}^n} (1 + |k|_2^2)^{s'} \, |c_k(f)|^2.$$

In case $s > s'$ one knows

$$\|f - S_N f \mid H_2^{s'}(\mathbb{T}^n)\| \leq C_6 \, N^{s'-s} \|f \mid B_{2,\infty}^s(\mathbb{T}^n)\|, \tag{3.2}$$

cf. e.g. [11]. This yields

$$\|L_N(f - S_N f) \mid L_2(\mathbb{T}^n)\| \leq C_7 \, N^{-s} \|f \mid B_{2,\infty}^s(\mathbb{T}^n)\|, \tag{3.3}$$

where again the constant C_7 does not depend on N and f. To finish the proof, observe that (3.1), (3.2) and (3.3) (applied with $s' = 0$) imply

$$\begin{aligned}
\|f - &L_N f \mid L_2(\mathbb{T}^n)\| \\
\leq \; & \|f - S_N f \mid L_2(\mathbb{T}^n)\| + \|S_N f - L_N(S_N f) \mid L_2(\mathbb{T}^n)\| \\
& + \|L_N(f - S_N f) \mid L_2(\mathbb{T}^n)\| \\
\leq \; & C \, N^{-s} \|f \mid B_{2,\infty}^s(\mathbb{T}^n)\|.
\end{aligned}$$

This proves the theorem. $\qquad\square$

Remark. We note that the most constants appearing in the proof only depend on the dimension n and on the smoothness s of the function to be interpolated. The dependency on the used fundamental interpolant Λ_N is reflected in the constants by the term $\|\{b_r\} \mid \ell_2(\mathbb{Z}^n)\|$ from the Strang–Fix conditions.

Remark. Recall, if X is a Banach space and W a subspace of X, then the linear N–width is defined as

$$\lambda_N(W, X) = \inf_{\substack{U_N \in Lin_N(X) \\ P \in \mathcal{L}(X, U_N)}} \sup_{f \in W} \|f - Pf \mid X\|,$$

where the infimum is taken over all subspaces U_N of X of finite dimension $\leq N$ and all linear operators P from X to U_N. Here we are interested in $X = L_2(\mathbb{T})$ and W the unit ball in the Nikol'skij–Besov space $B_{2,\infty}^s(\mathbb{T})$, denoted by $B_2^s(\mathbb{T})$. If $s > 0$, then

$$\lambda_N(B_2^s(\mathbb{T}), L_2(\mathbb{T})) \sim N^{-s},$$

cf. [9, Theorem 14.3.8]. In this sense, approximation of univariate functions with those interpolation operators L_N is nearly optimal (nearly optimal means the order of approximation is correct but may be not the constants). More details about widths may be found in [9, 27].

Remark.   An interesting limiting case has been observed by Pöplau [2, 14]. If Λ_N is a fundamental interpolant which satisfies the periodic Strang–Fix condition of order $m > n/2$, then there exists a constant C (independent of N) such that

$$\|f - L_N f \mid L_2(\mathbb{T}^n)\| \leq C\, N^{-m}\, \|f \mid B_{2,2}^m(\mathbb{T}^n)\|$$

holds for all $f \in B_{2,2}^m(\mathbb{T}^n)$. For a generalization in various directions, including different function spaces (defined by using decay properties of the Fourier coefficients), we refer to [23, 24].

Corollary 3.3.   *Let the univariate fundamental interpolant $\Lambda_N \in A(\mathbb{T})$ satisfy the periodic Strang–Fix conditions of order $m > 0$. Let $L_N \otimes L_N$ be the interpolation operator associated with the bivariate fundamental interpolant $\Lambda_N \otimes \Lambda_N$. Let $1 < s < m$. Then there exists a constant C (independent of N) such that*

$$\|f - (L_N \otimes L_N)f \mid L_2(\mathbb{T}^2)\| \leq C\, N^{-s}\, \|f \mid B_{2,\infty}^s(\mathbb{T}^2)\|$$

holds for all $f \in B_{2,\infty}^s(\mathbb{T}^2)$.

Proof.   Because $c_k(\Lambda_N \otimes \Lambda_N) = c_{k_1}(\Lambda_N)c_{k_2}(\Lambda_N)$, one proves easily that also the bivariate fundamental interpolant $\Lambda_N \otimes \Lambda_N$ satisfies periodic Strang–Fix conditions of order m. Then, Theorem 3.2 is applicable.   □

Example: B–Splines

As an example, we may use the interpolation by the 2π–periodized centered B–Spline $\mathcal{M}_{N,r}$ of order $r \in \mathbb{N}$. Its Fourier coefficients are known as

$$c_k(\mathcal{M}_{N,r}) = \frac{1}{N}\left(\operatorname{sinc}\frac{\pi k}{N}\right)^r, \qquad k \in \mathbb{Z},$$

with $\operatorname{sinc} t := \sin t / t$. The fundamental interpolant $\Lambda_{N,r}$ corresponding to the 2π–periodic centered B–spline of order r can be computed from

$$c_k(\Lambda_{N,r}) := \frac{c_k(\mathcal{M}_{N,r})}{N\, c_k^N(\mathcal{M}_{N,r})}, \qquad k \in \mathbb{Z}. \tag{3.4}$$

Then, $\Lambda_{N,r}$ satisfies the periodic Strang–Fix conditions of order r (cf. [14]) with the constants

$$b_0 = \begin{cases} \dfrac{1}{2^{r+1}} & \text{for } r \text{ odd,} \\[2ex] \dfrac{1}{2(2^r - 1)} & \text{for } r \text{ even,} \end{cases}$$

and

$$b_\ell = \frac{1}{\pi^r (2|\ell| - 1)^r} \begin{cases} 1 & \text{for } r = 1, \\[2ex] \dfrac{(r-1)!}{E_{(r-1)/2}} & \text{for } r > 1, \text{ odd,} \\[2ex] \dfrac{r!}{2^r (2^r - 1) B_{r/2}} & \text{for } r \text{ even,} \end{cases}$$

for $\ell \neq 0$. Here, B_s and E_s $(s \in \mathbb{N})$ denote the corresponding Bernoulli and Euler numbers.

Example: Trigonometric Interpolation

Another example is the trigonometric interpolation. The de la Vallée Poussin means $\mathcal{V}_N^K$ $(N, K \in \mathbb{N}, N > K)$ of the Dirichlet kernel are given by

$$\mathcal{V}_N^K(x) := \frac{1}{4KN} \sum_{\ell=N-K}^{N+K-1} \left(\sum_{k=-\ell}^{\ell} e^{ikx} \right).$$

They are fundamental interpolants for the grid $\mathcal{T}_{2N}$. So, we have a lot of different fundamental interpolants for the grids $\mathcal{T}_N$ (N even) belonging to different parameters K. We denote them by $\Lambda_{N,K} := \mathcal{V}_{N/2}^K$ for $N/2, K \in \mathbb{N}, N/2 > K$. Since the de la Vallée Poussin means are trigonometric polynomials they of course satisfy Strang–Fix conditions of arbitrary order. But the constants of the Strang–Fix conditions depend on the quotient of the parameters K and N. For $K = 1$, we obtain the best constants since $\Lambda_{N,1}$ is only a slight modification of the Dirichlet kernel whose Fourier coefficients are compared with the Fourier coefficients of the fundamental interpolant. For $K = N/2 - 1$, the corresponding de la Vallée Poussin mean is already very close to the Fejér kernel and the constants are much bigger (for details we refer to [23, 24]).

Example: Radial Basis Functions

A nice n–variate example can be found in [15]. Let the n–variate radial basis function φ be given by its Fourier coefficients

$$N^n c_k(\varphi) := |k|_2^{-\alpha}, \qquad k \in \mathbb{Z}^n \setminus \{0\},$$

for a fixed $\alpha > d$. In case $\alpha \in 2\mathbb{N}$, we obtain the periodized version of the cardinal polyharmonic splines [10]. The associated fundamental interpolant can be constructed analogously to the spline case (3.4) from its Fourier coefficients

$$N^n c_k(\Lambda_{N,\varphi}) := \begin{cases} \dfrac{|k|_2^{-\alpha}}{\sum_{\ell \in \mathbb{Z}^n} |k + \ell N|_2^{-\alpha}} & \text{for } k \in \mathbb{Z}^n \setminus N\,\mathbb{Z}^n, \\[2mm] 1 & \text{for } k = 0, \\[1mm] 0 & \text{for } k \in N\,\mathbb{Z}^n \setminus \{0\}. \end{cases}$$

Because of $\alpha > d$, this fundamental interpolant $\Lambda_{N,\varphi}$ belongs to the Wiener algebra. Furthermore, it satisfies the periodic Strang–Fix conditions of order α with the constants

$$b_0 = 2^\alpha \sum_{r=1}^{n} \binom{n}{r} \frac{1}{r^{\alpha/2}} \left(\frac{2\alpha - r}{\alpha - r} \right)$$

and

$$b_\ell = 2^\alpha |v(\ell)|_2^{-\alpha}, \qquad \ell \in \mathbb{Z}^n \setminus \{0\},$$

where the vector v has the components $v_r(\ell) = \delta_{0,\ell_r}(2|\ell_r| - 1)$ for $r = 1, \ldots, n$.

In addition to these examples, one can find more examples of bivariate functions in [13, 14] (3– and 4–direction box splines) satisfying periodic Strang–Fix conditions of certain order.

4 Interpolation on Sparse Grids

Now we want to define the interpolation operators for interpolation on sparse grids and give error estimates. The definition of the blending interpolation operator and its basic properties can be found e.g. in [1, 4]. This definition needs the notation of a chain of projectors.

The ordering relation $P \leq Q$ for projectors holds if $PQ = QP = P$. A family of projectors $\{P_j\}_{j=0}^{\infty}$ forms a chain if $P_j \leq P_{j+1}$, $j \in \mathbb{N}_0$. For two interpolation projectors L_K and L_N, the ordering $L_K \leq L_N$ holds if and only if the images $\operatorname{Im} L_K \subset \operatorname{Im} L_N$ as well as the grids $\mathcal{T}_K \subset \mathcal{T}_N$ are ordered.

Fix $d \in \mathbb{N}$. By the choice $N_j := d\, 2^j$, we immediately insure $\mathcal{T}_{N_j} \subset \mathcal{T}_{N_{j+1}}$. Furthermore, we assume

$$\mathrm{Im}\ L_{N_j} \subset \mathrm{Im}\ L_{N_{j+1}}. \tag{4.1}$$

This property has to be proved for every example by hand. Then, we have a chain

$$L_{N_0} \leq L_{N_1} \leq \cdots \leq L_{N_j} \leq L_{N_{j+1}} \leq \cdots \tag{4.2}$$

of interpolation operators.

Given a chain (4.2) of interpolation operators $L_{N_j}, j \in \mathbb{N}_0$, for univariate functions. For bivariate functions, we will consider the j-th order blending operator defined by the j-th order Boolean sum

$$B_j := \bigoplus_{r=0}^{j} L_{N_r} \otimes L_{N_{j-r}},$$

where $A \oplus B := A + B - AB$. The representation of B_j in terms of ordinary sums is known to be

$$B_j = \sum_{r=0}^{j} L_{N_r} \otimes L_{N_{j-r}} - \sum_{r=0}^{j-1} L_{N_r} \otimes L_{N_{j-r-1}}.$$

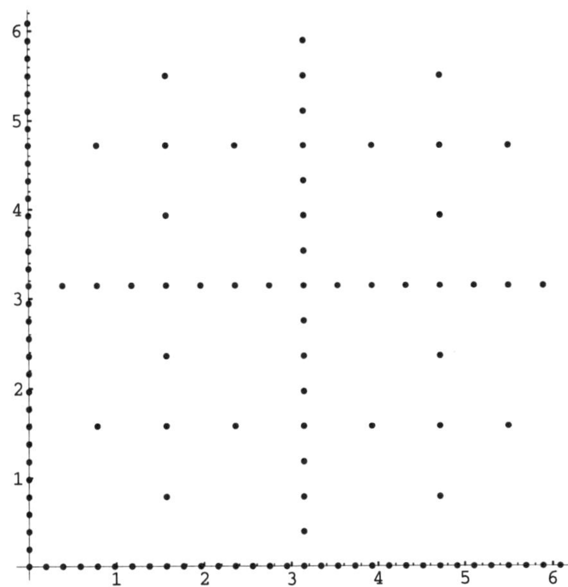

Figure 1: Sparse grid $\mathcal{T}_5^{\mathrm{B}}$ for $d = 1$.

The Boolean sums have the range $\operatorname{Im} B_j = \sum_{r=0}^{j} \operatorname{Im} L_{N_r} \otimes \operatorname{Im} L_{N_{j-r}}$. They interpolate on the sparse grid $\mathcal{T}_j^{\mathrm{B}} := \bigcup_{r=0}^{j} \mathcal{T}_{N_r} \times \mathcal{T}_{N_{j-r}}$ which has $d^2(j2^{j-1}+2^j)$ nodes which is essentially less than the $d^2\,2^{2j}$ nodes in the equidistant grid $\mathcal{T}_{N_j} \times \mathcal{T}_{N_j}$.

Theorem 4.1. *Suppose that the interpolation operators L_{N_j}, $j \in \mathbb{N}_0$, form a chain (4.2) and satisfy*

$$\sup_{j \in \mathbb{N}_0} N_j^{s_k} \|f - L_{N_j} f \mid L_2(\mathbb{T})\| \leq C_k \|f \mid B_{2,\infty}^{s_k}(\mathbb{T})\|$$

with constants C_k independent of f and for some fixed s_1, s_2 with $s_1, s_2 > 1/2$. Then in case $s_1 = s_2 = s$, we find

$$\|f - B_j f \mid L_2(\mathbb{T}^2)\| \leq C\,(j+1)\,N_j^{-s}\,\|f \mid \widetilde{S}_{2,\infty}^{s,s} B(\mathbb{T}^2)\|$$

for all $f \in \widetilde{S}_{2,\infty}^{s,s} B(\mathbb{T}^2)$, whereas in case $s_1 \neq s_2$, it holds that

$$\|f - B_j f \mid L_2(\mathbb{T}^2)\| \leq C\,N_j^{-\min(s_1,s_2)}\,\|f \mid \widetilde{S}_{2,\infty}^{s_1,s_2} B(\mathbb{T}^2)\|$$

for all $f \in \widetilde{S}_{2,\infty}^{s_1,s_2} B(\mathbb{T}^2)$. In both situations, C denotes a constant independent of j and f.

Proof. Because of $\widetilde{S}_{2,\infty}^{s_1,s_2} B(\mathbb{T}^2) \hookrightarrow A(\mathbb{T}^2) \hookrightarrow C(\mathbb{T}^2)$ for $s_1, s_2 > 1/2$, interpolation is well-defined. The remainder ($P^c := I - P$) of the blending interpolation has the representation

$$B_j^c = L_{N_j}^c \otimes I + I \otimes L_{N_j}^c - \sum_{r=0}^{j} L_{N_r}^c \otimes L_{N_{j-r}}^c + \sum_{r=0}^{j-1} L_{N_r}^c \otimes L_{N_{j-r-1}}^c,$$

cf. [4]. With this, the assertion follows from the triangle inequality, the uniformity of the norms (see (2.3)) and the assumption on the error for the univariate interpolation. $\qquad\square$

Corollary 4.2. *Let the 2π-periodic fundamental interpolants $\Lambda_{N_j} \in A(\mathbb{T})$ satisfy the periodic Strang–Fix conditions of order $m > 0$ with same sequence $\{b_\ell\}$ of constants. The corresponding interpolation operators L_{N_j}, $j \in \mathbb{N}_0$, form a chain (4.2). Let $1/2 < s_1, s_2 < m$.*

Then, in case $s_1 = s_2 = s$, we can estimate

$$\|f - B_j f \mid L_2(\mathbb{T}^2)\| \leq C\,(j+1)\,N_j^{-s}\,\|f \mid \widetilde{S}_{2,\infty}^{s,s} B(\mathbb{T}^2)\|,$$

for all $f \in \widetilde{S}_{2,\infty}^{s,s} B(\mathbb{T}^2)$.

In case $s_1 \neq s_2$, it holds that

$$\|f - B_j f \mid L_2(\mathbb{T}^2)\| \leq C \, N_j^{-\min(s_1,s_2)} \, \|f \mid \widetilde{S}_{2,\infty}^{s_1,s_2} B(\mathbb{T}^2)\|$$

for all $f \in \widetilde{S}_{2,\infty}^{s_1,s_2} B(\mathbb{T}^2)$. In both situations, C denotes a constant independent of j and f.

The same ideas as before yield the following estimate. It shows that the order of the interpolation error for equidistant grids does not improve for the smoother functions with dominating mixed smoothness properties in comparison to the isotropic case. For the functions with dominating mixed smoothness the error of interpolation on sparse grids is only by a logarithmic factor worse the result for equidistant grids.

Corollary 4.3.   *Let the univariate fundamental interpolant $\Lambda_N \in A(\mathbb{T})$ satisfy the periodic Strang–Fix conditions of order $m > 0$. Let $L_N \otimes L_N$ be the interpolation operator associated with the bivariate fundamental interpolant $\Lambda_N \otimes \Lambda_N$. Let $1/2 < s_1, s_2 < m$. Then there exists a constant C (independent of N) such that*

$$\|f - (L_N \otimes L_N)f \mid L_2(\mathbb{T}^2)\| \leq C \, N^{-\min(s_1,s_2)} \, \|f \mid \widetilde{S}_{2,\infty}^{s_1,s_2} B(\mathbb{T}^2)\|$$

holds for all $f \in \widetilde{S}_{2,\infty}^{s_1,s_2} B(\mathbb{T}^2)$.

Example: B–Splines

The fundamental interpolants $\Lambda_{N_j,r}$ belonging to the 2π–periodic centered B–spline of even order $r \in \mathbb{N}$ satisfy (4.1) automatically since at the step from j to $j+1$ only some new spline knots are added. Therefore, the corresponding interpolation operators form a chain (4.2). The constants for Strang–Fix conditions given in the previous section do not depend on N_j.

The fundamental interpolants $\Lambda_{N_j,r}$ belonging to the 2π–periodic centered B–spline of odd order $r \in \mathbb{N}$ do not satisfy (4.1). For splines for the grid $\mathcal{T}_{N_{j+1}}$ only totally new spline knots are used compared to the j–th grid.

Example: Trigonometric Interpolation

The de la Vallée Poussin means Λ_{N_j,K_j} satisfy the chain condition (4.1) only under certain restrictions on K_j and N_j. In [17], it was shown that for N_j as before and

$$K_j := \begin{cases} 2^{j-\kappa-1} & \text{for } j > \kappa, \\ 1 & \text{for } j \leq \kappa, \end{cases} \qquad \kappa \in \mathbb{N}, \ 3 \leq d \, 2^\kappa,$$

condition (4.1) is satisfied. The case $\kappa = \infty$ is allowed. With this choice of the parameters N_j and K_j, one can estimate the constants of the Strang–Fix conditions of order m uniformly by

$$b_\ell = \begin{cases} \dfrac{3^m}{2(2\pi)^m} & \text{for } \ell = -1, 0, 1, \\ 0 & \text{otherwise.} \end{cases}$$

Example: Radial Basis Functions

The fundamental interpolants $\Lambda_{N_j,\varphi}$ constructed from the radial basis function φ satisfy the periodic Strang–Fix conditions with constants not depending on N_j. Now we restrict ourselves to the univariate case. One can find constants a_k, $k = 0, \ldots, N_{j+1} - 1$, such that

$$c_{k+\ell N_{j+1}}(\Lambda_{N_j,\varphi}) = a_k \, c_{k+\ell N_{j+1}}(\Lambda_{N_{j+1},\varphi}), \qquad \ell \in \mathbb{Z}, \ k = 0, \ldots, N_{j+1} - 1.$$

These constants are $a_0 = 1$, $a_{N_j} = 0$, and $a_k = 1/2 \left(\sum_{\ell \in \mathbb{Z}} |k + 2\ell N_j|^{-\alpha} \right) / \left(\sum_{\ell \in \mathbb{Z}} |k + \ell N_j|^{-\alpha} \right)$, otherwise. This yields the chain property (4.1), cf. [12].

References

[1]   G. Baszenski, F.-J. Delvos: *A discrete Fourier transform scheme for Boolean sums of trigonometric operators*, in: *Multivariate Approximation Theory IV*, C. K. Chui, W. Schempp, and K. Zeller (eds.) ISNM **90**, 15–24, Birkhäuser, Basel 1989.

[2]   G. Brumme: *Error estimates for periodic interpolation by translates*, in: *Wavelets, Images, and Surface Fitting*, P. J. Laurent, A. Le Méhauté, and L. L. Schumaker (eds.), 75–82, AK Peters, Boston 1994.

[3]   F.-J. Delvos: *Approximation properties of periodic interpolation by translates of one function*, Math. Model. Numer. Anal. **28** (1994), 177–188.

[4]   F.-J. Delvos, W. Schempp: *Boolean Methods in Interpolation and Approximation*, Pitman Research Notes in Mathematics Series **230**, Longman Scientific & Technical, Harlow 1989.

[5]   R. A. DeVore, S. V. Konyagin, V. N. Temlyakov: *Hyperbolic wavelet approximation*, Constr. Approx. **14** (1998), 1–26.

[6]   K. Jetter: *Multivariate approximation from the cardinal point of view*, in: *Approximation Theory VII*, E. W. Cheney, C. K. Chui, and L. L. Schumaker (eds.), 131–161, Academic Press, New York 1992.

[7]   W. A. Light, E. W. Cheney: *Approximation Theory in Tensor Product Spaces*, Lecture Notes in Mathematics **1169**, Springer, Berlin 1985.

[8]   F. Locher: *Interpolation on uniform meshes by translates of one function and related attenuation factors*, Math. Comp. **37** (1981), 404–416.

[9]   G. G. Lorentz, M. von Golitschek, and Y. Makovoz: *Constructive Approximation: Advanced Problems*, Springer, Berlin 1996.

[10]   W. R. Madych, S. A. Nelson: *Polyharmonic cardinal splines*, J. Approx. Theory **60** (1990), 141–161.

[11]   A. Pietsch: *Approximation spaces*, J. Approx. Theory **32** (1981), 115–134.

[12]   G. Plonka, M. Tasche: *A unified approach to periodic wavelets*, in: *Wavelets: Theory, Algorithms, and Applications*, C. K. Chui, L. Montefusco, and L. Puccio (eds.), 137–151, Academic Press, 1994.

[13]   G. Pöplau: *Fast direct solvers for PDE's in shift–invariant spaces*, in: *Approximation Theory VIII, vol. 2: Wavelets and Multilevel Approximation*, C. K. Chui and L. L. Schumaker (eds.), 325–333, World Scientific Publishing, Singapore 1995.

[14]   G. Pöplau: *Multivariate periodische Interpolation durch Translate und Anwendungen*, Doctoral thesis, Universität Rostock 1995.

[15]   G. Pöplau: *Fast solvers for elliptic PDE's with radial basis functions*, Preprint 96/10, Universität Rostock 1996.

[16] G. Pöplau, F. Sprengel: *Some error estimates for periodic interpolation on full and sparse grids*, in: *Curves and Surfaces with Applications in CAGD*, A. Le Méhauté, C. Rabut, and L. L. Schumaker (eds.), 355–362, Vanderbilt University Press, Nashville 1997.

[17] J. Prestin, K. Selig: *Interpolatory and orthonormal trigonometric wavelets*, in: *Signal and Image Representation in Combined Spaces*, J. Zeevi and R. Coifman (eds.), 201–255, Academic Press, 1998.

[18] H.-J. Schmeißer: *An unconditional basis in periodic spaces with dominating mixed smoothness properties*, Anal. Math. **13** (1987), 153–168.

[19] H.-J. Schmeißer, H. Triebel: *Topics in Fourier Analysis and Function Spaces*, Akadem. Verlagsgesellschaft Geest & Portig, Leipzig 1987.

[20] I. J. Schoenberg: *Contributions to the problem of approximation of equidistant data by analytic functions. Part A: On the problem of smoothing or graduation. A first class of analytic formulae*, Quart. Appl. Math **IV** (1946), 45–99.

[21] W. Sickel: *Tensor products of Besov spaces*, Jenaer Schriften zur Mathematik und Informatik **99/04**, 1999.

[22] W. Sickel, F. Sprengel: *Interpolation on sparse grids and tensor products of Nikol'skij–Besov spaces*, J. Comput. Anal. Appl. **1** (1999), to appear.

[23] F. Sprengel: *Interpolation und Waveletzerlegung multivariater periodischer Funktionen*, Doctoral thesis, Universität Rostock 1997.

[24] F. Sprengel: *A class of function spaces and interpolation on sparse grids*, Preprint 98/13, Universität Rostock 1998.

[25] F. Sprengel: *A tool for approximation in bivariate periodic Sobolev spaces*, in: *Approximation Theory IX, vol. 2: Computational Aspects*, C. K. Chui and L. L. Schumaker (eds.), 319–326, Vanderbilt University Press, Nashville 1998.

[26] G. Strang, G. Fix: *A Fourier analysis of the finite element variational method*, in: *Constructive Aspects of Functional Analysis*, G. Geymont (ed.), 793–840, C.I.M.E., 1973.

[27]   V. M. Tichomirov: *Approximation theory*, in: *Encyclopedia of Math. Science: Analysis II*, vol. 14, R. V. Gamkrelidze (ed.), 93–244, Springer, Berlin 1990.

[28]   C. Zenger: *Sparse grids*, in: *Notes on Numerical Fluid Mechanics*, vol. 31, W. Hackbusch (ed.), 297–301, Vieweg, Braunschweig 1991.

Addresses:

WINFRIED SICKEL
Mathematisches Institut
Friedrich–Schiller–Universität Jena
D–07740 Jena
Germany

FRAUKE SPRENGEL
Centrum voor Wiskunde en Informatica
P.O.Box 94079
NL–1090 GB Amsterdam
The Netherlands

The Uniform Error of Hyperinterpolation on the Sphere

Ian H. Sloan and Robert S. Womersley

Dedicated to Manfred Reimer on the occasion of his 65th birthday

Abstract

This paper considers the problem of approximation of a continuous function on the unit sphere $S^{r-1} \subseteq \mathbb{R}^r$ by a spherical polynomial from the space $\mathbb{P}_n$ of all spherical polynomials of degree $\leq n$. For $r = 3$ it was shown in [16] that the hyperinterpolation approximation $L_n f$ (which is the linear projection obtained by approximating the Fourier coefficients in the exact L_2 orthogonal projection by a positive–weight quadrature rule that integrates exactly all polynomials of degree $\leq 2n$) has the exact order $\|L_n\| \asymp n^{1/2}$ for its uniform norm, provided the underlying quadrature rule satisfies an additional 'quadrature regularity' assumption. This rate of growth is the same as that of $\|P_n\|$, where $P_n f$ is the L_2 orthogonal projection, and is optimal among all linear projections on $\mathbb{P}_n$ for $r = 3$. For $r \geq 3$ an upper bound on $\|L_n\|$ of non–optimal asymptotic order $O(n^{(r-1)/2})$ also holds, without any special assumption on the quadrature rule. This paper first surveys these recent theoretical results, and then for the first time presents numerical results, both on the growth of the uniform norm of the hyperinterpolation operator up to degree 200, and on the uniform error in the approximation for a set of test functions.

1 Introduction

This paper surveys recent results on the construction of good approximations to continuous functions on the unit sphere $S^{r-1} \subseteq \mathbb{R}^r$ from the space $\mathbb{P}_n$ of polynomials of degree $\leq n$ restricted to the sphere S^{r-1}.

In particular, this paper is concerned with the 'hyperinterpolation' approximation $L_n f$, which may be described as an approximation to the L_2 orthogonal projection $P_n f$ (or the partial sum of the Laplace series for f) in which the exact L_2 inner products are approximated by a quadrature rule, with the quadrature rule being required to have all weights positive and to integrate all polynomials of degree $\leq 2n$ exactly.

Advances in Multivariate Approximation; W. Haußmann, K. Jetter and M. Reimer (eds.)
Mathematical Research, Vol. 107, pp. 289–306, ISBN 3-527-40236-5
© WILEY-VCH, Berlin 1999

Recently it was shown, in [16], that for $r = 3$ the uniform norm of L_n, defined by

$$\|L_n\| = \sup_{f \in \mathbb{P}_n, f \neq 0} \frac{\|L_n f\|_\infty}{\|f\|_\infty}, \tag{1.1}$$

has the optimal rate of growth with n, namely the exact order $n^{1/2}$, among all linear projections onto $\mathbb{P}_n$, provided only that the family of quadrature rules satisfies a mild 'quadrature regularity' assumption. The quadrature regularity assumption (explained below) is satisfied by all of the common quadrature rules.

It was also shown in [16], for all $r \geq 3$ and with no quadrature regularity assumption, that $\|L_n\| \leq d_n^{1/2}$, where d_n is the dimension of the space $\mathbb{P}_n$. In the case $r = 3$ the dimension is $d_n = (n + 1)^2$, thus the bound in this case is $\|L_n\| \leq n + 1$.

This paper surveys these theoretical results, and for the first time presents numerical results for hyperinterpolation. Hyperinterpolation was introduced in the paper [14], and shown there to have the optimal value for its norm in the setting in which the error is observed in the L_2 norm. Specifically, it was shown there that $\|L_n\|_{C \to L_2} = |S^{r-1}|^{1/2}$, where $|S^{r-1}|$ is the surface area of the sphere S^{r-1}. (That this is optimal follows by observing that the function $f \equiv 1$ on S^{r-1} has $\|f\|_\infty = 1$, yet gives $\|L_n f\|_2 = \|f\|_2 = |S^{r-1}|^{1/2}$.) Subsequently it was shown in [15] that for classical interpolation the same norm is strictly larger than $|S^{r-1}|^{1/2}$ if $r \geq 3$ and $n \geq 3$, no matter how the interpolation points may be chosen. Reimer [13] has extensively studied classical polynomial interpolation on the sphere, while Freeden et al. [7] cover both polynomial and other types of approximation.

For simplicity, and because the most important of the results above is available only for $r = 3$, this presentation is restricted to the case $r = 3$.

2   Approximation Schemes and Background

With r taken to be 3, a convenient basis for $\mathbb{P}_n$ is provided by the spherical harmonics of degree $\leq n$,

$$\{Y_{\ell,k} : 0 \leq \ell \leq n, 1 \leq k \leq 2\ell + 1\},$$

which we assume to be orthonormal, i.e.

$$\int_{S^{r-1}} Y_{\ell,k}(x) Y_{\ell',k'}(x) dx = \delta_{\ell\ell'} \delta_{kk'},$$

where dx is the surface measure of S^{r-1}. Then the L_2 orthogonal projection of f onto $\mathbb{P}_n$, or the partial sum of the Laplace series, is defined by

$$P_n f = \sum_{\ell=0}^{n} \sum_{k=1}^{2\ell+1} (f, Y_{\ell,k}) \, Y_{\ell,k},$$

where

$$(f, g) = \int_{S^{r-1}} f(x) g(x) dx.$$

An equivalent definition is

$$P_n f \in \mathbb{P}_n, \quad (P_n f, p) = (f, p) \quad \forall \, p \in \mathbb{P}_n.$$

The hyperinterpolation approximation may be defined in a similar way, but with the exact inner products replaced by a suitable quadrature rule. For $g \in C(S^{r-1})$, let $Q_m g$ denote an m–point quadrature rule, i.e.

$$Q_m g = \sum_{j=1}^{m} w_j g(t_j),$$

where

$$w_j > 0, \quad t_j \in S^{r-1}, \quad j = 1, \ldots, m,$$

and where Q_m has precision $2n$, i.e.

$$Q_m p = \int_{S^{r-1}} p(x) dx \quad \forall \, p \in \mathbb{P}_{2n}.$$

Then $L_n f$ may be defined by

$$L_n f = \sum_{\ell=0}^{n} \sum_{k=1}^{2\ell+1} (f, Y_{\ell,k})_m \, Y_{\ell,k}, \tag{2.1}$$

or alternatively by

$$L_n f \in \mathbb{P}_n, \quad (L_n f, p)_m = (f, p)_m \quad \forall \, p \in \mathbb{P}_{2n},$$

where

$$(f, g)_m = \sum_{j=1}^{m} w_j f(t_j) g(t_j) = Q_m(fg).$$

In practice it is often simpler to express $L_n f$ differently, by introducing the 'reproducing kernel'

$$
\begin{aligned}
G_n(x, y) &= \sum_{\ell=0}^{n} \sum_{k=1}^{2\ell+1} Y_{\ell,k}(x) Y_{\ell,k}(y) \\
&= \sum_{\ell=0}^{n} \frac{2\ell+1}{4\pi} P_\ell(x \cdot y),
\end{aligned}
\tag{2.2}
$$

where in the last step we used the 'addition theorem' for spherical harmonics, and P_ℓ is the Legendre polynomial of degree ℓ.

By rearranging the summation in (2.1) (remembering that $(\cdot, \cdot)_m$ is itself a quadrature sum), $L_n f$ can now be written in the form

$$
L_n f(x) = \sum_{j=1}^{m} w_j G_n(t_j, x) f(t_j),
\tag{2.3}
$$

in which form the approximation is very easy to implement numerically. In a similar vein we can write

$$
P_n f(x) = \int_{S^2} G_n(t, x) f(t) dt.
\tag{2.4}
$$

The approximation schemes described here are projection methods: that is, they take the form $T_n f$ where T_n is a projection on $\mathbb{P}_n$, i.e.

$$
\begin{aligned}
&T_n : C(S^2) \longrightarrow \mathbb{P}_n \text{ is linear,} \\
&\mathrm{range}(T_n) = \mathbb{P}_n, \\
&T_n^2 = T_n.
\end{aligned}
$$

As is well known, projection methods have error bounds closely related to the error of best approximation, in that for arbitrary $p \in \mathbb{P}_n$,

$$
\begin{aligned}
\|f - T_n f\|_\infty &= \|(f - p) - T_n(f - p)\|_\infty \\
&\leq (1 + \|T_n\|) \|f - p\|_\infty,
\end{aligned}
$$

and hence

$$
\|f - T_n f\|_\infty \leq (1 + \|T_n\|) \inf_{p \in \mathbb{P}_n} \|f - p\|_\infty.
$$

This is the motivation for wanting the norms $\|T_n\|$ to be small.

3   Norms – Theoretical Results

It is known that the projection method with range $\mathbb{P}_n$ that has the least value for its uniform norm is the L_2 orthogonal projection P_n. This possibly surprising fact was proved by Berman [1] for $r = 2$ and by Daugavet [3] for $r \geq 3$. The precise behaviour of P_n for $r = 3$ was established by Gronwall [9], who showed that

$$\lim_{n \to \infty} n^{-1/2}\|P_n\| = 2\sqrt{2/\pi}.$$

Thus we know that

$$\|P_n\| \asymp n^{1/2},$$

where $a_n \asymp b_n$ means $c_1 b_n \leq a_n \leq c_2 b_n$ for some positive constants c_1, c_2. And we also know that this is the ideal behaviour for a projection method with range $\mathbb{P}_n$.

The orthogonal projection is of course expensive to compute; indeed it is in principle impossible to compute $P_n f$ exactly if only point values of f are available. We therefore look to the hyperinterpolation approximation $L_n f$ as a realistic alternative. It loses essentially nothing compared to the orthogonal projection if we can prove $\|L_n\| \asymp n^{1/2}$. We shall see that this is indeed possible, under an additional mild assumption on the quadrature rule.

A concrete expression for the norm of $\|L_n\|$, obtained in [16] from (2.3) and the norm definition (1.1) by an easy argument, is

$$
\begin{aligned}
\|L_n\| &= \max_{x \in S^2} \sum_{j=1}^{m} w_j |G_n(t_j, x)| \\
&= \sum_{j=1}^{m} w_j |G_n(t_j, x_0)|,
\end{aligned}
\tag{3.1}
$$

for some (unknown) point $x_0 \in S^2$.

From this and the Cauchy–Schwarz inequality follows, without any additional assumption on the quadrature rule,

$$
\begin{aligned}
\|L_n\| &\leq \left(\sum_{j=1}^{m} w_j \right)^{1/2} \left(\sum_{j=1}^{m} w_j G_n(t_j, x_0)^2 \right)^{1/2} \\
&= (4\pi)^{1/2} \left(\int_{S^2} G_n(x, x_0)^2 dx \right)^{1/2}
\end{aligned}
$$

$$\begin{aligned} &= (4\pi)^{1/2} \left(G_n(x_0, x_0)^2 \right)^{1/2} \\ &= n + 1, \end{aligned}$$

where we used the reproducing kernel property $(G_n(\cdot, u), G_n(\cdot, v)) = G_n(u, v)$, which follows from (2.2), together with $P_n(1) = 1$. This is Theorem 5.2 of [16] in the special case $r = 3$.

The improved result $\|L_n\| \asymp n^{1/2}$ is proved in [16] directly from (3.1) under an additional 'quadrature regularity' assumption:

Assumption [Quadrature regularity] 3.1. *Let $r = 3$. The infinite family of positive–weight m–point quadrature rules $\{Q_m\}$ is said to satisfy the quadrature regularity assumption if there exists $c > 0$, with c independent of m, such that for every spherical cap A of spherical radius $1/\sqrt{m}$ we have*

$$\sum_{t_j \in A} w_j \leq c|A|, \tag{3.2}$$

where $|A|$ is the surface area of A.

A spherical cap of spherical radius α is the closed portion of the unit sphere cut out by a cone of half angle α; the axis of the cap is the axis of the cone.

The area of the spherical cap of spherical radius $1/\sqrt{m}$ is approximately π/m when m is large. Given that the expected number of points which lie in such a cap is approximately $m(\pi/m)/(4\pi) = 1/4$, it may seem a reasonable restriction that the sum of the weights corresponding to points lying in the cap should not grow unboundedly in relation to its area. In any event, it is shown in [16] that most quadrature rules, including standard tensor product rules, do satisfy the quadrature regularity property.

The following theorem is the main result in [16].

Theorem 3.2. *Let $r = 3$. For each $n \geq 0$ let $m = m_n$ satisfy $m \geq (n + 1)^2$. Moreover let $\{Q_m\}$ be a family of positive–weight m–point quadrature rules satisfying the quadrature regularity assumption. Then there exists $c > 0$, with c independent of n, such that*

$$\|L_n\| \leq cn^{1/2}.$$

We do not repeat the lengthy proof, but instead sketch the main ideas. The first step is to express the reproducing kernel (2.2) in the closed form

$$G_n(x, y) = \frac{n + 1}{4\pi} P_n^{(1,0)}(x \cdot y)$$

using results from Szegö [19]. Here $P_n^{(\alpha,\beta)}$ is the Jacobi polynomial corresponding to the weight function $(1-x)^\alpha(1+x)^\beta$ on the interval $[-1,1]$. Since $P_n^{(1,0)}(-z) = (-1)^n P_n^{(0,1)}(z)$, this allows us to write, from (3.1),

$$\|L_n\| \le \frac{n+1}{4\pi} \sum_{\substack{j=1 \\ z_j \ge 0}}^{m} w_j \left(|P_n^{(1,0)}(z_j)| + |P_n^{(0,1)}(z_j)| \right),$$

where we define

$$z_j = t_j \cdot x_0 = \cos\theta_j, \quad 0 \le \theta_j \le \frac{\pi}{2}.$$

Now Szegö shows that for $0 \le \theta \le \pi$ both $|P_n^{(1,0)}(\cos\theta)|$ and $|P_n^{(0,1)}(\cos\theta)|$ are bounded by $P_n^{(1,0)}(1) = n+1$, and also by $cn^{-1/2}\theta^{-3/2}$ for some constant $c > 0$, thus we may deduce

$$\|L_n\| \le \frac{n+1}{4\pi} \sum_{\substack{j=1 \\ z_j \ge 0}}^{m} w_j u_n(z_j), \tag{3.3}$$

where u_n is the monotone function on $[0,1]$ defined by

$$u_n(\cos\theta) = \min\left(2(n+1), cn^{-1/2}\theta^{-3/2}\right), \quad 0 \le \theta \le \frac{\pi}{2}.$$

The strategy for bounding (3.3) consists of slicing the hemisphere given by $z = x \cdot x_0 \ge 0$ by planes perpendicular to the axis x_0; then for each 'spherical collar' bounding $u_n(z)$ by its value on its upper circular edge (using monotonicity of u_n); then bounding the sum of the weights corresponding to points within a given spherical collar by a constant times the surface area of the collar, after establishing that the quadrature regularity property extends to spherical collars that are not too thin; and then controlling the major part of the resulting sum by recognising it as the Riemann sum for a monotone increasing function, which can therefore be bounded by an integral if the corresponding slices are of uniform thickness. Some fine tuning of the slicing near the 'north pole' is needed to make sure that there are no contributions which grow faster than $n^{1/2}$. For the rest of the hemisphere it turns out to be sufficient to use slices of uniform thickness $1/n$. For further details we refer to [16].

4 Numerical Results

This section presents numerical results on the norms of hyperinterpolation operators compared with the norm of the orthogonal projection, and on the uniform error in hyperinterpolation approximations applied to a set of test functions. For these calculations we use three different quadrature rules, each of tensor product form, and each of precision $2n$ or more.

4.1 Tensor Product Quadrature Rules

With the sphere S^2 parametrised by $z = \cos\theta$ and ϕ, where θ and ϕ are the polar and azimuthal angles, a tensor product quadrature rule has the form

$$\sum_k \mu_k \sum_j \nu_j F(z_j, \phi_k) = q_\phi q_z F, \tag{4.1}$$

where

$$q_\phi g := \sum_k \mu_k g(\phi_k), \quad q_z h := \sum_j \nu_j h(z_j), \tag{4.2}$$

for appropriate choices of the 1–dimensional rules q_ϕ and q_z, and with both sums finite. For the azimuthal integration a sensible choice for the rule q_ϕ, and the only one considered here, is the rectangle rule with spacing $\pi/(n+1)$,

$$q_\phi g := \frac{\pi}{n+1} \sum_{k=0}^{2n+1} g\left(\frac{k\pi}{n+1}\right), \tag{4.3}$$

because this rule is exact for all trigonometric polynomials of degree $\leq 2n+1$.

For the integration rule q_z three examples are considered here, all satisfying the property that the rule is of algebraic degree of precision at least $2n$. It was proved in [16] that all three of these rules satisfy the quadrature regularity assumption.

Example [Gauss–Legendre] 4.1. *Here we choose q_z to be the $(n+1)$–point Gauss-Legendre rule (see, for example, [18, 8]).*

This choice gives $m = 2(n+1)^2$ for the total number of points.

Example [Clenshaw–Curtis] 4.2. *Next choose q_z to be the Clenshaw–Curtis rule [2],*

$$q_z h = \sum_{j=0}^{2n} \nu_j h\left(\cos\frac{j\pi}{2n}\right).$$

This is an interpolatory rule, in which (as pointed out by Imhof [10]), the weights can be written explicitly as

$$\nu_0 = \nu_{2n} = \frac{1}{n} \sum_{k=0}^{n}{}'' \frac{1}{1 - 4k^2}$$

$$\nu_j = \nu_{2n-j} = \frac{2}{n} \sum_{k=0}^{n}{}'' \frac{1}{1 - 4k^2} \cos \frac{kj\pi}{n}, \quad j = 1, \ldots, n.$$

(The double prime on the sum indicates that the first and last terms are to be halved.)

This is a positive–weight rule (see [10]) with degree of precision $2n + 1$. The total number of points is $m = (2n-1)2(n+1)+2 = 4n^2+2n$, as on the sphere there is only one point with $z = +1$, and one with $z = -1$.

Example [Fejér] 4.3. *In 1933 Fejér [5] discussed an interpolatory quadrature formula based on the 'Filippi' points, which are the Clenshaw–Curtis quadrature points excluding the two end–points. The rule was rediscovered recently by [4, Section 2.5.5]. The Fejér rule of the appropriate precision is*

$$q_z h = \sum_{j=1}^{2n+1} \nu_j h \left(\cos \frac{j\pi}{2n + 2} \right), \tag{4.4}$$

where

$$\nu_j = \frac{2}{n + 1} \sin \left(\frac{j\pi}{2n + 2} \right) \sum_{\ell=1}^{n+1} \frac{1}{2\ell - 1} \sin \left(\frac{(2\ell - 1)j\pi}{2n + 2} \right), \quad j = 1, \ldots, 2n + 1.$$

It was shown by [5] that $\nu_j > 0$ for $j = 1, \ldots, 2n + 1$. The hyperinterpolation approximation with this choice of quadrature rule was in effect discussed by [12], and $\|L_n\|$ shown there to be of order $O(n^{1/2})$. With this rule the total number of points is $m = (2n + 1)(2n + 2) = 4n^2 + 6n + 2$.

4.2 Norms – Numerical Results

The norm $\|L_n\|$ of the hyperinterpolation operator for a given quadrature rule Q_m can be estimated by using a large number of points on the sphere to estimate the maximum in (3.1), or equivalently in

$$\|L_n\| = \max_{x \in S^2} \sum_{j=1}^{m} w_j |g_n(t_j \cdot x)|, \tag{4.5}$$

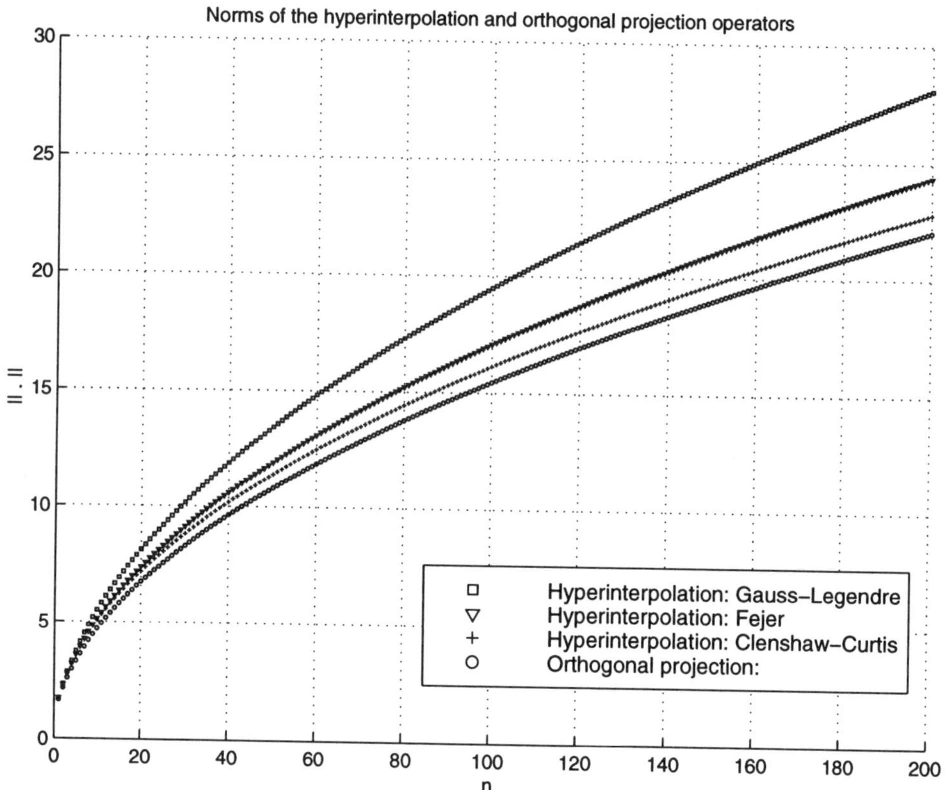

Figure 1: Norms for hyperinterpolation $\|L_n\|$ and orthogonal projection $\|P_n\|$

where

$$g_n(z) = \sum_{\ell=0}^{n} \frac{2\ell+1}{4\pi} P_\ell(z).$$

The norm of the orthogonal projection is

$$\|P_n\| = 2\pi \int_{-1}^{1} |g_n(z)|\, dz. \tag{4.6}$$

Figure 1 gives the growth of $\|L_n\|$ as a function of n for the Gauss–Legendre, Clenshaw–Curtis and Fejér tensor product rules, and for $\|P_n\|$. The norm (4.5) was initially estimated using up to $101, 105$ points, corresponding to $\ell = 282$, in the formulae

$$\begin{aligned}
&\theta_i = i\pi/\ell, \quad i = 0,\ldots,\ell \\
&n_0 = 1, \ \phi_{0,1} = 0, \quad n_\ell = 1, \ \phi_{\ell,1} = 0 \\
&n_i = \lfloor (2\pi)/\cos^{-1}\left((\cos(\pi/\ell) - \cos^2\theta_i)/\sin^2\theta_i\right)\rfloor, \quad i = 1,\ldots,\ell-1 \\
&\phi_{ij} = (j - 1/2)(2\pi/n_i), \quad j = 1,\ldots,n_j, \quad i = 1,\ldots,\ell-1
\end{aligned} \tag{4.7}$$

(see Freeden et al. [7, Example 7.1.9]). Empirically, for the Gauss–Legendre and Fejér points, the maximum in (4.5) is attained at the north and south poles. For the Clenshaw–Curtis points the maximum in (4.5) is attained within a spherical collar around the equator, in particular for $n \geq 2$ at $\overline{\phi}_j = (2j+1)\pi/(2n+2)$ for $j = 0,\ldots,2n+1$, and close to $\overline{\theta}_j = \pi/2 \pm \xi_n\pi/(4n)$ where $\xi_n = (n-2)/(n-1)$ for even $n \geq 4$ and $\xi_n = (2n-4)/(2n-3)$ for odd $n \geq 5$. The maximum was then obtained accurately by a quasi–Newton optimization routine starting from one of the points $\overline{\theta}_j, \overline{\phi}_j$.

In each case the observed results can be fitted rather accurately by a curve of the form $a + bn^{1/2}$, for the constants a and b in Table 1. For orthogonal projection the asymptotic value of b is $2\sqrt{2/\pi} \approx 1.596$.

Method	a	b
Gauss–Legendre	−1.05	2.057
Fejér	−0.43	1.752
Clenshaw–Curtis	−0.03	1.615
Orthogonal projection	−0.31	1.577

Table 1: Least squares fit of $a + bn^{1/2}$ to $\|L_n\|$, $\|P_n\|$

The striking feature of the results in Figure 1 is how close the norms of the hyperinterpolation operators are to those for the optimal orthogonal projection. This is especially true for the Clenshaw–Curtis rule, for which the norm $\|L_n\|$ is only a few percent larger than $\|P_n\|$, but even for the worst case (the Gauss rule), $\|L_n\|$ is about 25% larger than $\|P_n\|$. (However, for a given n the Gauss rule uses only about half the number of points of the Clenshaw–Curtis rule.)

The good numerical results for the norms encourage us to expect that the hyperinterpolation approximation $L_n f$ is of essentially the same quality as the orthogonal projection $P_n f$, especially for the Clenshaw–Curtis and Fejér cases.

4.3 Uniform Errors on Test Functions

We now consider the uniform error of the hyperinterpolation approximation for each of the quadrature rules on a set of test functions. The test functions used are

$$
\begin{aligned}
f_1(x) &= x_1 x_2 x_3, \\
f_2(x) &= e^{x_1}, \\
f_3(x) &= \tfrac{1}{10} e^{x_1 + x_2 + x_3}, \\
f_4(x) &= -5 \sin(1 + 10 x_3), \\
f_5(x) &= 1/(101 - 100 x_3), \\
f_6(x) &= \|x\|_1 / 10, \\
f_7(x) &= \sin^2(1 + \|x\|_1)/10, \\
f_8(x) &= 1/\|x\|_1, \\
f_9(x) &= \begin{cases} h_0 \cos^2(\tfrac{\pi}{2} \mathrm{dist}(x, x_0)/R) & \text{if } \mathrm{dist}(x, x_0) < R \\ 0 & \text{if } \mathrm{dist}(x, x_0) \geq R, \end{cases} \\
f_{10}(x) &= \sum_{i=1}^{5} \alpha_i e^{-\beta_i \mathrm{dist}(x, y_i)^{2 p_i}},
\end{aligned}
$$

$$(4.8)$$

$$(4.9)$$

where the great circle distance between two points x, y on the unit sphere is

$$
\mathrm{dist}(x, y) = \cos^{-1}(x \cdot y).
$$

i	$y_{i,1}$	$y_{i,2}$	$y_{i,3}$	α_i	β_i	p_i
1	0	0	1	2	5	1
2	0.932039	0	0.362358	0.5	7	1
3	-0.362154	0.619228	0.696707	-2	6	2
4	0.904035	0.279651	-0.323290	-2	5	1
5	-0.0479317	-0.424684	-0.904072	0.2	2.1	1

Table 2: Parameters for test function f_{10}

The functions $f_1, \ldots, f_8$ were used by Fliege and Maier [6] to test the quality of their numerical integration scheme, which is based on integration of the

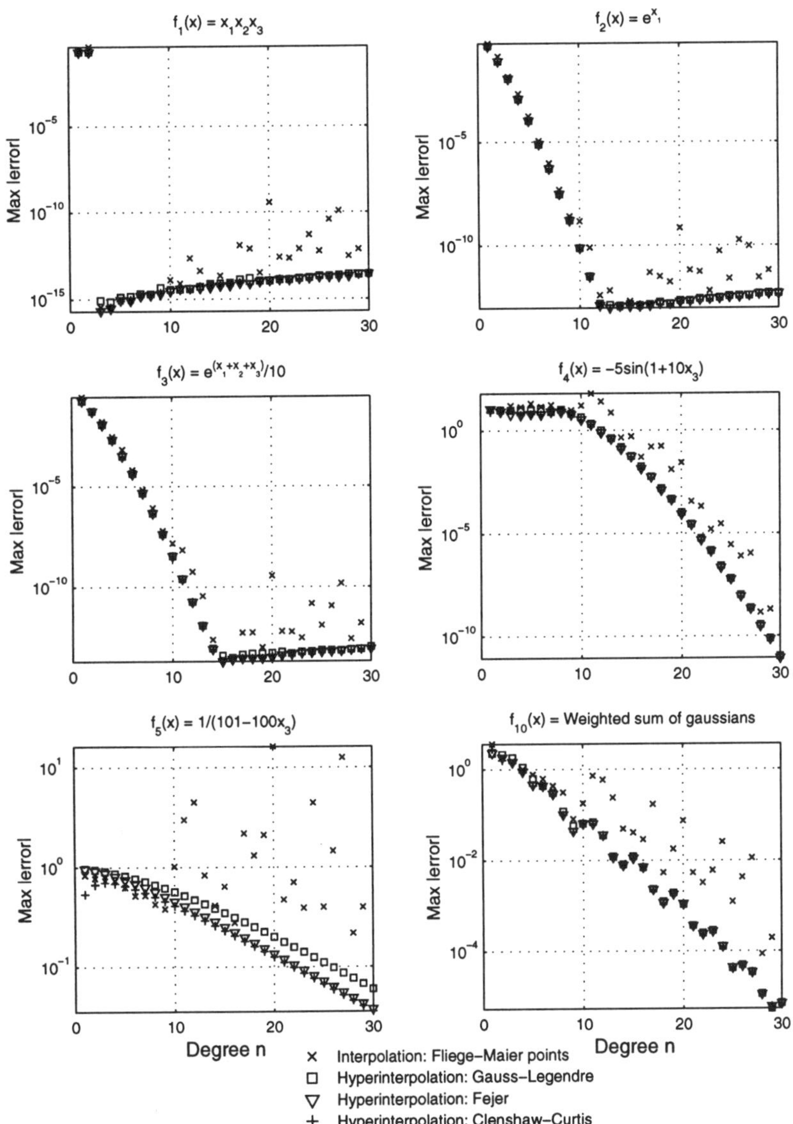

Figure 2: $\|f - L_n f\|_\infty$ estimated using 101,105 points

polynomial interpolant through their calculated points. The function f_1, a polynomial of low degree, is useful for code checking. Functions $f_2, \ldots, f_5$ are analytic over S^2 (although f_5 has a pole just off the surface of the sphere at $x = [0\ 0\ 1.01]$), while the functions $f_6, \ldots, f_8$ are only C^0 (in particular they are not continuously differentiable at points where any component of x is zero).

The cosine cap f_9, defined in (4.8), is part of a standard test set [20] for numerical approximations to the shallow water equations in spherical geometry. The point $x_0 \in S^2$ is the centre of the cap with radius R and height h_0. The results in Figure 3 used a centre x_0 with polar coordinates $(\theta_0, \phi_0) = (\pi/4, 5\pi/4)$, radius $R = 1/3$ and amplitude $h_0 = 1$. Other parameters can easily be used. The cosine cap has local support and is only once continuously differentiable at a distance R from the centre of the cap.

The function f_{10}, defined in (4.9), introduced by Jetter, Stöckler and Ward [11], is a weighted sum of five Gaussian–like functions with the parameters α_i, β_i, p_i and points $y_i \in S^2$ given in Table 2. The function $f_{10}(x)$ is analytic, with a global maximum near y_1 and minima near y_3 and y_4. The component centred at y_3 decays radially like $e^{-\xi^4}$, and is the hardest to approximate accurately.

The uniform norm of the errors is estimated by

$$\|f - L_n f\|_\infty \approx \max_{x \in X} |f(x) - L_n f(x)|, \tag{4.10}$$

where X is a large, but finite, set of well distributed points over the sphere, for instance the points (4.7).

Errors for each of the test functions, and for each of the three quadrature rules in Examples 4.1 to 4.3, are shown in Figures 2 and 3. In Figure 2 the test functions are analytic functions over the sphere, so should be well approximated by polynomials. The exception is f_5 which has a pole just off the sphere. The functions f_6, f_7, f_8 in Figure 3 are only C^0, while the cosine cap f_{10} is C^1.

To give something to compare with in these figures, we also show the estimated uniform errors for polynomial interpolation at the sets of points found by Fliege and Maier [6]. (Fliege and Maier recommend these points for quadrature, not interpolation. However, because the quadrature approximation for a function f is just the exact integral of the polynomial interpolant, it seems not unreasonable to look at the quality of the interpolant itself.)

The hyperinterpolation results are generally very satisfactory, bearing in mind the difficulty of some of the test functions, especially those in Figure 3. For the easier test functions in Figure 2 the three hyperinterpolation results cannot be distinguished. (Of course machine rounding errors dominate the results for the larger values of n for the functions f_1, f_2 and f_3.) For the harder functions there

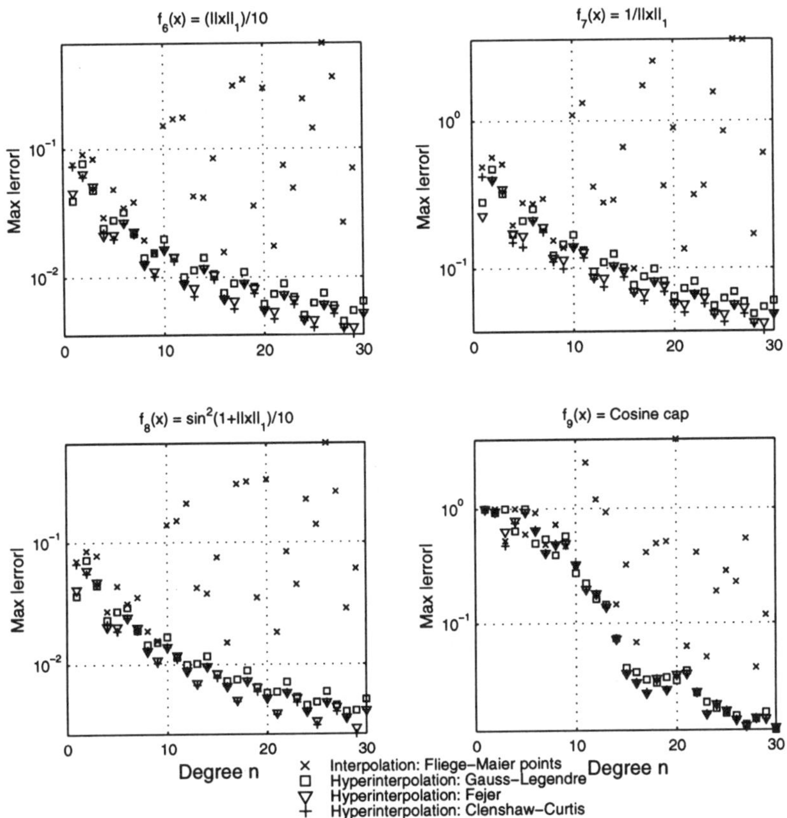

Figure 3: $\|f - L_n f\|_\infty$ estimated using 101,105 points

is little to choose between the three hyperinterpolation schemes (as might be expected from Figure 1), though the Clenshaw–Curtis results are consistently the best.

The hyperinterpolation results are in stark contrast to interpolation using the points from Fliege and Maier [6], which in many cases seem not even to be converging as n increases. The relatively poor performance for non–smooth functions f was already noted in [6]. Of course the Fliege–Maier points, as noted, were proposed for quadrature over the sphere, not interpolation.

Note also that interpolation uses only $m = (n + 1)^2$ points, while the Gauss–Legendre quadrature rule uses $m = 2(n + 1)^2$, and the Clenshaw–Curtis and Fejér rules use $m = 4n^2 + 2n$ and $m = 4n^2 + 6n + 2$ points respectively. One major advantage of the hyperinterpolation approximation is that the points and weights can be generated with very little work, while choosing good interpolation points is a significant computational task for large n.

Evaluating the hyperinterpolation approximation of degree n at N separate points, using an $m-$point quadrature rule, requires $6nmN + O(mN)$ floating point operations (flops), with the standard recurrence for Legendre polynomials. Interpolation requires (if we interpret m as $(n + 1)^2$) the same number of flops, plus the solution of an m by m symmetric positive definite linear system needing $(1/3)m^3 + O(m^2)$ flops. A simple Matlab implementation on a 400 MHz Pentium II PC computed $N = 5002$ hyperinterpolation values for degree $n = 10$ and the Clenshaw Curtis rule (and therefore with $m = 420$) in 17 seconds.

We shall return to the problem of polynomial interpolation in a subsequent paper [17], with the aim of showing that interpolation can produce a much better approximation than would appear from Figures 2 and 3.

Acknowledgements. The support of the Australian Research Council is gratefully acknowledged, along with programming assistance by D. Dowsett.

References

[1] D. L. Berman: *On a class of linear operators* (Russian), Dokl. Akad. Nauk SSSR **85** (1952), 13–16; Math. Reviews **14**, 57.

[2] C. W. Clenshaw, A. R. Curtis: *A method for numerical integration on an automatic computer*, Numer. Math. **2** (1960), 197–205.

[3] I. K. Daugavet: *Some applications of the Marcinkiewicz–Berman identity*, Vestnik Leningrad Univ. Math. **1** (1974), 321–327.

[4] J. R. Driscoll, D. M. Healy, Jr.: *Computing Fourier transforms and convolutions on the 2–sphere*, Adv. Appl. Math. **15** (1994), 202–250.

[5] L. Fejér: *Mechanische Quadraturen mit positiven Cotesschen Zahlen*, Math. Z. **87** (1933), 287–308.

[6] J. Fliege, U. Maier: *The distribution of points on the sphere and corresponding cubature formulae*, IMA J. Num. Anal., to appear 1999.
http://www.mathematik.uni-dortmund.de/lsx/fliege/nodes.html

[7] W. Freeden, T. Gervens, M. Schreiner: *Constructive Approximation on the Sphere*, Clarendon Press, Oxford 1998.

[8] G. H. Golub, J. H. Welsch: *Calculation of Gauss quadrature rules*, Math. Comp. **23** (1969), 221–230.

[9] T. H. Gronwall: *On the degree of convergence of Laplace's series*, Trans. Amer. Math. Soc. **15** (1914), 1–30.

[10] J. P. Imhof: *On the method for numerical integration of Clenshaw and Curtis*, Numer. Math. **5** (1963), 138–141.

[11] K. Jetter, J. Stöckler, J. D. Ward: *Error estimates for scattered data interpolation on spheres*, Math. Comp., to appear.

[12] A. K. Kushpel, J. Levesley: *Radial quasi–interpolation on S^2*, J. Approx. Theory, to appear.

[13] M. Reimer: *Constructive Theory of Multivariate Functions*, BI Wissenschaftsverlag, Mannhein–Wien–Zürich 1990.

[14] I. H. Sloan: *Polynomial interpolation and hyperinterpolation over general regions*, J. Approx. Theory **83** (1995), 238–254.

[15] I. H. Sloan: *Interpolation and hyperinterpolation on the sphere*, in: *Multivariate Approximation: Recent Trends and Results*, 255–268, W. Haußmann, K. Jetter, and H. Reimer (eds.), Akademie Verlag, Berlin 1997.

[16] I. H. Sloan, R. S. Womersley: *Constructive polynomial approximation on the sphere*, Tech. Rep. **19**, School of Mathematics, University of New South Wales, Sydney, Australia, 1998.
http://www.maths.unsw.edu.au/~rsw/sphere.ps

[17] I. H. Sloan, R. S. Womersley: *How good can polynomial interpolation on the sphere be?*, School of Mathematics, University of New South Wales, Sydney, Australia, 1999, in preparation.

[18] A. H. Stroud: *Approximate Calculation of Multiple Integrals*, Prentice–Hall, Englewood Cliffs 1971.

[19] G. Szegö: *Orthogonal Polynomials*, Amer. Math. Soc. Coll. Publ. Vol. **23**, 4th ed., Amer. Math. Soc., Providence, Rhode Island, 1975.

[20] D. L. Williamson, J. B. Brake, J. J. Hack, R. Jakob, P. N. Swarztrauber: *A standard test set for numerical approximations to the shallow water equations in spherical geometry*, J. Comp. Phys. **102** (1992), 211–224.

Address:

IAN H. SLOAN, ROBERT S. WOMERLSEY
School of Mathematics
University of New South Wales
Sydney, NSW, 2052
Australia

Simultaneous Approximation in the Dirichlet Space

Arne Stray

Dedicated to Manfred Reimer on the occasion of his 65th birthday

Abstract

Some general problems in joint approximation, originally raised by Lee Rubel, are discussed. The emphasis is on spaces of analytic functions in the unit disc D. Besides giving a survey of known results for the Hardy spaces and the space $H(D)$ consisting of all analytic functions in D, we announce some new results for the Dirichlet space $\mathcal{D}$ consisting of all analytic functions in D having finite Dirichlet integral.

Introduction

Consider topological function spaces E_1 defined on a set D_1 and E_2 defined on a set D_2. Let E consist of all functions f on $D_1 \cup D_2$ such that $f|_{D_i} \in E_i$ for $i = 1, 2$. Suppose there is a subset X of E such that $X|_{D_i}$ is dense in E_i for $i = 1, 2$. Given $g \in E$, is there a net $\{p_\alpha\} \subset X$ such that $p_\alpha|_{D_i} \to g|_{D_i}$ for $i = 1, 2$?

Suppose for example $E_i = C(K_i)$, $i = 1, 2$, where $C(K_i)$ denotes the continuous complex valued functions on a compact simple curve K_i. By Mergelyan's Theorem (see [23, p. 68]), the set P of all polynomials is dense in E_i for $i = 1, 2$. But P is dense in $C(K_1 \cup K_2)$ only if $K_1 \cup K_2$ does not divide the plane.

Let us now be more concrete and consider a linear space A of analytic functions in the unit disc $D = \{z : |z| < 1\}$ containing the set of polynomials as a dense subset. We also fix a relatively closed subset F of D.

If g is a function defined on the set B, we shall use the notation

$$\|g\|_B = \sup \{|g(z)| : z \in B\}.$$

Definition 0.1. *A relatively closed subset F of D is called a Farrell set for A if for any $f \in A$ there are polynomials p_n such that $p_n \to f$ in A and $\|p_n\|_F \to \|f\|_F$, as $n \to \infty$.*

Advances in Multivariate Approximation; W. Haußmann, K. Jetter and M. Reimer (eds.)
Mathematical Research, Vol. 107, pp. 307–319, ISBN 3-527-40236-5
© WILEY-VCH, Berlin 1999

Since we know a priori that the polynomials are dense in A, the problem is to get the right bound on $||p_n||_F$ given that f is bounded on F. If we have additional information about f on F, we may require more from $\{p_n\}$:

Definition 0.2 *A relatively closed subset F of D is called a Mergelyan set for A if for any $f \in A$ with $f|_F$ being uniformly continuous, there are polynomials p_n such that $p_n \to f$ in A and $||p_n - f||_F \to 0$, as $n \to \infty$.*

Let $H(D)$ denote the space of all analytic functions in D equipped with the topology of uniform convergence on compact subsets of D. Also let $H^p(D)$, $0 < p \leq \infty$, denote the classical Hardy spaces in D. For these spaces the classes of Farrell and Mergelyan sets have been characterized completely in terms of geometric or topological properties. We give an overview of these results in Section 1. The main new results we announce, concern analytic functions having finite Dirichlet integral in D. These results are given in Section 2.

1   Farrell and Mergelyan Sets for $H(D)$ and $H^p(D)$

Mergelyan sets were first introduced for the space $H(D)$ in a paper by L. Rubel, A. L. Shields and B. A. Taylor [17]. A complete characterization was later found in [19]. Rubel had conjectured that Mergelyan sets should be described in terms of the polynomial hull of $\overline{F} \cup K$ for arbitrary compact subsets K of D. This turned out to be the case.

If $M \subset D$ is relatively closed, let

$$\widetilde{M} = \{z \in D : |f(z)| \leq ||f||_M, f \in H(D)\}$$

denote the $H(D)$-hull of M, and let

$$\widehat{M} = \{z \in D : |p(z)| \leq ||p||_M, p \in P\}$$

denote the polynomial hull of $\overline{M}$ intersected with D. Then we have:

Theorem 1.1. *The classes of Farrell sets and Mergelyan sets for $H(D)$ are identical. They consist of all relatively closed subsets F of D such that*

$$\widetilde{F \cup K} = \widehat{F \cup K}$$

for any compact subset K of D. Moreover, $\widetilde{F \cup K}$ consists of all $z \in D\backslash(F \cup K)$ that can not be connected to the unit circle T by an arc in $D\backslash(F \cup K)$.

From Theorem 1.1 it follows that any countable relatively closed subset F_1 of D not clustering at all of T is a Mergelyan set for $H(D)$. On the other hand, if $F_2 = \{z : |z - \frac{1}{2}| = \frac{1}{2}\} \cap D$, then F_2 is not a Mergelyan set. This is easily seen noting that for general M, $\widehat{M}$ consists of M and the bounded components of $\mathbb{C} \setminus \overline{M}$.

For the Hardy spaces the situation is quite different. If $0 < p < \infty$, let $H^p(D)$ denote the set of all analytic functions f in D such that

$$||f||_p^p = \int_0^{2\pi} |f(re^{i\theta})|^p d\theta$$

is bounded as $r \to 1$. Moreover, $d(f, g) = ||f - g||_p^p$ makes $H^p(D)$ a complete metric space containing the polynomials P as a dense subset. For $p \geq 1$, $||f||_p$ defines a norm on $H^p(D)$. If $p = \infty$, $H^p(D)$ consists of all bounded analytic functions in D and $||f||_\infty$ denotes the sup–norm of $|f|$ on D. In this norm, the polynomials are not dense, but P is dense in $H^\infty(D)$ with respect to pointwise bounded convergence in D. (For more about different topologies on $H^\infty(D)$ corresponding to this notion of convergence, see [16].)

The growth restriction on functions in $H^p(D)$ implies the existence of nontangential limits

$$f(e^{i\theta}) = \lim f(z), \quad z \to e^{i\theta}, \quad z \in \Gamma_\alpha,$$

where

$$\Gamma_\alpha = \{z \in D : |z - e^{i\theta}| < \alpha(1 - |z|)\}$$

and $\alpha > 0$. (See [11] for this result of Fatou and more about Hardy spaces.)

To formulate the main result about Farrell and Mergelyan sets for $H^p(D)$, we need a concept related to the existence of nontangential limits for $f \in H^p$. We define the nontangential closure of F as

$$F_{nt} = \{e^{i\theta} : \exists \alpha > 0 \ \text{such that} \ e^{i\theta} \in \overline{F \cap \Gamma_\alpha}\}.$$

The tangential closure F_t of F is now given by

$$F_t = \overline{F} \cap T \backslash F_{nt}.$$

We now have

Theorem 1.2. *The classes of Mergelyan and Farrell sets are the same for any of the Hardy spaces $H^p(D)$, and are characterized by the property that the linear measure $|F_t|$ of F_t is zero.*

This theorem was proved in several steps. The Farrell sets for H^∞ were first described by A. M. Davie using methods from functional analysis. At the same time, J. Detraz [10] proved that $|F_t| = 0$ was necessary for F being a Mergelyan set for $H^\infty(D)$. Then A. Stray found a constructive proof of Davies' result and observed that Davies argument also gave that $|F_t| = 0$ was sufficient for F being a Mergelyan set for $H^\infty(D)$ [20]. Details about the further development can be found in [13].

We can now observe that the classes of Farrell and Mergelyan sets for $H(D)$ and $H^p(D)$ are completely different. Consider the sets F_1 and F_2 defined in the remarks following Theorem 1.1. It is evident that F_2 is a Mergelyan set for $H^p(D)$. On the other hand, if $F_1 = \{z_\nu\}$ is infinite and $\sum_\nu (1 - |z_\nu|) < \infty$, then $|F_{nt}| = 0$ and hence F_1 is not Mergelyan for $H^p(D)$ if $|\overline{F} \cap T| > 0$.

2 Farrell and Mergelyan Sets for the Dirichlet Space

We consider now the space $\mathcal{D}$ consisting of all analytic functions f in D such that the Dirichlet integral

$$D(f) = \int \int |f'|^2 dx dy$$

is finite. In [6, page 362], we raised the question of describing the Farrell and Mergelyan sets for $\mathcal{D}$. We obtain a geometric description of the Farrell sets for $\mathcal{D}$, and show that any Farrell set for $\mathcal{D}$ also is a Mergelyan set for $\mathcal{D}$. It seems to be an open problem whether any Mergelyan set also is a Farrell set for $\mathcal{D}$.

In the following, we identify the unit circle T with the interval $[-\pi, \pi)$. If B is a measurable subset of the real line $\mathbb{R}$, the capacity of B is given by

$$\operatorname{cap}(B) = C_{\frac{1}{2},2}(B),$$

where $C_{\frac{1}{2},2}(B)$ denotes the Bessel capacity of B as defined in [1, page 21]. For details see (2.4) below. The linear measure of B will be denoted by $|B|$.

The set B is said to be thick at $x \in \mathbb{R}$ if

$$\int_0^1 \frac{\operatorname{cap}[B \cap (x - t, x + t)]}{t} dt = \infty.$$

Otherwise B is called thin at x. See [1, p. 166] for more about this concept.

To formulate our main result about Farrell sets for $\mathcal{D}$, we adopt the definition of F_t and F_{nt} from Section 1. In addition, if $z \in D$ and $z \neq 0$, let $I_z = \{w \in T : |w - z| < 2(1 - |z|)\}$. Then define

$$S_m(F) = \bigcup_{z \in F, |z| \geq 1 - \frac{1}{m}} I_z.$$

We can now state

Theorem 2.1. *A relatively closed subset F of D is a Farrell set for $\mathcal{D}$ if and only if $F_{nt} \cup S_m(F)$ is thick at all $z \in \overline{F}$ except for a set of capacity zero, for $m=1,2,\dots$.*

A detailed proof of Theorem 2.1 can be found in [21]. Here we shall only outline the main ideas. We start with

Lemma 2.2. *If $f \in \mathcal{D}$, the limit*

$$f(e^{i\theta}) = \lim f(z), \quad z \to e^{i\theta}, \quad z \in \Gamma_\alpha,$$

exists for almost all $e^{i\theta} \in T$ apart from a set of zero capacity.

Proof. See [4] or [22, p. 344]. More general results can be found in [1, p. 161] or [9, p. 55]. $\square$

We shall use the fact that the Dirichlet integral can be expressed as a boundary integral

$$D(f) = \int_{-\pi}^{\pi} \int_{-\pi}^{\pi} \frac{|\widetilde{f}(s - t) - \widetilde{f}(s)|^2}{|e^{it} - 1|^2} \, ds \, dt, \tag{2.1}$$

where $\widetilde{f}$ denotes the 2π–periodic function corresponding to f by $\widetilde{f}(t) = f(e^{it})$, $t \in \mathbb{R}$.

We also consider the Bessel potential space $L^{\frac{1}{2},2}(\mathbb{R})$ and recall that a function g belongs to $L^{\frac{1}{2},2}(\mathbb{R})$ if and only if

$$\int_{-\infty}^{\infty} \int_{-\infty}^{\infty} \frac{|g(x - y) - g(x)|^2}{y^2} \, dx \, dy \tag{2.2}$$

is finite. (For details about (2.1) see [2] and for (2.2) we refer to [18].) If $\phi \in C^\infty(\mathbb{R})$ has compact support, it follows directly from (2.2) that $\phi g \in L^{\frac{1}{2},2}(\mathbb{R})$ whenever $g \in L^{\frac{1}{2},2}(\mathbb{R})$. Our next result is certainly well known and easy to prove.

Lemma 2.3. *Let $\phi \in C^\infty(\mathbb{R})$ have compact support on $(-\pi, \pi)$.*

(a) *If g is defined on $\mathbb{R}$ by $g = \tilde{f}$ for some $f \in \mathcal{D}$, then $\phi g \in L^{\frac{1}{2},2}(\mathbb{R})$.*

(b) *Let $h \in L^{\frac{1}{2},2}(\mathbb{R})$ be real valued. Define $u = \phi h$ on $(-\pi, \pi)$. Then $u = \mathrm{Re}\, \tilde{f}$ for some $f \in \mathcal{D}$.*

Proof. Since $\left|\frac{e^{it}-1}{t}\right| \to 1$ as $t \to 0$, the integrals (2.1) and (2.2) are simultaneously bounded for ϕg. The same argument applies to $u = \phi h$ as defined in (b) and then Douglas' formula (see [2]) gives

$$\int\int_D |\nabla u|^2 dx dy < \infty, \tag{2.3}$$

where u is extended to D by Poisson's formula. If v is a harmonic conjugate to u in D, the Cauchy–Riemann equations give $u + iv \in \mathcal{D}$.   □

Before going into details about the proof of Theorem 2.1, let us be a little more specific about the Bessel capacity $\mathrm{cap}(B)$ for a Borel set $B \subset \mathbb{R}$. Let $G_{\frac{1}{2}}$ denote the Bessel kernel on $\mathbb{R}$. (The explicit expression for $G_{\frac{1}{2}}$ is not important here, see [1, pages 10–11] for details.) Then $L^{\frac{1}{2},2}(\mathbb{R})$ consist of all functions g on $\mathbb{R}$ of the form

$$g(x) = G_{\frac{1}{2}} \star f = \int_{-\infty}^{\infty} G_{\frac{1}{2}}(y - x) f(x) dx,$$

where $f \in L^2(\mathbb{R})$. Now the capacity of B is given as

$$\mathrm{cap}(B) = \inf \left\{ \|f\|_2^2 : G_{\frac{1}{2}} \star f \geq 1 \ \text{ on } \ B \right\}, \tag{2.4}$$

where $\|f\|_2^2 = \int_{-\infty}^{\infty} |f|^2 dx$. (For full details about Bessel capacities see [1, chapter 2].) The proof of Theorem 2.1 requires the following lemma:

Lemma 2.4. *Let F be a subset of the unit disc D. The following statements are equivalent:*

(a) *F is a Farrell set for $\mathcal{D}$.*

(b) *If $f \in \mathcal{D}$, then $|f| \leq \|f\|_F$ on $\overline{F} \cap T$ with the exception of a set of capacity zero.*

We refer to [21] for the proof of Lemma 2.4.

We shall need one more concept from potential theory. Suppose A and B are subsets of $\mathbb{R}$ and that $A \subset B$. Then A is said to be representative for B if

$\text{cap}(A \cap \Delta) = \text{cap}(B \cap \Delta)$ for any open disc Δ. We refer to Theorem 11.4.2 and Remark 11.4 on page 327 in [1] for more about such sets.

In particular, Theorem 11.4.2 in [1] shows that the thickness condition in Theorem 2.1 can be restated as follows: $S_m(F) \cup F_{nt}$ is representative for $S_m(F) \cup (\overline{F} \cap T)$ for any $m \geq 1$.

Let us now assume that F is a Farrell set for $\mathcal{D}$. Let $A = S_m(F) \cup F_{nt}$ and $B = S_m(F) \cup (\overline{F} \cap T)$. We assume A fails to be representative for B for some integer m, and shall from this get a contradiction. We choose a disc Δ such that

$$\text{cap}\,(A \cap \Delta) < \text{cap}\,(B \cap \Delta)$$

and from the definition (2.4) of capacity we deduce that there is $u \in L^{\frac{1}{2},2}(\mathbb{R})$ such that $u \geq 1$ on $A \cap \Delta$ and $u \leq t < 1$ on a set $K \subset B \cap \Delta$ with $\text{cap}(K) > 0$. We may assume K is compact. Now choose $\phi \in C_0^\infty(\Delta)$ such that $0 \leq \phi \leq 1$, $\phi = 1$ near K, and define v by $v = \phi(1 - u)$. Then $v \geq 1 - t$ on K. Replacing v by $\min\{v, 1 - t\}$ if necessary, we may assume $v \leq (1 - t)$ everywhere and $v \leq 0$ on $S_m(F)$. By Lemma 2.3, w defined by

$$w(e^{i\theta}) = v(\theta), -\pi \leq \theta < \pi,$$

extends by Poisson's integral formula to a harmonic function in D having finite Dirichlet integral. It is evident that $\|w\|_F < 1 - t$ since if $z \in F, |z| > 1 - \frac{1}{m}$, the contribution to $w(z)$ coming from integrating the Poisson kernel over $S_m(F)$ is significant. If $f = e^{w+i\widetilde{w}}$, where $\widetilde{w}$ is a harmonic conjugate to w in D, then $f \in \mathcal{D}$ and $|f| > \|f\|_F$ on K. By Lemma 2.4, F is not a Farrell set for $\mathcal{D}$, contradicting our assumption.

To prove the converse in Theorem 2.1, we assume $A = S_m(F) \cup F_{nt}$ is representative for $B = S_m(F) \cup (\overline{F} \cap T)$. If F fails to be a Farrell set for $\mathcal{D}$, there is, by Lemma 2.4, an $f \in \mathcal{D}$ such that $\|f\|_F < 1$ while $|f| \geq 1$ on a set $K \subset \overline{F}$ of positive capacity. The function f admits a factorization $f = Ig$ (see Carleson [8]) where $|I| \leq 1$ in D and g is given by

$$g(z) = \exp\{\int_{-\pi}^{\pi} \frac{e^{i\theta} + z}{e^{i\theta} - z} u(\theta)d\theta\}.$$

Here $u(\theta) = \log|f(e^{i\theta})|$. It follows by Carleson's representation formula for the Dirichlet integral (see [8]) that if we replace u by $u_0 = \min\{u, 0\}$ and define

$$g_0(z) = \exp\{\int_{-\pi}^{\pi} \frac{e^{i\theta} + z}{e^{i\theta} - z} u_0(\theta)d\theta\}$$

then the function f_0 defined by $f_0 = Ig_0$ is in $\mathcal{D}$ and satisfies $|f_0| \leq |f|$ pointwise in D. Moreover, $|f_0| = 1$ on K, while $\|f_0\|_F < 1$. Replacing f_0 by f_0^N if necessary, we may assume $\|f_0\|_F < \varepsilon$ if $\varepsilon > 0$ is given in advance. If $u = \mathrm{Re}(\frac{1}{2}(1 + e^{i\alpha}f_0))$, we can choose α real such that $u \geq 1 - \frac{\varepsilon}{2}$ on a subset K' of K with $\mathrm{cap}(K') > 0$. We also have $\|u\|_F < \frac{1}{2}(1 + \varepsilon)$ and $0 \leq u \leq 1$ on $[-\pi, \pi)$.

If $z \in F$, it follows that there is a subset E_z of I_z with $|E_z| > c \cdot |I_z|$ for some $c > 0$ independent of z such that $u \leq (1 - \varepsilon)$ on E_z. For $m = 1, 2, \ldots$ there is a subset D_m of K' with $\mathrm{cap}(D_m) = 0$, such that $S_m(F) \cup F_{nt}$ is thick at any $z_0 \in K' \backslash D_m$. Let $D = \bigcup_{m \geq 1} D_m$. Also let

$$E_m = \bigcup_{z \in F, |z| \geq 1 - m^{-1}} E_z.$$

The desired contradiction is obtained as follows: Choose $z_0 \in K' \backslash D$. One proves that there is a subset E' of E_m such that $E' \cup F_{nt}$ is thick at z_0. If u is finely continuous at z_0, as we may well assume (u is finely continuous outside a set of capacity zero, [1, p. 177]), it follows that $u(z_0) \leq 1 - \varepsilon$. But on the other hand, $u(z_0) > 1 - \frac{\varepsilon}{2}$ since $z_0 \in K'$.

For the full details about this proof, we refer to [21].

3   Farrell Sets for $\mathcal{D}$ are Mergelyan Sets for $\mathcal{D}$

As mentioned in Section 1, the classes of Farrell and Mergelyan sets for $H(D)$ and for $H^p(D)$, $0 < p < \infty$, are identical. We do not know if there is a similar result for the Dirichlet space $\mathcal{D}$, but at least the following is true:

Theorem 3.1. *The following statements are equivalent for a subset F of D:*

(a) *If $f \in \mathcal{D}$, there are polynomials p_n such that $\|p_n - f\| \to 0$ and $\|p_n\|_F \to \|f\|_F$ as $n \to \infty$.*

(b) *If $f \in \mathcal{D}$ and u is a uniformly continuous function on F, there are polynomials p_n such that $\|p_n - f\| \to 0$ and $\|p_n - u\|_F \to \|f - u\|_F$ as $n \to \infty$.*

We shall not give the proof of Theorem 3.1 here, but only indicate the main ideas: First one observes that only $(a) \Rightarrow (b)$ needs to be proved. Then it is a trivial matter to verify that (b) follows from (a) whenever u is a constant function.

Using Lemma 2.4, one shows that being a Farrell set for $\mathcal{D}$ is a local property: F is a Farrell set for $\mathcal{D}$ if and only if for any disc Δ with $\Delta \cap F \neq \emptyset$, $F \cap \Delta$ is a Farrell set for $\mathcal{D}$.

We then select a finite covering $\{\Delta_i\}$ of F with discs Δ_i and select constants u_i such that $u \approx u_i$ on $\Delta_i \cap F$ for each i.

Then a local problem is solved: For each i we find polynomials $p_i^k, k = 1, 2, ...,$ such that $p_i^k \to f - u_i$ in $\mathcal{D}$ and $||p_i^k||_{F \cap \Delta_i} \to ||f - u_i||_{F \cap \Delta_i}$ as $k \to \infty$.

From these polynomials one can then build polynomials solving the global problem on F. Passing to finer and finer coverings $\{\Delta_i\}$ of F, one obtains (b) from (a). Full details about the proof are in [21] and a similar argument works for the Hardy spaces as shown in [13].

In particular if f has a uniformly continuous restriction $f|_F$ to F, we may choose $u = f|_F$ in (b) in Theorem 3.1 and the conclusion is that any Farrell set for $\mathcal{D}$ is also a Mergelyan set for $\mathcal{D}$.

4   Sets of Determination for $\mathcal{D}$

Consider again a general space A of functions on D. A subset F of D is called a set of determination for A if

$$||f||_F = ||f||_D$$

for all $f \in A$.

Suppose now A is closed with respect to uniform convergence on D and that convergence in A implies pointwise convergence in D. In addition, we assume that the function f_r given by

$$f_r(z) = f(rz), z \in D,$$

is in A for $0 < r < 1$ when $f \in A$. Then it is easy to see that F is a set of determination for A if and only if $\overline{F} \supset T$ and F is a Farrell set for A. We have the following characterization of sets of determination for $\mathcal{D}$:

Theorem 4.1. *Let F be a subset of D. The following statements are equivalent:*

(a) *For all $f \in \mathcal{D}$ the supremum of $|f|$ on F equals the supremum of $|f|$ on D.*

(b) $\mathrm{cap}\,(T) = \mathrm{cap}\,(T \backslash F_t).$

Proof. If $\overline{F} \supset T$, we have $S_m(F) \cup F_{nt} = T \backslash F_m$ where $F_m = F_t \backslash S_m(F)$. By Theorem 2.1 and Hedberg [15, Theorem 13], we only have to prove

$$\{\mathrm{cap}(T) = \mathrm{cap}(T \backslash F_m), m = 1, 2, ...\} \Rightarrow \mathrm{cap}(T) = \mathrm{cap}(T \backslash F_t).$$

To get a contradiction, assume $\mathrm{cap}(T \backslash F_t) < \mathrm{cap}(T)$ while $\mathrm{cap}(T \backslash F_m) = \mathrm{cap}(T)$ for $m = 1, 2, ...$ Then by regularity properties of capacity, there is a compact subset K of F_t such that

$$\mathrm{cap}(T \backslash K) < \mathrm{cap}(T). \tag{4.1}$$

By a theorem of L. V. Ahlfors and A. Beurling [3, Theorem 14], (4.1) holds if and only if there is a nonconstant function h_K analytic in $\mathbb{C} \backslash K$ and having finite Dirichlet integral there. In the following we assume K is minimal in the sense that h_K does not extend to be analytic near any $z \in K$. This minimal set is still denoted by K and (4.1) will remain valid because of the Ahlfors–Beurling theorem. Let $K_m = K \cap F_m$. Since F_m is closed relative to F_t, each K_m is compact. Let $z \in K$ and assume there is a disc Δ containing z such that $\Delta \cap K \subset K_m$. Since no K_m can contain an interval (if it did, we would have $\mathrm{cap}(T \backslash K_m) < \mathrm{cap}\, T$), we may assume the boundary of Δ does not intersect K. By Theorem 5 in [3], h_K extends to be analytic near $K \cap \Delta$, contradicting the minimality of K. We conclude that $K \backslash K_m$ is dense in K for all m. Since $\bigcap_{m=1}^{\infty} K \backslash K_m = \emptyset$, this contradicts the Baire category theorem, and Theorem 4.1 is proved. $\square$

Sets of determination have been studied recently for many function spaces. The first result we are aware of is the theorem of L. Brown, A. L. Shields and K. Zeller [7], stating that F is a set of determination for $H^{\infty}(D)$ (the bounded analytic functions in D) if and only if $\overline{F} \supset T$ and $|F_t| = 0$. More recently, F. Bonsall [5] found that the same condition characterizes sets of determination for the bounded harmonic functions in D. For the space consisting of differences of positive harmonic functions in D, W. K. Hayman and T. J. Lyons [14] described sets of determination based on a minimum principle for positive harmonic functions going back to A. Beurling. More recently, Stephen J. Gardiner [12] extended both the work of Bonsall and Hayman–Lyons to harmonic functions in several variables.

Let us finally show by an example that the presence of the set $S_m(F)$ in Theorem 2.1 can not be avoided. So the Farrell sets for $\mathcal{D}$ can not be described in terms of F_t and F_{nt} only. Let K denote a compact totally disconnected subset of T such that $\mathrm{cap}(T)=\mathrm{cap}(T\backslash K)$ and still $|K| > 0$. Such sets exist (see Adams–Hedberg [1, Remark 2 on page 314] or [3, page 128]). Let

$$T\backslash K = \bigcup I_\nu,$$

where the open arcs I_ν are disjoint. Let

$$I_\nu = \{e^{i\theta} : \alpha_\nu < \theta < \beta_\nu\}$$

and define arcs $J_\nu \subset D$, $\nu = 1,2,\dots$. by

$$J_\nu = \{r_\nu(\theta)e^{i\theta} : \alpha_\nu < \theta < \beta_\nu\},$$

where the functions $r_\nu(\theta)$ are chosen such that r_ν map I_ν into $(0,1)$ and

$$r_\nu(\theta) \to 1 \; if \; \theta \to \alpha_\nu \; or \; \theta \to \beta_\nu$$

from within J_ν. We may arrange it so that F given by

$$F = \bigcup_{\nu=1}^{\infty} J_\nu$$

has the following properties:

$$F_{nt} = \emptyset$$

and

$$S_m(F) = (\cup I_\nu)\backslash L_m$$

for some compact subset L_m of $\cup I_\nu$ for $m = 1, 2, \dots$.

Since $T\backslash K$ is assumed to be representative for T, it follows that $S_m(F)$ is representative for $S_m(F) \cup K = S_m(F) \cup (\overline{F} \cap T)$ for all m. By Theorem 2.1, F is a Farrell set for $\mathcal{D}$ even if $F_{nt} = \emptyset$ and $|F_t| > 0$.

References

[1]   D. R. Adams, L. I. Hedberg: *Function Spaces and Potential Theory*, Springer, Berlin–Heidelberg–New York 1996.

[2]   L. V. Ahlfors: *Conformal Invariants*, McGraw–Hill, New York 1973.

[3]   L. V. Ahlfors, A. Beurling: *Conformal invariants and function theoretic null sets*, Acta Math. **83** (1950), 101–129.

[4]   A. Beurling: *Ensembles exceptionelles*, Acta Math. **72** (1940), 1–13.

[5]   F. F. Bonsall: *Domination of the supremum of a bounded harmonic function by its supremum over a countable subset*, Proc. Edinburgh Math. Soc. **32** (1987), 471–477.

[6]   D. A. Brannan, J. G. Clunie: *Aspects of Contemporary Complex Analysis*, Academic Press, London–New York 1980.

[7]   L. Brown, A. L. Shields, K. Zeller: *On absolutely convergent exponential sums*, Trans. Amer. Math. Soc. **96** (1960), 162–183.

[8]   L. Carleson: *A representation formula for the Dirichlet integral*, Math. Z. **56** (1960), 190–196.

[9]   L. Carleson: *Selected Problems on Exceptional Sets*, Van Nordstrand, Princeton, N. J., 1967.

[10]   J. Detraz: *Algèbres de fonctions analytiques dans le disque*, Ann. Sci. École Norm. Sup. **4** (1970), 313–352.

[11]   P. Duren: *Theory of H^p-spaces*, Academic Press, New York 1970.

[12]   S. J. Gardiner: *Sets of determination for harmonic functions*, Trans. Amer. Math. Soc. **338** (1993), 233–243.

[13]   F. Pérez–González, A. Stray: *Farrell and Mergelyan sets for H^p-spaces*, Michigan Math. J. **36** (1989), 379–386.

[14]   W. K. Hayman, T. J. Lyons: *Bases for positive continuous functions*, J. London Math. Soc. **42** (1990), 292–308.

[15]   L. I. Hedberg: *Removable singularities and condenser capacities*, Arkiv för Matematik **12** (1974), 181–201.

[16]   L. A. Rubel: *Bounded convergence of analytic functions*, Bull. Amer. Math. Soc. **77** (1971), 13–20.

[17]   L. A. Rubel, A. L. Shields, B. A. Taylor: *Mergelyan sets and the modulus of continuity*, J. Approx. Theory **15** (1975), 23–40.

[18]   E. M. Stein: *The characterization of functions arising as potentials*, Bull. Amer. Math. Soc. **67** (1961), 102–104.

[19]   A. Stray: *Characterization of Mergelyan sets*, Proc. Amer. Math. Soc. **44** (1974), 347–352.

[20]   A. Stray: *Pointwise bounded approximation by functions satisfying a side condition*, Pacific J. Math. **51** (1974), 301–305.

[21]   A. Stray: *Simultaneous approximation in the Dirichlet space*, to appear.

[22]   M. Tsuji: *Potential Theory in Modern Function Theory*, Maruzen, Tokyo 1959.

[23]   L. Zalcman: *Analytic Capacity and Rational Approximation*, Springer, Berlin–Heidelberg–New York 1968.

Address:

ARNE STRAY
Department of Mathematics
University of Bergen
Johannes Bruns gt. 12
5008 Bergen
Norway

The Equivalence between Weighted K–Functionals and Moduli of Smoothness on a Simplex and their Applications

Pei–Cai Xuan

To Prof. M. Reimer on the occasion of his 65th birthday

Abstract

In this paper we consider the equivalence between a class of weighted K–functionals and moduli of smoothness first, and then we obtain a characterization of multidimensional Bernstein operators on the simplex in weighted approximation with multidimensional Jacobi weights.

1 Introduction

Since J. Peetre [2] introduced the K–functional and interpolation space theory, there have appeared many papers studying the K–functional theory and its applications. Recently, Z. Ditzian and V. Totik [3] obtained the equivalence between a class of K–functionals and moduli of smoothness, then easily gave the characterizations of approximation order by exponential operators. After that time, H. Berens and Y. Hu [1] extended these results to the multidimensional case. In the 1990s, Z. Ditzian and V. Totik [4] obtained another equivalence between a weighted K–functional and moduli of smoothness:

Theorem 1.1. *Let $w(x) = x^\beta (1-x)^\eta$, $0 < \beta$, $\eta < 1$, $wf \in L_p[0,1]$. Then*

$$K_\varphi^r(f, t^r)_{w,p} \sim \omega_\varphi^r(f, t)_{w,p}, \qquad t \to 0,$$

where

$$K_\varphi^r(f, t^r)_{w,p} = \inf_{g^{(r-1)} \in AC_{loc}} \{\|w(f-g)\|_p + t^r \|w\varphi^r g^{(r)}\|_p\},$$

$$\omega_\varphi^r(f, t)_{w,p} = \sup_{0 < h \leq t} \|w\Delta_{h\varphi}^r f\|_p, \quad \varphi(x) = \sqrt{x(1-x)}.$$

Advances in Multivariate Approximation; W. Haußmann, K. Jetter and M. Reimer (eds.)
Mathematical Research, Vol. 107, pp. 321–334, ISBN 3–527–40236–5
© WILEY–VCH, Berlin 1999

Here we use the notation $g^{(r-1)} \in AC_{loc}$ if and only if g is $(r-1)$–times differentiable and $g^{(r-1)}$ is absolutely continuous in every closed interval $[c,d] \subset [0,1]$.

By using this equivalence, we can consider the weighted approximation on some linear positive operators $[4] - [8]$. In this paper, we will extend the above theorem to the multidimensional case. We give an equivalence between a class of weighted K–functionals and moduli of smoothness on the d–dimensional simplex first. Then, using this theory, we obtain the characterization in weighted approximation of the Bernstein operator, though it is unbounded in the usual weighted norm.

2 Equivalence between Weighted K–Functionals and Moduli of Smoothness on a d–Dimensional Simplex

Let $\vec{x} = (x_1, x_2, \ldots, x_d) \in \mathbb{R}^d$, and let S be the following d–dimensional simplex

$$S = \{\vec{x} \in \mathbb{R}^d : 0 \le x_i \le 1,\ 0 \le i \le d,\ 1 - |\vec{x}| \ge 0\}.$$

Let $|\vec{x}| = \sum_{i=1}^{d} x_i$, $\quad \vec{x}^{\vec{k}} = x_1^{k_1} x_2^{k_2} \ldots x_d^{k_d}$, $\quad \vec{k}! = k_1! k_2! \ldots k_d!$, and let $\vec{k} = (k_1, k_2, \ldots, k_d) \in \mathbb{N}_0^d$. Furthermore, let

$$\binom{n}{\vec{k}} = \frac{n!}{\vec{k}!(n - |\vec{k}|)!}.$$

Then the Bernstein operator on the d–dimensional simplex can be written as

$$B_{n,d}(f; \vec{x}) = \sum_{|\vec{k}| \le n} f\left(\frac{\vec{k}}{n}\right) P_{n,\vec{k}}(\vec{x}), \tag{2.1}$$

where

$$P_{n,\vec{k}}(\vec{x}) = \binom{n}{\vec{k}} \vec{x}^{|\vec{k}|}(1 - |\vec{x}|)^{n - |\vec{k}|},\ n \in \mathbb{N}_0.$$

For $1 \le p \le \infty$, let $L_p(S)$ denote the space of Lebesgue measurable functions, for which $\|f\|_p^p = \int_S |f|^p$ is finite. Further let

$$\varphi_i(\vec{x}) = \varphi_{ii}(\vec{x}) = \sqrt{x_i(1 - |\vec{x}|)}, \qquad\qquad 1 \le i \le d,$$

$$\varphi_{ij}(\vec{x}) = \sqrt{x_i x_j}, \qquad\qquad\qquad 1 \le i < j \le d,$$

$$D_i = D_{ii} = \frac{\partial}{\partial x_i}, \qquad\qquad\qquad 1 \le i \le d,$$

$$D_{ij} = D_i - D_j, \qquad\qquad\qquad 1 \le i < j \le d,$$

$$D_{ij}^r = D_{ij}(D_{ij}^{r-1}), \quad D^{\vec{k}} = D_1^{k_1} D_2^{k_2} \cdots D_d^{k_d}, \qquad\qquad \vec{k} \in \mathbb{N}_0^d,$$

where $D_{ij}^0 f = f$. For $1 \le p < \infty$, we define the following weighted Sobolev space:

$$W_{\varphi,w}^{r,p}(S) = \{wf \in L_p(S) \ : \ D^{\vec{k}} f \in L_{loc}(S), \ |\vec{k}| \le r, \ \vec{k} \in \mathbb{N}^d,$$

$$\text{and } w\varphi_{ij}^r D_{ij}^r f \in L_p(S), \ 1 \le i \le j \le d\}.$$

For $p = \infty$, we define:

$$C_{\varphi,w}^r(S) = \{wf \in C(S) \ : \ f \in C^r(\overset{\circ}{S}),$$

$$\text{and } w\varphi_{ij}^r D_{ij}^r f \in C(S), 1 \le i \le j \le d\},$$

where $\overset{\circ}{S} = S \setminus \partial S$, and where $w(\vec{x})$ is a multivariate Jacobi weight

$$w(\vec{x}) = \vec{x}^{\vec{\beta}}(1 - |\vec{x}|)^\eta,$$

with $\vec{\beta} = (\beta_1, \beta_2, \cdots, \beta_d) \in \mathbb{R}^d, 0 < \beta_i < 1, \ 1 \le i \le d$ and $0 < \eta < 1$.

Let $K_\varphi^r(f, t^r)_{w,p}$ be the weighted K–functional on the simplex S:

$$K_\varphi^r(f, t^r) = \inf_g \{\|w(f - g)\|_p + t^r \cdot \sum_{1 \le i \le j \le d} \|w\varphi_{ij}^r D_{ij}^r g\|_p\},$$

where $g \in W_{\varphi,w}^{r,p}(S)$ for $1 \le p < \infty$, and $g \in C_{\varphi,w}^r(S)$ for $p = \infty$.

Let $\vec{e_i} \in \mathbb{R}^d$ be the unit vector

$$\vec{e_i} = (0, \cdots, 0, \overset{i}{1}, 0, \cdots, 0),$$

and $\vec{e_{ij}} = \vec{e_i} - \vec{e_j}$.

For any vector $\vec{e} \in \mathbb{R}^d$, we write for the r–th symmetric difference of a function f in the direction of $\vec{e}$

$$
\Delta_{h\vec{e}}^{r} f(\vec{x}) = \begin{cases} \sum_{i=0}^{r} \binom{r}{k}(-1)^{k} f(\vec{x} + (\tfrac{r}{2} - k)h\vec{e}), & \vec{x} + \tfrac{hr}{2}\vec{e} \in S, \\[2em] 0, & \text{otherwise.} \end{cases}
$$

Based on this we define the weighted modulus of smoothness on the d–dimensional simplex as follows:

$$
\omega_{\varphi}^{r}(f,t)_{w,p} = \sup_{0 < h \le t} \sum_{1 \le i \le j \le d} \| w \Delta_{h\varphi_{ij}\vec{e}_{ij}}^{r} f \|_{p}
$$

with $wf \in L_p(S)$ for $1 \le p < \infty$, and $wf \in C(S)$ for $p = \infty$.

Then we have

Theorem 2.1. *There exist positive constants M_1, M_2, such that*

$$
M_1 \omega_{\varphi}^{r}(f,t)_{w,p} \le K_{\varphi}^{r}(f,t^{r})_{w,p} \le M_2 \omega_{\varphi}^{r}(f,t)_{w,p}, \tag{2.2}
$$

where, for $1 \le p < \infty$, $wf \in L_p(S)$, and $wf \in C(S)$ for $p = \infty$.

Proof. For $d = 1$, our statement is a theorem by Ditzian–Totik [4, p. 11, Theorem A]. So, we only discuss the case $d > 1$. Let $\vec{x_1} = (x_2, \cdots, x_d)$, $S_1 = \{\vec{x_1} : \vec{x} = (x_1, \vec{x_1}) \in S\}$, and $x_1 = (1 - |\vec{x}|)z$, $0 \le z \le 1$, $F(z) = F(z, \vec{x_1}) = f((1 - |\vec{x_1}|)z, \vec{x_1})$. Then

$$
\varphi_1(\vec{x}) = (1 - |\vec{x_1}|)\varphi(z), \qquad (D_1^{r} f)(\vec{x}) = (1 - |\vec{x_1}|)^{-r} F^{(r)}(z),
$$

$$
\Delta_{h\varphi_1 \vec{e_1}}^{r} f(\vec{x}) = \Delta_{h\varphi}^{r} F(z), \qquad w(\vec{x}) = \vec{x_1}^{\vec{\beta_1}}(1 - |\vec{x_1}|)^{\beta_1 + \eta} w(z),
$$

where $\vec{\beta_1} = (\beta_2, \beta_3, \cdots, \beta_d)$ and $0 < \eta < 1$. Hence, by [3, p. 58],

$$
w(\vec{x}) \Delta_{h\varphi_1 \vec{e_1}}^{r} f(\vec{x}) = \vec{x_1}^{\vec{\beta_1}}(1 - |\vec{x_1}|)^{\beta_1 + \eta} w(z) \Delta_{h\varphi}^{r} F(z),
$$

for $1 \leq p < \infty$. If $wf \in L_p(S)$, then we have

$$\|w(\vec{x})\Delta^r_{h\varphi_1 \vec{e_1}} f(\vec{x})\|^p_p = \int_{S_1} \int_0^{1-|\vec{x_1}|} |w\Delta^r_{h\varphi_1 \vec{e_1}} f(\vec{x})|^p dx_1 d\vec{x_1}$$

$$= \int_{S_1} (1 - |\vec{x_1}|)(\vec{x_1}^{\vec{\beta_1}} (1 - |\vec{x_1}|)^{\beta_1 + \eta})^p \int_0^1 |w(z)\Delta^r_{h\varphi} F(z)|^p dz \, d\vec{x_1}$$

$$\leq M \int_{S_1} (1 - |\vec{x_1}|)(\vec{x_1}^{\vec{\beta_1}} (1 - |\vec{x_1}|)^{\beta_1 + \eta})^p \int_0^1 |w(z) F(z)|^p dz \, d\vec{x_1}$$

$$= M \int_{S_1} \int_0^{1-|\vec{x_1}|} w(z) \vec{x_1}^{\vec{\beta_1}} (1 - |\vec{x_1}|)^{\beta_1 + \eta} f(x_1, \vec{x_1})|^p dx_1 \, d\vec{x_1}$$

$$= M \cdot \|wf\|^p_p.$$

where M is a constant. Similarly, for $f \in W^{r,p}_{\varphi,w}(S)$, by [3, p. 58], we have

$$\|w(\vec{x})\Delta^r_{h\varphi_1 \vec{e_1}} f(\vec{x})\|^p_p$$

$$= \int_{S_1} (1 - |\vec{x_1}|)(\vec{x_1}^{\vec{\beta_1}} (1 - |\vec{x_1}|)^{\beta_1 + \eta})^p \int_0^1 |w(z)\Delta^r_{h\varphi} F(z)|^p dz \, d\vec{x_1}$$

$$\leq M \int_{S_1} (1 - |\vec{x_1}|)(\vec{x_1}^{\vec{\beta_1}} (1 - |\vec{x_1}|)^{\beta_1 + \eta})^p (h\varphi(z))^{rp} \int_0^1 |w(z)\varphi^r(z) F^{(r)}(z)|^p dz \, d\vec{x_1}$$

$$= M(h\varphi(z))^{rp} \int_{S_1} \int_0^{1-|\vec{x_1}|} |w(z)\vec{x_1}^{\vec{\beta_1}} (1 - |\vec{x_1}|)^{\beta_1 + \eta} \varphi^r(z) F^{(r)}(z)|^p dz \, d\vec{x_1}$$

$$\leq M(h\varphi_1(\vec{x}))^{rp} \int_{S_1} \int_0^{1-|\vec{x_1}|} |w(\vec{x})\varphi_1^r(\vec{x}) D_1^r f(x_1, \vec{x_1})|^p dx_1 d\vec{x_1}$$

$$\leq M h^{rp} \|w\varphi_1^r D_1^r f(\vec{x})\|^p_p.$$

$$(2.3)$$

For $p = \infty$, $wf \in C(S)$, we have

$$\|w(\vec{x})\Delta^r_{h\varphi_1 \vec{e_1}} f(\vec{x})\|_\infty \leq M \sup_{\vec{x} \in S} |\vec{x_1}^{\vec{\beta_1}} (1 - |\vec{x_1}|)^{\beta_1 + \eta} w(z) F(z)|$$

$$\leq M \sup_{\vec{x} \in S} |w(\vec{x}) f(\vec{x})| = M \|wf\|_\infty.$$

$$(2.4)$$

If $f \in C^r_{\varphi,w}(S)$, then we have

$$\|w(\vec{x})\Delta^r_{h\varphi_1 \vec{e_1}} f(\vec{x})\|_\infty \le M \sup_{\vec{x}\in S} |\vec{x_1}^{\vec{\beta_1}}(1-|\vec{x_1}|)^{\beta_1+\eta} w(z)(h\varphi(z))^r \varphi^r(z) F^{(r)}(z)|$$

$$\le M \sup_{\vec{x}\in S} |w(\vec{x}) D^r_1 f(\vec{x}) \varphi^r_1(\vec{x}) (h\varphi_1(\vec{x}))^r| \le M(h\varphi_1(\vec{x}))^r \|w\varphi^r_1 D^r_1 f(\vec{x})\|_\infty$$

$$\le M h^r \|w\varphi_1 D^r_1 f(\vec{x})\|_\infty.$$

$$(2.5)$$

Similarly, using the transformation $T_i : S \to S$, $x_i \to 1 - |\vec{x_i}|$, $x_j \to x_i$ for $j \ne i$, $D_{ij}f = D_j(f \circ T_i)$, we have

$$\|w\Delta^r_{h\varphi_{ij}\vec{e_{ij}}} f(\vec{x})\|_p \le M \begin{cases} \|wf\|_p, & wf \in L_p(S), \\ h^r \|w\varphi^r_{ij} D^r_{ij} f\|_p, & f \in W^{p,r}_{\varphi,w}(S) \end{cases} \qquad (2.6)$$

$$\|w\Delta^r_{h\varphi_{ij}\vec{e_{ij}}} f(\vec{x})\|_\infty \le M \begin{cases} \|wf\|_\infty, & wf \in C(S), \\ h^r \|w\varphi^r_{ij} D^r_{ij} f\|_\infty, & f \in C^r_{\varphi,w}(S). \end{cases} \qquad (2.7)$$

Hence by (2.3)–(2.8),we have for $1 \le p \le \infty$

$$\|w\Delta^r_{h\varphi_{ij}\vec{e_{ij}}} f(\vec{x})\|_p \le \|w\Delta^r_{h\varphi_{ij}\vec{e_{ij}}} (f(\vec{x}) - g(\vec{x}))\|_p + \|w\Delta^r_{h\varphi_{ij}\vec{e_{ij}}} g(\vec{x})\|_p$$

$$\le M \left\{ \|w(f-g)\|_p + t^r \sum_{1 \le i,j \le d} \|w\varphi^r_{ij} D^r_{ij} g\|_p \right\}$$

where $g \in W^{r,p}_{\varphi,w}(S)$ resp. $g \in C^r_{\varphi,w}(S)$. Then we have

$$\|w\Delta^r_{h\varphi_{ij}\vec{e_{ij}}} f(\vec{x})\|_p \le M \inf_g \left\{ \|w(f-g)\|_p + t^r \sum_{1 \le i,j \le d} \|w\varphi^r_{ij} D^r_{ij} g\|_p \right\}$$

$$= M K^r_\varphi(f, t^r)_{w,p}.$$

With $M_1 = \frac{1}{M}$, we have

$$M_1 \omega^r_\varphi(f,t)_{w,p} \le K^r_\varphi(f, t^r)_{w,p}. \qquad (2.8)$$

Thus we have proved first estimate in (2.2). By [3, p. 18] we know that for fixed $\vec{x_1}$ there is a $G_t(z) = G_t((1-|\vec{x_1}|)z, \vec{x_1}) \in W^{r,p}_{\varphi,w}(S)$, $(1 \le p < \infty)$, such that

$$\|w(z)(F - G_t)(z)\|^p_{p,} \le \frac{M}{t} \int_0^t \|w\Delta^r_{h\varphi} f\|^p_p du$$

and

$$t^{rp}\|w(z)\varphi^r(z)G_t^{(r)}(z)\|_p^p \le \frac{M}{t}\int_0^t \|w\Delta_{h\varphi}^r f\|_p^p du.$$

From this we have

$$\|w(\vec{x})(f-g)(\vec{x})\|_p^p$$

$$= \int_{S_1}(1-|\vec{x_1}|)\int_0^1 |\vec{x_1}|^{\vec{\beta_1}}(1-|\vec{x_1}|)^{\beta_1+\eta}w(z)(F(z)-G_t(z))|^p dz\, d\vec{x_1}$$

$$\le M\int_{S_1}(1-|\vec{x_1}|)(\vec{x_1}^{\vec{\beta_1}}(1-|\vec{x}|)^{\beta_1+\eta})^p\frac{1}{t}\int_0^t\int_0^1 |w(z)-\Delta_{h\varphi}^r F(z)|^p dz du\, d\vec{x_1}$$

$$\le M\frac{1}{t}\int_0^t\int_{S_1}\int_0^{1-|\vec{x_1}|} |w(\vec{x})\Delta_{u\varphi_1\vec{e_1}}^r f(x_1,\vec{x_1})|^p dx\, d\vec{x_1}du$$

$$\le M\frac{1}{t}\int_0^t \|w\Delta_{u\varphi_1\vec{e_1}}^r f(\vec{x})\|_p^p du,$$

$$(2.9)$$

where $g(\vec{x}) = G_t\left((1-|\vec{x_1}|)z,\vec{x_1}\right)$. Similarly, we get

$$t^{rp}\|w(\vec{x})\varphi_1^r(\vec{x})D_1^r g(\vec{x})\|_p^p$$

$$= t^{rp}\int_{S_1}(1-|\vec{x_1}|)(\vec{x_1}^{\vec{\beta_1}}(1-|\vec{x_1}|)^{\beta_1+\eta})^p\int_0^1 |w(z)\varphi^r(z)G_r^{(r)}(z)|^p dz d\vec{x_1}$$

$$\le M\frac{1}{t}\int_0^t\int_{S_1}(1-|\vec{x_1}|)\int_0^1 |w(\vec{x_1})\Delta_{h\varphi_1\vec{e_1}}^r g(\vec{x})|^p dx_1\, d\vec{x_1}du$$

$$\le M\frac{1}{t}\int_0^t \|w(\vec{x})\Delta_{h\varphi_1\vec{e_1}}^r g(\vec{x})\|_p^p du.$$

$$(2.10)$$

Similarly, by using the transformation $T_i : S \to S, x_i \to 1-|\vec{x}|, x_j \to x_i$, $i \ne j$, $D_{ij}f = D_i(f \circ T_i)$, we get

$$\|w(f-g)\|_p^p + t^{rp}\|w\varphi_{ij}^r D_{ij}g\|_p^p \le M\frac{1}{t}\int_0^t \|w\Delta_{h\varphi_{ij}\vec{e_{ij}}}^r g\|_p^p du. \qquad (2.11)$$

Combining (2.10)–(2.12), we get

$$\|w(f-g)\|_p + t^r \sum_{1\le i\le j\le d}\|w\varphi_{ij}^r D_{ij}^r g\|_p \le M\omega_\varphi^r(f,t)_{w,p}.$$

Hence, for $1 \le p < \infty$, we have

$$K_\varphi^r(f,t^r)_{w,p} \le M_2\omega_\varphi^r(f,t)_{w,p}.$$

Similarly, we can prove the case $p = \infty$, and we have proved the second estimate in (2.2). $\qquad\square$

3 The Convergence Rate of Multivariate Bernstein Polynomials in Weighted Approximation on a Simplex

Recently, Ding–Xuan Zhou [7, 8, 9] considered the rate of convergence of the Bernstein polynomials in weighted approximation, where the weights are Jacobi weights. In this section we shall discuss the multidimensional case by K-functional. At first we point out that the two–dimensional Bernstein operator $B_{n,2}$ is unbounded in the weighted uniform norm $\|wf\|_\infty$. Here we have $w(x,y) = x^\beta y^r(1-x-y)^\eta, (x,y) \in S$, where

$$S = \{(x,y) : x \geq 0, y \geq 0, x + y \leq 1\}$$

is a simplex. Let

$$P_{n,k,m}(x,y) = \frac{n!}{k!m!(n-k-m)!}x^k y^m(1-x-y)^{n-k-m} = P_{n,k}(x)P_{n-k,m}\left(\frac{y}{1-x}\right),$$

where $P_{n,k}(x) = \binom{n}{k}x^k(1-x)^{n-k}$. Here we have the following

Lemma 3.1. *If $wf \in C(S)$, then*

$$\left| w(x,y) \cdot \sum_{\substack{k,m>0 \\ k+m<n}} P_{n,k,m}(x,y)f\left(\frac{k}{n},\frac{m}{n}\right) \right| \leq M \cdot \|wf\|_\infty.$$

Proof. It is evident that

$$\sum_{k=0}^{n} n^m P_{n,k}(x)(k+1)^{-m} \leq m!x^{-m},$$

$$\sum_{k=0}^{n} n^m P_{n,k}(x)(n-k+1)^{-m} \leq (m!)(1-x)^{-m}.$$

Hence we have

$$\left| w(x,y) \sum_{\substack{k,m>0 \\ k+m<n}} P_{n,k,m}(x,y) f(\tfrac{k}{n}, \tfrac{m}{n}) \right|$$

$$\leq \|wf\|_\infty x^\beta \sum_{k=1}^{n-1} P_{n,k}(x) y^r (1-x-y)^\eta (\tfrac{k}{n})^{-\beta} (\tfrac{n-k}{n})^{-r-\eta} \times$$

$$\times \left\{ 2^{r+\eta} \sum_{m=1}^{n-k} \left[(\tfrac{n-k}{m+1})^r + (\tfrac{n-k}{n-k-m+1})^\eta \right] P_{n-k,m}(\tfrac{y}{1-x}) \right\} \leq 2^{2(r+\beta+\eta)} \|wf\|_\infty . \quad \square$$

Lemma 3.2. *The Bernstein operator $B_{n,2}$ is unbounded in $(C(S), \|wf\|_\infty)$.*

Proof. For $p \geq 0$, $f_p(x,y) = \dfrac{x+y}{e^{-p} + x^\beta y^r (1-x-y)^\eta}$, $0 < \beta, r, \eta < 1$.

We have

$$\|wf_p\|_\infty \leq 1,$$

and by Lemma 3.1 we get

$$\|w(x,y) B_{n,2}(f_p; x, y)\|_\infty$$

$$\geq \left\| w(x,y) \sum_{\substack{k=0 \\ m \neq 0,n}} P_{n,k,m}(x,y) f_p(\tfrac{k}{n}, \tfrac{m}{n}) \right\|_\infty - 4^{\beta+r+\eta} \|wf_p\|_\infty$$

$$\geq e^p \left\| w(x,y) \sum_{m=1}^{n-1} P_{n,0,m}(x,y) m(n-m) n^{-2} \right\|_\infty - 4^{\beta+r+\eta}$$

$$\geq e^p \binom{n-1}{m-1} (\tfrac{1}{4})^{3n-m+\beta+2r+\eta} - 4^{\beta+r+\eta} \to \infty, \ p \to \infty.$$

This shows that $B_{n,2}$ is unbounded in $(C(S), \|wf\|_\infty)$. $\quad \square$

Now we let $f(x,y) \in C_0(S)$, where

$$C_0(S) = \{ f \in C(S), f|_{x=0} = 0, f|_{y=0} = 0, f|_{x+y=1} = 0 \}.$$

Lemma 3.3. *The Bernstein operator $B_{n,2}$ is bounded in $(C_0(S), \|wf\|_\infty)$.*

Proof. This result is immediate by Lemma 3.1. $\quad \square$

Lemma 3.4. *If $f \in C_0(S)$, then*

$$\|w\varphi_i^2 D_i^2 B_{n,2}(f)\|_\infty \leq Mn\|wf\|_\infty, \quad i = 1, 2, 3,$$

where $\varphi_1^2 = x(1 - x)$, $\varphi_2^2 = y(1 - y)$, $\varphi_3^2 = xy = \varphi_{12}^2$, $D_3^2 = D_{12}^2$.

Proof. Let $f_1(x, y) = f(1 - x - y, y)$. Then

$$(B_{n,2}(f))_1(x, y) = (B_{n,2}(f_1))(x, y).$$

We only need the following estimate from [8, p. 82]:

$$\|w\varphi_i^2 D_i^2 B_{n,2}(f)\|_{L_\infty(S \cap \{x+y \leq \frac{3}{4}\})} \leq Mn\|wf\|_\infty, \quad i = 1, 2, 3.$$

Hence we have

$$
\begin{aligned}
|D_1^2 B_{n,2}(f)| \;\leq\; & x^{-2}(1 - x - y)^{-2}\|wf\|_\infty \Big(\sum_{0<k+m<n} (\tfrac{k}{n})^{-3} A(x,y) \Big)^{\beta/3} \\[4pt]
& \times \Big(\sum_{0<k+m<n} (\tfrac{m}{n})^{-3} A(x,y) \Big)^{r/3} \cdot \Big(\sum_{0<k+m<n} A(x,y) \Big)^{1-\frac{\beta+r+\eta}{3}} \\[4pt]
& \times \Big(\sum_{0<k+m<n} (\tfrac{n-k-m}{n})^{-3} A(x,y) \Big)^{\eta/3},
\end{aligned}
$$

where

$$A(x,y) =$$

$$= P_{n,k,m}(x,y)\{[k(1 - x - y) - (n - k - m)x]^2 + k(1 - x - y)^2 + (n - k - m)x^2\}$$

and

$$
\begin{aligned}
\sum (\tfrac{k}{n})^{-3} A(x,y) \;\leq\; & 4^3 x^{-3} \sum P_{n+3,k+3,m}(x,y)\{[(k + 3)(1 - x - y) - ((n + 3) \\[4pt]
& -(k + 3) - m)x]^2 + 16k(1 - x - y)^2 + 7(n - k - m)x\} \\[4pt]
\leq\; & 4^3 x^{-3}\{(n + 3)x(1 - y) + 23(n + 3)x(1 - x - y)\} \\[4pt]
\leq\; & 4^6(n + 3)x^{-2}, \\[8pt]
\sum A(x,y) \;\leq\; & B_{n,2}(n^2(u(1 - x - y) - (1 - u - v)x)^2 + nu(1 - x - y)^2 \\[4pt]
& +n(1 - u - v)x^2; x, y) \leq 5nx.
\end{aligned}
$$

This yields

$$\|w\varphi_1^2 D_1^2 B_{n,2}(f)\|_{L_\infty(S\cap\{x+y\le\frac{3}{4}\})} \le M n\|wf\|_\infty. \tag{3.1}$$

Similarly we get

$$\|w\varphi_2^2 D_2^2 B_{n,2}(f)\|_{L_\infty(S\cap\{x+y\le\frac{3}{4}\})} \le M n\|wf\|_\infty. \tag{3.2}$$

Noticing that

$$|(xy)^{-1}\{km(1-x-y)^2-(ky+mx)(n-k-m)(1-x-y)+(n-k-m)^2xy\}|$$

$$\le x^{-2}[k(1-x-y)-(n-k-m)x]^2+y^{-2}[m(1-x-y)-(n-k-m)y]^2,$$

we have

$$\|w(x,y)xyD_3^2 B_{n,2}(f)\|_{L_\infty(S\cap\{x+y\le\frac{3}{4}\})} \le M n\|wf\|_\infty. \tag{3.3}$$

Combining (3.1)–(3.3), we have proved the lemma. $\qquad\square$

Lemma 3.5.   *Let $f \in C^2_{\varphi,w}(S)$, then*

$$\|w(B_{n,2}(f) - f)\|_\infty \le M\frac{1}{n}(\|wf\|_\infty + \sum_{i=1}^{3}\|w\varphi_i^2 D_i^2 f\|_\infty).$$

Proof. By [8], we only need to estimate $\|w(B_{n,2}(f) - f)\|_{L_\infty(S\cap\{x+y\le\frac{3}{4}\})}$. Let $f_x(y) = f(x,y)$ for fixed x. Then we have

$$w(x,y)(B_{n,2}(f(u,v) - f(x,y))$$

$$= w(x,y)(\sum_{k=0}^{n} P_{n,k}(x) \sum_{m=0}^{n-k} P_{n-k,m}(\frac{y}{1-x})f_{\frac{k}{n}}((1-\frac{k}{n})\frac{m}{n-k}) - f_x(y))$$

$$= w(x,y)(\sum_{k=0}^{n} P_{n,k}(x)B_{n-k}(f_{\frac{k}{n}}((1-\frac{k}{n})u; \frac{y}{1-x}) - f_x((1-x)\frac{y}{1-x}))$$

$$= x^\beta(1-x)^{n+r} \sum_{k=0}^{n} P_{n,k}(x)\{(\frac{y}{1-x})^r(1-\frac{y}{1-x})^\eta(B_{n-k}(f_{\frac{k}{n}}(1-\frac{k}{n})u)$$

$$-f_x((1-x)\tfrac{y}{1-x}); \tfrac{y}{1-x})\}$$

$$= x^\beta(1-x)^{n+r} \sum_{k=0}^{n} P_{n,k}(x)\{w(z)B_{n-k}(f_{\frac{k}{n}}((1-\frac{k}{n})u) - f_x((1-x)z); z)\}.$$

Next we get from [3]

$$|w(B_{n,2}(f) - f)|_{S\cap\{x+y\le\frac{3}{4}\}}$$

$$\le M\{\sum_{k=0}^{n} P_{n,k}(x)\frac{1}{n-k}\sup_{z}\{x^{\beta}(1-x)^{\eta+r}(|w(z)f_x((1-x)z)|$$

$$+|w(z)\varphi^2(z)f_x(1-x)z)|)\}\}_{S\cap\{x+y\le\frac{3}{4}\}}$$

$$\le M\{\sum_{k=0}^{\infty} P_{n,k}(x)\frac{1}{n-k}(\|w(x,y)f(x,y)\|_{\infty}$$

$$+ \quad \|w(x,y)\varphi_1^2(x,y)D_1^2F(x,y)\|_{\infty}\}_{S\cap\{x+y\le\frac{3}{4}\}}$$

$$= \frac{M}{n}(\|wf\|_{\infty} + \|w\varphi_1^2 D_1^2 f\|_{\infty})\{B_n(\frac{1}{1-u};x)\}_{S\cap\{x+y\le\frac{3}{4}\}}$$

$$\le \frac{4M}{n}(\|wf\|_{\infty} + \|w\varphi_1^2 D_1^2 f\|_{\infty})$$

$$\le \frac{4M}{n}(\|wf\|_{\infty} + \sum_{i=1}^{3}\|w\varphi_i^2 D_0^2 f\|_{\infty}).$$

This completes the proof of Lemma 3.5.   □

Theorem 3.6. *Let $f \in C_0(S)$, $0 < \alpha < 1$. Then the following statements are equivalent :*

(i) $\|w(B_{n,2}(f) - f)\|_{\infty} = O(n^{-\alpha})$,

(ii) $K_{\varphi}^2(f,t)_{w,\infty} = O(t^{\alpha})$,

(iii) $\omega_{\varphi}^2(f,t)_{w,\infty} = O(t^{2\alpha})$.

Proof. (i) $\Longrightarrow$ (ii). Let $g \in C_{\varphi,w}^2(S)$. From Lemma 3.4 we get

$$K_{\varphi}^2(f,t)_{w,\infty} \le \|w(B_{n,2}(f) - f)\|_{\infty} + t(\sum_{i=1}^{3}\|w\varphi_i^2 D_i^2 B_{n,2}(f)\|_{\infty})$$

$$\le \|w(f - B_{n,2}(f))\|_{\infty} + t(\sum_{i=1}^{3}\|w\varphi_i^2 D_i^2 B_{n,2}(f - g)\|_{\infty} + \sum_{i=1}^{3}\|w\varphi_i^2 D_i^2(y)\|_{\infty})$$

$$\le M_1 n^{-\alpha} + Mt(n\|w(f - g)\|_{\infty} + \sum_{i=1}^{3}\|w\varphi_i^2 D_i^2 g\|_{\infty}),$$

and hence we have

$$K_\varphi^2(f,t)_{w,\infty} \leq Mn^{-\alpha} + MtnK_\varphi^2(f,\tfrac{1}{n})_{w,\infty}.$$

By a result of Berens–Lorentz [3, p. 122], (ii) is valid.

(ii)$\Longrightarrow$(i). Take $g \in C_{\varphi,w}^2(S)$, by Lemmas 3.1 and 3.5, we obtain

$$\|w(B_{n,2}(f) - f)\|_\infty \leq \|w(B_{n,2}(f) - g)\|_\infty + \|w(B_{n,2}(g) - g)\|_\infty + \|w(g - f)\|_\infty$$

$$\leq M_3\|w(f - g)\|_\infty + M_4\tfrac{1}{n}\Big(\sum_{i=1}^3 \|w\varphi_i^2 D_i^2 g\| + \|wg\|_\infty\Big)$$

$$\leq M_5\|w(f - g)\|_\infty + \tfrac{1}{n}\Big(\sum_{i=1}^3 \|w\varphi_i^2 D_i^2 g\|\Big) + M_4\tfrac{1}{n}\|wg\|_\infty.$$

Hence

$$\|w(B_{n,2}(f) - f)\|_\infty \leq M_6 K_\varphi^2(f,\frac{1}{n})_{w,\infty} + M_4\frac{1}{n}\|wg\|_\infty \leq Mn^{-\alpha}.$$

Hence (ii)$\Longrightarrow$(i) is proved. By Theorem 2.1, we know (ii)$\Longleftrightarrow$(iii). Thus the theorem is proved. $\qquad\qquad\Box$

Acknowledgments. The author thanks W. Haußmann, K. Jetter and D. X. Zhou for their help. In addition, the author thanks the referee for his many useful suggestions.

References

[1] H. Berens, Y. Hu: *K-moduli, moduli of smoothness, and Bernstein polynomials on a simplex,* Indag. Math. N. S. **2** (4) (1991), 411–421.

[2] J. Bergh, J. Löfström: *Interpolation Spaces,* Berlin–Heidelberg–New York, Springer, 1976.

[3] Z. Ditzian, V. Totik: *Moduli of Smoothness,* Berlin–Heidelberg–New York–London–Paris, Springer 1987.

[4] Z. Ditzian, V. Totik: *K-functionals and weighted moduli of smoothness,* J. Approx. Theory **63** (1–2) (1990), 3–29.

[5]   P.–C. Xuan: *Weighted approximation by Baskakov type operators,* Chin. Ann. Math. **16A** (1995), 614–624.

[6]   P.–C. Xuan: *Weighted approximation for a class of multidimensional linear positive operators in L^p,* Acta Math. Appl. **19** (3) (1996), 369–384.

[7]   D.–X. Zhou: *Rate of convergence for Bernstein operators with Jacobi weights,* Acta. Math. Sinica **35** (3) (1992), 331–338.

[8]   D.–X. Zhou: *Inverse theorems for multidimensional Benstein-Durrmeyer operators in L^p,* J. Approx. Theory **70** (1) (1992), 68–93.

[9]   D.–X. Zhou: *Weighted approximation by multidimensional Bernstein operators,* J. Approx. Theory **76** (3) (1994), 403–422.

Address:

PEI–CAI XUAN
Department of Computer Sciences
Shaoxing College of Arts and Sciences
Shaoxing, Zhejiang 312000
P. R. China

WILEY-VCH